LONGBOW
A SOCIAL AND MILITARY HISTORY

ROBERT HARDY

Cover image: *Entrainement des archers gallois*: Psalm 79, from The Luttrell Psalter.

First published 1976 by
Patrick Stephens, an imprint of Haynes Publishing

First published by Sutton Publishing 2006
This edition first published 2024

The History Press
97 St George's Place, Cheltenham,
Gloucestershire, GL50 3QB
www.thehistorypress.co.uk

© Robert Hardy, 1976, 2006, 2024

The right of Robert Hardy to be identified as the Author
of this work has been asserted in accordance with the
Copyright, Designs and Patents Act 1988.

All rights reserved. No part of this book may be reprinted
or reproduced or utilised in any form or by any electronic,
mechanical or other means, now known or hereafter invented,
including photocopying and recording, or in any information
storage or retrieval system, without the permission in writing
from the Publishers.

British Library Cataloguing in Publication Data.
A catalogue record for this book is available from the British Library.

ISBN 978 1 80399 802 2

Typesetting and origination by The History Press
Printed and bound in Great Britain by TJ Books, Padstow, Cornwall

Trees for Life

Contents

Author's Note, 2006		5
Author's Preface, 1992		7
Foreword by Bernard Cornwell		9
Acknowledgements		11
Introduction		13
Chapter 1	The Beginnings	15
Chapter 2	The Longbow into Britain	38
Chapter 3	From Edward I in Wales to Edward III in France – the Longbow Comes of Age	56
Chapter 4	The Archer at Sea, in a River, and on the Downs at Crécy-en-Ponthieu	78
Chapter 5	The Men of the Bow	101
Chapter 6	The Long Road to Agincourt	128
Chapter 7	From Joan of Arc to Roger Ascham	149
Chapter 8	Playing Bows and Arrows	170
Chapter 9	The American Way	197
Chapter 10	The Weapon and the Hunt, the Wood and the Making	218
Chapter 11	Forward to the Past	245
	Testing the Bows	260
	What Sort of Men? Some Conclusions	270

Appendices   Some Technical Considerations
              1. The Design and Materials of the Bow, by
                 P. H. Blyth   277
              2. The Arrow, by P. L. Pratt   287
              3. The Target, by Peter Jones   298

Bibliography   307
Index   310

AUTHOR'S NOTE, 2006

The advances in the technology of printing are beyond me, but in the case of this new edition of *Longbow*, they mean it is not possible to edit or in any sense re-write the original text. That was what I wanted to do, when first I heard that a 30th year edition was to be published, but since then I have come to believe the publishers are right not to let me interfere. A study that against the odds has survived from its first appearance in 1976 should be allowed stand as it is.

Since *Longbow* first saw the light of day much research has revealed new facts and figures which are reflected in many new books and articles, and a powerful force of scholarly examination, led by Professor Ann Curry, Dr Andrew Ayton, Sir Philip Preston, and my colleague Professor Matthew Strickland has reinvaded the history of the Hundred Years War in marvellous and revealing detail. What readers do not find in *Longbow* they are likely to come across in Strickland and Hardy's *Great War Bow*.

Longbow is a sort of bird's eye view of the world history of the simple wooden, unadorned archers' weapon of the hunt, of defence and of war. Later studies have tended to focus on the rise and fall of military archery from somewhere around the time of Hastings to the sinking of the *Mary Rose* warship in 1545, the year that Roger Ascham presented his treatise on archery to Henry VIII, to whose son Edward, and daughter Elizabeth he was tutor, in the study of classical languages and the use of the bow and arrow.

The greatest revelations have come, on the academic level, from new and concentrated study of records contemporary to the peak of the longbow's military use, and the fascination of the subject is proved by the fact that in the last two years there have been no less than seven new published books on the campaigns of Crécy and Agincourt, each of which battles vies with the other as the moment of military archery's greatest achievements.

On the practical side great advances have also been made. Developing from tests and studies carried out during the writing of *Longbow*, and as a result of finding a truly remarkable archer in Simon Stanley, together with the generous scientific assistance of Professor Anna Crowley of the Royal Military Collect of Science and the skill of fine bowyers, the understanding of the medieval weapon of war, as distinct from the sporting longbow still in use today, has deepened immeasurably. There is still much practical work to be done: for instance, the next series of tests for penetration and damage will involve targets made from a steel that can be accepted as representing the average quality of late medieval metal armours, to be attacked at distances from 300 yards (274m) to 20 yards (18m), with heavy arrows of known medieval designs, shot from bows of the great strengths which those recovered from the wreck of the *Mary Rose* prove were in military use until the end of the longbow's dominance as a weapon of war.

I am grateful indeed to my publishers who thirty years ago had faith in a work which might be thought to attract a very narrow interest, who year by year have kept that faith and found it justified, and who are now offering this special edition. These same publishers, whose name and banner may have varied, but are now the great house of Sutton, are responsible for Book Two, as it were, *The Great War Bow*. I dare say that those who read both will have a pretty broad idea of the history of one of the most fascinating, in its day most lethal weapons of war, and companions of peace. At least they are stepping stones in the long journey towards the truth of the warfare of the Middle Ages.

AUTHOR'S PREFACE, 1992

This is the third edition of *Longbow*, first published in 1976. The first edition contained no more mention of the *Mary Rose* and her longbows than was known then, and which in this new edition remains as it was in the original text. The second edition which appeared in 1986 had a substantial new introduction which spoke in broad terms of the longbows and arrows and other equipment for the military archer of the early 16th century that had been brought up in miraculously good condition from the wreck site, by Dr Margaret Rule and her team. This new, third edition has an added chapter, at the end of the original book and before the appendices, which attempts to set into context, with more detail than hitherto, the importance of the *Mary Rose* finds and of what we have learned from them. The alternative was to rewrite piecemeal. But this was thought to be an almost impossible undertaking, and was finally abandoned in view of the fact that I am preparing, together with Dr Matthew Strickland of Cambridge University, a wholly new book on the Great War Bow of the Middle Ages which will base its technical conclusions firmly on what we have learned from the *Mary Rose* experience.

The answer, for the present, seemed to be an added chapter, with references back to the original text where necessary. This chapter has been written by the author, and his colleague in the conservation and examination of the *Mary Rose* archery equipment, Professor Peter Pratt, Professor of Crystal Physics at the Imperial College of Science and Technology. It is a chapter which will explain itself, and will help to throw light into an area of dark uncertainty, of murky assumptions and ill-illumined guesswork. From the gloom and mud

Longbow

of the silted bed of the Solent during 1979, 1980 and 1981, there were borne tenderly into the gleam of day military longbows and arrows of 1545; objects of beauty, objects of use, objects whose examination, though it still leaves us with uncertainties, offers us many answers to questions long asked.

Robert Hardy

Robert Hardy
Upper Bolney, 1992

NB: a draw length of 28 inches is equivalent to 71.12 centimetres
a draw length of 30 inches is equivalent to 76.2 centimetres
a draw length of 36 inches is equivalent to 91.44 centimetres

FOREWORD BY BERNARD CORNWELL

Over 6,000 names of those who fought at Agincourt in 1415 are recorded on muster rolls kept in the National Archives. It is a list of sturdy English surnames (though well salted with Welshmen) and among those names, sandwiched between Thomas Hardewyn and John Hardys is John Hardy, man at arms, and between John Hardegate and John Hardyng is William Hardy, archer.

I like to think Robert Hardy is descended from John or William, two men who fought for their king on a cold, damp day in the cloying mud of a field in Picardy. The army in which they marched was sick, tired, hungry and vastly outnumbered by the French. Yet the English had the longbow, and at day's end it was the French who were beaten.

The English war bow is the most extraordinary weapon. The Duke of Wellington, no fool, wanted to raise a Corps of Longbowmen to fight Napoleon, but alas, there were not enough trained archers in England. Benjamin Franklin, another wise man, believed that if the American rebels had been armed with the longbow then the revolution would have been over in a matter of months rather than years. The longbow was far more accurate than any musket and had a much quicker rate of shot. It was responsible for that astonishing victory at Agincourt, just as it won Crécy and Poitiers, Halidon Hill and Neville's Cross. The list of its victories could fill a page.

The bow has always fascinated me. When I wrote some novels set during the Hundred Years' War I consulted all sorts of sources, and was thus exposed to many historians' opinions and prejudices, and I would frequently find that those eminent historians disagreed. Yet I also learned that Robert Hardy's sturdy English sense usually cut through the confusion and I came to trust his judgement above all others. It is extraordinary that a man who is one of our greatest and most compelling actors should also be the foremost authority on the war bow, but so he is, and we are fortunate that he is also a superb storyteller. The longbow's story is, in large part, England's story, and Robert Hardy tells it wonderfully. I think that Will Hardy, his fingers sore from loosing arrows at Agincourt, would be quietly proud.

ACKNOWLEDGEMENTS

To acknowledge my debt to everyone who has helped me in the preparation of this book is a practical impossibility. There are many who, in spite of their encouragement, their direct or indirect assistance, or their work in this or in related fields, inevitably remain unthanked. I thank all who have earned my gratitude yet are not mentioned here, and hope to have earned their pardon.

My grateful, and in many cases affectionate acknowledgment of assistance and inspiration is offered to Count Gregers Ahlefeldt-Laurvig-Bille, whose personal contribution to this study will be seen to have been both vital and vivid, and whose experience and unstinting help and advice have amply justified my determination not to publish without involving him to the greatest extent his busy life would allow; to the members of the Medieval Society, in particular to John Waller, who has shot all our tests and whose wisdom and instinct for the longbow and its history have given indispensable guidance; and to Ray Monnery who has contributed so much hard work and skill in forging all the arrowheads which have made the tests possible.

My debt to Professor P. L. Pratt, Professor of Crystal Physics, Imperial College of Science & Technology, to Peter Jones of the Royal Armaments Research and Development Establishment, and to Henry Blyth, Leverhume Research Fellow in Classics & Technology at Reading University can be gauged from the importance of their joint contribution in providing the technical appendices at the end of the book. E. G. Heath, well known to archer and historian alike, has given invaluable help in every direction, especially with the illustrations, and Lt Cdr W. F. Paterson, RN, of the Society of Archer Antiquaries has afforded kind guidance.

My gratitude is also due to Dr Maurice Keen of Baliol College, Oxford, for pointing me the way to discover more about the medieval longbowman himself; to Mr William Reid and the staff of the National Army Museum, London; to Mr Russell Robinson, Keeper of Armour at the Tower of London; to Professor Dimbleby of the Institute of Archaeology, University of London; to the Librarian of the National Library of Wales, Aberystwyth; to the staff of the Wildfowl Trust at Slimbridge, Gloucestershire. To Nadine, Countess of Shrewsbury; to Vicomte A. de Châbot-Tramecourt; and to Mr Mostyn Owen Jodrell for kind help and information concerning their families' involvement in medieval history. To Sir Cennydd Traherne, KG, and the Company of the Freemen of Llantrisant; to Mr Graham Deane of the Nuffield Orthopaedic Centre; to Mr P. S. Leathart, MBE, of the Royal Forestry Society; to the Librarians of the Forestry Commission and the Common-wealth Forestry Institute; to Mr T. G. H. James and the staff of the Department of Egyptian Antiquities, British Museum; to Mr J. Cherry of the Medieval & Later Department of the British Museum; to the Curator of the Brecknock Museum; to Fred Lake, archivist of the Royal Toxophilite Society; to the Wakefield Metropolitan District Council; to Dennis Holman, author of *The Elephant People* and to his publishers, Messrs John Murray; to G. P. Putnams Sons, New York, the Trustees of the Elmer Estate and Alfred A. Knopf Inc, New York, and Charles Scribner's Sons, New York. To all those museums, libraries, publishing houses and private owners who have given permission for the reproduction of illustrations in their collections, and which are separately acknowledged in the appropriate captions; and to the following for permission to use the illustrations that appear at the beginning of each chapter: Editions Arthaud, Paris (photograph by G. Franceschi) (Chapter 1); British Museum (Chapter 3); Bibliothèque Nationale, Paris (Chapters 4 and 7); BBC (Chapters 5 and 6); Radio Times Hulton Picture Library (Chapter 8); E. G. Heath (Chapter 9); Count Ahlefeldt-Laurvig-Bille (Chapter 10.)

Last, but by no means least, my research assistant, Mrs W. M. Garcin, and I offer our thanks to the staff of the Reading Room and the European Manuscripts Room of the British Museum, of the Public Record Office and of the many museums and libraries both in Britain and elsewhere whose assistance has always proved invaluable.

INTRODUCTION

A longbow is only a bow that is long rather than short. It has come to mean a particular kind of bow, the type still used today and presented quite precisely in the rules of the British Long Bow Society.

'A long bow is defined as the traditional type with stacked belly, horn nocks, and limbs made of wood only. All surfaces shall be convex.'

That description already excludes many kinds of bow that for the purposes of this book will be called longbows. But the Society exists to perpetuate the use of the traditional longbow which has ancient and honourable antecedents. It derives from the bows that were used at Crécy and Agincourt, through the decline of the military weapon and its rise as a weapon of sport. During that slow transition the bow underwent gradual changes. As the desire for maximum strength lessened, it was replaced by the need for ease of use and for moderate power in return for minimum exertion. For target shooting, the longbow must, without tiring the archer, be capable of casting arrows accurately and consistently. So the design of the limbs developed in ways that would have made the weapon too critical, too liable to damage under conditions of campaign and battle.

'The length between nocks [the slots in the horn-tips to which the string is attached when the bow is "braced", or "armed"], measured along the back of the bow, shall not be less than 5 ft for arrows up to 26 inches long, and not less than 5 ft 6 inches for arrows over 26 inches long.'

These measurements, agreed in 1975, are from 3 to 6 inches shorter than those in common use when all target shooting was with longbows. The length of arrow is determined by the height and proportion of the archer, the distance between the hand that holds the bow and the one that draws the string when he has learned to draw comfortably to his best length. Between the arrow length and the sensible length of bow there is a direct relation. The bow should be long enough to bend safely, allowing the arrow to be drawn to its head, and at a given draw-weight, or strength, to drive that arrow with sufficient force and speed. There is no upper limit to the length of a bow. The longer the bow, the safer it is to bend it but, within certain limits, the longer a bow the weaker it is at a given length of draw.

'The thickness [depth] of limbs, measured from belly [the inside bend of the drawn bow, facing the archer] to back [the outside of the bent bow, facing the target] shall at no point be less than three quarters of the overall width of the limb at the same point.'

In other words, the traditional longbow is a deep, D-sectioned wooden spring, not a flat wooden one. That is what is meant by 'stacked'. The ratio between depth and width was formerly seldom less than four to five, sometimes higher than six to seven; then, extra length was essential for safety in a very 'high stacked' bow.

'At the arrow-plate [where the shaft of the arrow flies past the bow, on release] the bow shall not be narrower than at any other place on the top limb.'

There are no concessions to advances or alterations in design, which in modern bows or ancient short bows allow the arrow to fly from the centre or nearly the centre of the vertical plane of the limbs, by an acute narrowing to the handle.

'There shall be no arrow rest built into or attached to the bow. The stele [shaft] of the arrow shall be of wood, the fletchings of feathers.'

If you shoot a longbow with the Long Bow Society now, you must shoot with the simple bow that is a direct descendant of the medieval wooden weapon which described a near semi-circle when drawn, and which sent the arrows hissing with appalling and destructive force, 'so thick that it seemed snow', at the advancing enemy.

The purpose of this book is to trace the evolution of the longbow from the mists of pre-history, through its military glory, to the highly polished weapon that is slipped from a green baize cover at a longbow tournament today. It is a story of rigours and skills, of tenacity, and delight.

CHAPTER 1

THE BEGINNINGS

There are monkeys who drop things from trees, and a few even who go through some sort of throwing motion, but mankind alone among the animal creation has been obsessed for uncountable ages with the idea of propelling things through the air, both aggressively and defensively, and sometimes out of sheer exuberance. It is known that nearly half a million years ago he used wooden spears, their points hardened by partial burning. Before that he must have thrown sticks and stones. It is so deeply rooted, so ancient a human instinct, that one of the first things a child will do, on a gravel path, or a country lane, or by the water's edge, is to sort out a throwable stone, and hurl it with tiny force a few inches beyond his foot, or even, lacking future co-ordination, behind him. Time and growth allows the child to overcome the early problems of stone throwing, until it becomes a dangerous achievement, and when used as a concerted demonstration of disapproval calls from opposed authority the deployment of highly sophisticated weapons and tactics in retaliation.

The step from throwing with the hand to throwing with some sort of aid, such as a throwing stick, is a long one, and perhaps resulted from man's watchful eye noting the strength in a wind-bent sapling, or that a taller man with a longer arm could throw further. Once that lesson was learned, the whip of the sapling whisking back to the upright from a bent position must have obsessed the inventive man. Whatever alternatives presented themselves and were put to use, the boomerang, the weighted throwing-spear or the knob-kiri, the use of the sapling's strength rode his mind until he took the sapling out of the earth, chose a centre point, made the limbs on either side of that centre roughly equal and harnessed the force of the bent wood with some kind of a string attached to the two ends. Then he held in his hand the bow. Whether he first chose to throw stones or sticks from his bow, it was the thrown stick that lasted, and developed, and was responsible for the foundation and the tearing down of dynasties, potentates and empires across the world.

The invention of the bow was an enormous advance for man. When he had achieved the ability to store the slow energy of his muscles and release it suddenly and efficiently, he was able to hunt game with success and comparative safety, and to strike at his human enemies at a distance. The success, the safety and the distance grew as man's understanding of the bow increased, and as he learned to use sharpened stones and flints, animal teeth and sharpened horn and bone, on the ends of his arrow-sticks. No doubt the spear came before the bow, and no doubt spears were often fitted with sharp stone or bone heads, but the moment the bow and arrow emerged the archer's advantage was enormous. Not only could he attack his adversary from beyond spear-throwing range, he could also carry many more light arrows than his enemy could carry spears. When the first archers shot their arrows from their bows they started the use of a weapon which lasted unchallenged for tens of thousands of years. But when did those first archers shoot, and where?

Bows are basically of two kinds, wooden and composite. The wooden bow explains itself; the composite bow is one which is made either of wood, reinforced with horn or sinew, or entirely of materials other than wood. Whatever bows were made of, all the materials were organic, and so in the passing of thousands of years have perished and disappeared. Our only guides to date and place are arrowheads of stone, flint and obsidian which have survived time, and pictures painted on rock or carved into the stone. The earliest arrowheads were found at Bir-El-Ater in Tunisia, and also in Algeria, in Morocco and in the Sahara. Some of them date from about 50,000 years ago when the ice of the last Glacial period covered great parts of Europe. Fifty thousand years ago is only a short time in mankind's existence, but those heads represent a comparatively sophisticated state in the advance from the sharpened stick, and suggest that the bow and arrow was well established, at least in warm Africa, long before that period.

The arrowheads are all carefully pointed, some of the later ones tanged to fit into the end of the arrowshaft. Some are perhaps too large to have been anything but spearheads, but there is no doubt at all that of those which are incontestably arrowheads, expertly shaped from obsidian and flint, chalcedony and jasper, many date from the Upper-Paleolithic, or Old Stone Age, before 10,000 years BC. Before the stone head and superseding the sharpened

The Beginnings

stick there were no doubt centuries, perhaps millennia, of bone and horn, tooth and thorn arrowheads, but none can have survived. Sharpened hardwood arrowheads survive from periods long after the existence of flint heads, for instance in Egypt from the early dynasties to the New Kingdom, in the Caribbean, certainly until the 17th century AD, and in parts of South America and Africa until the present day.

What kinds of bow may have been used 50,000 years ago can only be guessed. We can only be sure that there *were* bows, that where at that period there grew suitable timber they were probably made entirely or partly of wood, and that where timber could not be had they were made of horn, bone, sinew and gut in various combinations. That is an important distinction to make because the longbow is almost always a wooden weapon, and the existence of trees is subject to climate.

In the Sahara, for instance, the desert is a comparatively recent occurrence, probably not complete until about 500 BC, the gradual desiccation having taken something like 5,000 years. Remnants of charcoal from old camp fires, and pollen samples, suggest that pine and holm oak, alder and lime trees grew there among other timber, so if we were to guess at wooden bows as companions of the earliest arrowheads we should probably be right. During the slow drying out of the Sahara region, through the Mesolithic (20,000–7,500 BC)

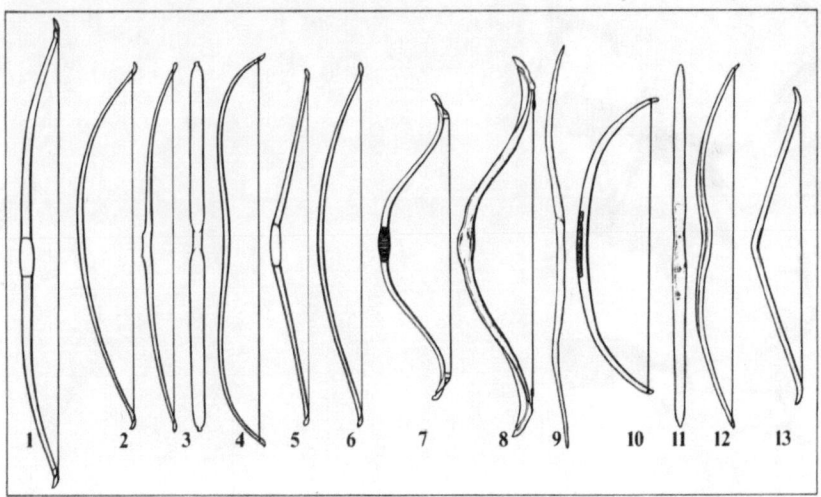

Simple bows **(1–6)** *and composite bows* **(7–13)**. **1** *Longbow.* **2** *Asymmetric bow.* **3** *Flat bow – Holmegaard type.* **4** *Doubly convex Egyptian type.* **5** *Joined angular bow.* **6** *Simple segment bow.* **7** *Chinese.* **8** *Turkish.* **9** *Siberian (made of antlers).* **10** *Hurrite.* **11 and 12** *Hittite.* **13** *Assyrian angular* (Drawing by Clifford Anscombe).

and Neolithic (7,500–3,500 BC) periods, the Middle and New Stone Ages, many different kinds of bow were beautifully illustrated on the cave walls and rock faces of the area, as well as in parts of Europe, where the retreating ice had allowed timber to spread northwards from the warmer lands of the south. These paintings show longbows and short bows of 'simple segment', which means that when drawn they describe roughly the segment of a circle, and such bows were almost certainly made of wood; but they also show bows of more complicated shapes, deflexed and reflexed, which suggest composite manufacture. As timber became scarce men invented substitutes, but not only that: in some parts of the world the composite bow seems to have arrived not as a development of the simple bow, but alone, as an unaided achievement of man's genius. Where that was true, the inventive skill is staggering. That an efficient bow could have been evolved from strips of horn and sinew, glued or bound together, is almost incredible. Where there had been wooden bows that were, as timber failed, adapted by the addition of horn and sinew is more understandable.

The true composite bows were mainly to be found in north-eastern Europe and in Asia. In Africa there seems to have been timber enough for the needs of archers from the earliest days, and probably the surviving wooden bows in use today among certain tribes in Kenya, in Lesotho, among the bushmen in other

Above, left *Neolithic Spanish bowmen with segment and doubly convex bows. Rock painting from Cueva de los Caballos, Castellon.*

Above, right *Archer from Jabbaren. Bovidian period. Tassili rock frescoes* (Exclusive to Editions Arthaud, Paris, photograph by G. Franceschi).

parts of Africa, and among the bowmen of South America are much the same as have been used for thousands of years. It may well be true of the wooden bows of India and Ceylon and of the Andaman and Solomon Islands. Only the sub-continent of Australia never produced an aboriginal bow of any kind.

We are concerned with wooden bows, the finest and most famous example of which was the late medieval war bow of the English, but before we trace the antecedents of the European longbow we must give some attention to the composite bow of the East. Its design rests on the facts that animal horn will compress and that animal sinew is elastic. Where timber was available, the core of the composite bow was made of wood. On the belly side, the surface of the bow facing the archer, horn was glued; on the back, the surface away from the archer, sinew. When the bow was drawn the horn compressed and the sinew lengthened; when the bow was released the compressed horn expanded quickly to its original state at the same time as the elastic sinew swiftly shortened. In the best designed and most developed bows of this kind the speed of the limbs' return to their normal position is greater than that of any kind of timber, no matter how skilful the design or the execution of the bow made from it. This is why the wooden bow has been abandoned now by most modern archers in favour of weapons that rely to a great extent on the principles of the old composite bows. In the modern bow, man-made plastics and fibres of brilliant efficiency replace the horn and sinew of its predecessors. It was fortunate for the English warriors of the Hundred Years' War with France that the Eastern composite bow had not at that time penetrated to Western Europe, even though it was the composite bow that drove so many of the inhabitants of Europe, centuries earlier, from their dwelling places further east and north.

The composite bow was the outcome of necessity. The obvious material for bowmaking, wood, was not available. The long history of archery is not alone in showing that man will invent and manufacture with more skill when driven by necessity, than when he has easy access to materials, ways and means that do not stretch his ingenuity to the limit. In some cases the developed composite bow seems to have arrived without the stimulus of need, but in those cases it almost certainly represented a derivation from earlier forms of composites made out of necessity, and exhibiting such outstanding mechanical results that later bowmakers abandoned the natural efficiency of woods that had already been proved good for bowmaking.

By the time the ice had retreated to approximately its present areas both composite and wooden bows were used, and their use was spreading through the world. There were wooden bows in western Europe, in Iceland, through the greater part of Africa, all India, and the island groups of the Indian and

Pacific oceans, and also through the greater parts of both north and south America. Composite bows of one kind or another lived with wooden bows in some parts of eastern Europe and north America, in isolated areas of south America, and seem to have been exclusively used through all of what is now Russia and China, the northern half of north America and in the coastal parts of Greenland.

These were times before static civilisations had established themselves, when groups and tribes wandered the world in search of survival, following the migration of game and the unsure fertility that good seasons brought, fleeing drought and flood and cold, as well as more powerful enemies seeking the same survival. It is not until some patterns of stability began to emerge in the geographical locations of major groups of peoples that there begins also to be seen a pattern in the use of weapons, and the many thousands of years of dark, unrecorded history allow us moments of insight, facts and indications, like brief flashes of lightning:

 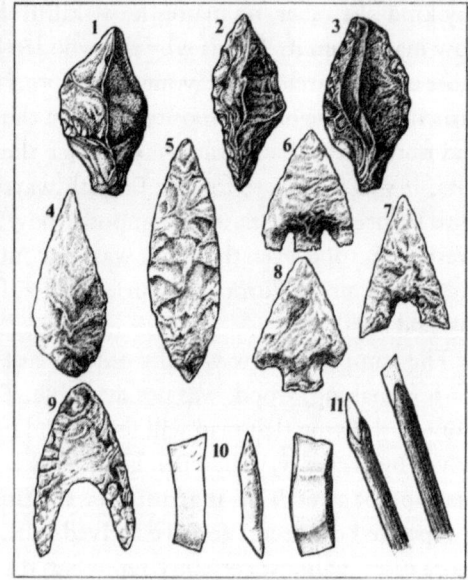

Above, left *Ti-n-Tazarift archers with bows of various patterns. Tassili rock frescoes* (Exclusive to Editions Arthaud, Paris, photograph by G. Franceschi).

Above, right *Flint and stone arrowheads.* **1, 2 and 3** *Middle Paleolithic.* **4 and 5** *Neolithic leaf-shaped.* **6** *Bronze Age barbed and tanged.* **7 and 8** *Mesolithic.* **9** *Pre-Pharaonic Egyptian 'knapped' jasper swallowtail.* **10 and 11** *Ertebolle culture (Denmark) chisel-shaped flint arrowhead, loose and set in the arrowshaft* (Drawing by Clifford Anscombe).

The Beginnings

In China in the second century BC workshops existed, which have been unearthed, for the making of bows and arrows, which were used from horseback and from chariots in time of war.

Homer describes the bow of Pandarus: 'His polished bow was made from the horns of an ibex that he had shot. They measured sixteen hand spans. A bowmaker worked on them, joined them together, smoothed them and set a nock of gold at each end. Pandarus strung the bow, opened his quiver, chose an arrow and fitted it to the string. He held the arrow nock on the string of sinew and drew back the string to his breast until the iron arrowhead touched the bow. When he had drawn the great bow to full compass, he loosed the string, which sang the sharp arrow into the air.' Homer's Greeks were the spear and bow warriors from the north and east who drove the earlier peoples out of their mainland and their islands.

In the Lake Bajkal area of Russia, early Hun burials of the fourth century AD show the importance of archery: warriors' bows were buried with them. One Siberian warrior was buried, his wife and child, both shot full of arrows, beside him.

The Andaman Islanders still make a flat, paddle-shaped bow with a longer upper limb, a proportion which suggests an ancient origin before the sapling was cut from the earth, when perhaps a string was tied to the top and to the bottom near the ground and a spear or an arrow could have been shot from it well below the mid-point, as was the practice of the Dyaks of North Borneo.

African bushmen and south American Indians still shoot with wooden bows which cannot have changed much since the Stone Age. So do the Veddahs of Ceylon.

As an aside, there was an instance of 'primitive' archery only a year or two ago, in England. An archer without his bow was at a field shoot, where the competitors were using the most sophisticated modern equipment, shooting at targets of varying ranges. Unable to bear watching others while he did nothing, the archer cut a hazel stick from a hedge, put a string to it, and using modern target arrows shot a complete round, ending with a score half way up the card.

From the mists of legend and occasional fact, there do at least emerge, one by one, undeniable proofs of the derivation of the longbow.

Two fragments of bows of the Mesolithic period were found at the Stellmore site near Hamburg, the longest piece being less than ten inches (24.8 cm). They are made from the heartwood of pine, a timber which lasted into the present century as bowmaking material in Siberia, and their shape, in section, is very like the longbow of today, with a flat back and a deeply arched, or 'D' section belly. Since it is likely that no pine grew in the area, which then

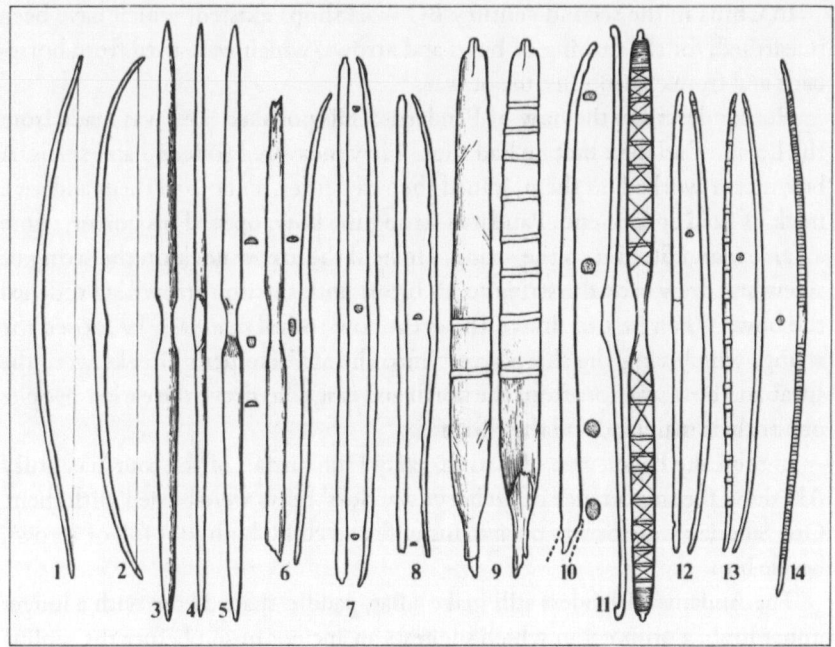

Above *Prehistoric bows.* **1 and 2** *Neolithic.* **3, 4 and 5** *Holmegaard.* **6, 7 and 8** *Neolithic flat bows.* **9** *The Meare Heath bow.* **10** *The Ashcott Heath bow.* **11** *Meare Heath bow reconstruction.* **12** *A Vimose bow.* **13 and 14** *Nydam bows* (Drawing by Clifford Anscombe).

Opposite *Limbs of Meare Heath bow and Ashcott Heath bow* (University Museum of Archaeology & Ethnology, Cambridge).

was tree-less tundra, it seems that the bows were made elsewhere and brought by hunters following migrating game or warriors in pursuit of or pursued by their enemies.

There are rock paintings in Spain of the early Neolithic period at Cueva de los Caballos, at Los Lavaderos de Tello and at Cogul. The first of these represents deer being driven towards a waiting line of archers. The arrows that have already been shot have struck exactly where they would most swiftly kill the animals. The bows are of two kinds: simple and double-convex bows, which suggest not a composite construction but staves that have been heated to the point where they can be permanently bent. It is hard to guess why wooden bows should have been bent towards the draw in this way, but it is a recurrent question which must be answered.

The Beginnings

From the later Stone Age in central Europe, at sites in Switzerland and Germany, more than 20 bows or fragments of bows have been found. Lengths have been estimated from fragments, or parts of bows, and the complete bows measured. The longest of these is 69 in (175 cm), though another of 67 in would probably have been 70 in originally, and another well over 80 in. The shortest, apart from a child's toy bow, is likely in complete condition to have measured about 61.5 in (155.5 cm). All the bows are of yew wood, that finest of all timbers for making bows, but they are all cut from the core of the log, or are made from branches, as their makers had not yet discovered the magic that lies in the use of the sapwood and the heartwood of yew together. They are all of a deep section, none is a flat bow, and most of them are of the 'D' section which is characteristic of the later longbows; one only has a slightly concave back, some have a pronounced keel along the belly, which would have been left to stiffen the stave, one so pronounced a keel as to be almost triangular in section. Even the bows cut from branches have a flattened back, showing the natural contour of the wood in the sides and the belly. All the shapes are worked smoothly except for a single example, the fragments of which are hexagonal, and so, possibly, unfinished. Any of these weapons, handed to an English bowman of the 15th century AD, or to a member of the British Long Bow Society today, would be instantly understandable and familiar.

From northern Europe and the British Isles in the same late Stone Age come some bows of similar construction, and some others. The two earliest are from the Mesolithic-Neolithic Transition period. They were found, one in a complete state, at Holmegaard, on Zealand, Denmark in 1944. They are flat bows, of a shallow 'D' section, and of considerable sophistication, with a cut-away handle allowing for an almost 'centre-shot' passage of the arrow. They are made of elm, a timber used for bows 4,000 years later in Wales, and from the narrowness of the year-rings it is probably elm grown in the shade, which in north Europe was the next best bowtimber to yew. One complete bow is 60.5 in (154 cm) long, the other was probably 64 in

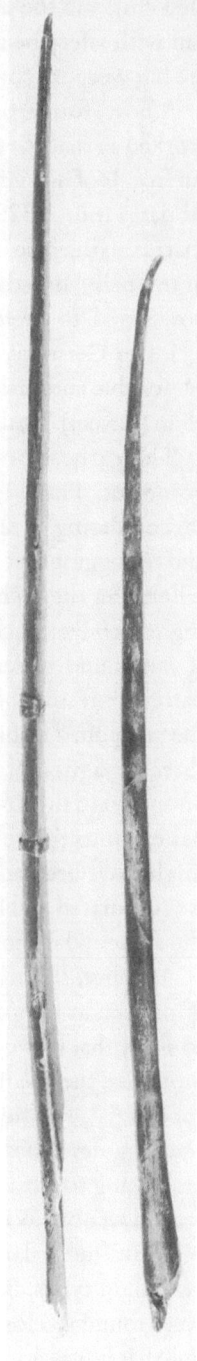

(163 cm), and the more powerful of the two. Bearing in mind that a flat bow can with safety be made shorter, for the same strength, than a longbow, these are big weapons for their type.

A bow, found on Zealand in 1959, also of shade elm, and most carefully worked so that the grain was not cut into, and knots left 'proud' of the general surface, is of a type between true longbow and flat bow. Radio-carbon analysis dates it to 2,820 BC. One curious fact about these three Danish bows is that, as in the case of bows made from branches, the outside of the log is used in the belly, not the back, which is the opposite of the natural way to bend bow wood, to stretch the outer and compress the inner wood of the log.

From Germany and Holland come half a dozen yew longbows of the Mesolithic and early Neolithic periods, one of them surprisingly short, about 45 in (115 cm). Radio-carbon datings range from 2,400 BC to 1,550 BC.

The earliest bow yet found in England comes from Meare Heath, in Somerset. There is only one limb, but that was sufficient to give a radio-carbon dating of 2,690 BC, with an allowed error either way of 120 years, and to suggest that it was once over six feet long, or up to 190 cm, a good deal taller than the archer who would have shot in it. It is a flat yew bow with a slightly convex back and an almost flat belly, a little keeled towards the handle. It was bound, along the limbs, with leather thongs, in an intricate criss-cross pattern, presumably to prevent the wood fibres starting, and so weakening the weapon, a fashion used by much later, Iron Age Danes. But, remarkably, there are signs that the thonging was only finally cut to shape after it had been bound on to the limbs, and in some places the flint knife used for the trimming has cut into the surface of the bow's back, which suggests that the binding might have been added after the time of the bow's useful life, when perhaps it was decorated and laid up as a tribute to the size, strength and achievements of the archer who shot so powerful a weapon.

Another, slightly younger, yew bow, dated 2,665 BC, was also found in Somerset, at Ashcott Heath. Again only half the bow remains, but enough to show that it was originally about 63 in long, or 160 cm, and that it is much more like the medieval weapon than its companion from Meare Heath. The back is flat, the sides and belly rounded and deeply 'stacked'. High or deep stacking describes the Roman or even Gothic arch section of the bow, the arch being towards the belly. Low or shallow stacking describes a more flatly arched section. What can we deduce from these two bows of the third millennium BC in England? Certainly that there already existed fine weapons of the two main types, flat longbow and high stacked longbow. The fact that they were found so close together suggests that the two types existed side by side, but that it was not yet recognised by bowmakers that high stacking combined with length produced a safe, but more powerful weapon than the broader, flatter bow.

It is clear that in Mesolithic and Neolithic times, as far as the northward shrinking of the ice allowed, wooden bows, many of which we can legitimately call longbows, were most efficiently made, most efficiently used – the number of animal skeletons with embedded arrowheads is proof enough – and greatly cherished, since many of the old hunters and warriors were buried with their bows beside them.

As we have seen, where timber was not to be had, composite bows were often achieved. This was true in the northern parts of Bronze Age Scandinavia. But vegetation stretched slowly further northwards, first small plants, then little trees, arctic willow and dwarf birch, followed by juniper and crowberry, large birch and aspen, and pine. Both bows and arrow shafts can be made from these woods, and no doubt they were. Hazel had reached the British Isles by about 6,000 BC. Elm followed, and oak, lime and ash were not long behind. Yew was well established by 5,000 BC at the latest, and probably much more widespread than it is today. Indeed in the Interglacial period it had existed in Britain, and in Pre-Glacial times was well distributed over the eastern parts of the country. After the last ice age, once re-established, it gave its ancient name to many places across Europe and Scandinavia. Youghal means yew wood, Dromanure, yew hill, Glenure, yew glen and Mayo is yew field. The reason for its comparative scarcity now may well be that, though yew grows readily in a temperate climate, and survives remarkably in difficult conditions, it offers so much to so many living creatures. It has been cut for weapons, tools, furniture and fuel, and in the winter has provided green eating for cattle, deer, hares and rabbits. Its poisonous properties were never as great a protection for it as prickles were to the holly.

The marvellous potential of yew wood for bowmaking was realised very early though it was probably not used in the most effective way until after the general renaissance of archery that followed a curious decline.

Archaeological traces suggest that the wooden bow in Europe was widespread and general from the tenth millennium BC, and that it was a predominant weapon. From about 1,500 BC it seems to have become less and less important. There are some bows dating from the first and second millennia BC. One found at Edington Burtle in Somerset, a branch-wood yew bow of 'D' section has been dated at 1,320 BC.

A fascinating find was made in Denmark, when a longbow from the late Stone Age, dated between 1,500 and 2,000 BC and made of oak, was dug out of a peat bog. The latest research (February 1976) suggests even a considerably earlier date. Count Ahlefeldt-Laurvig-Bille, whose knowledge and wisdom

Longbow

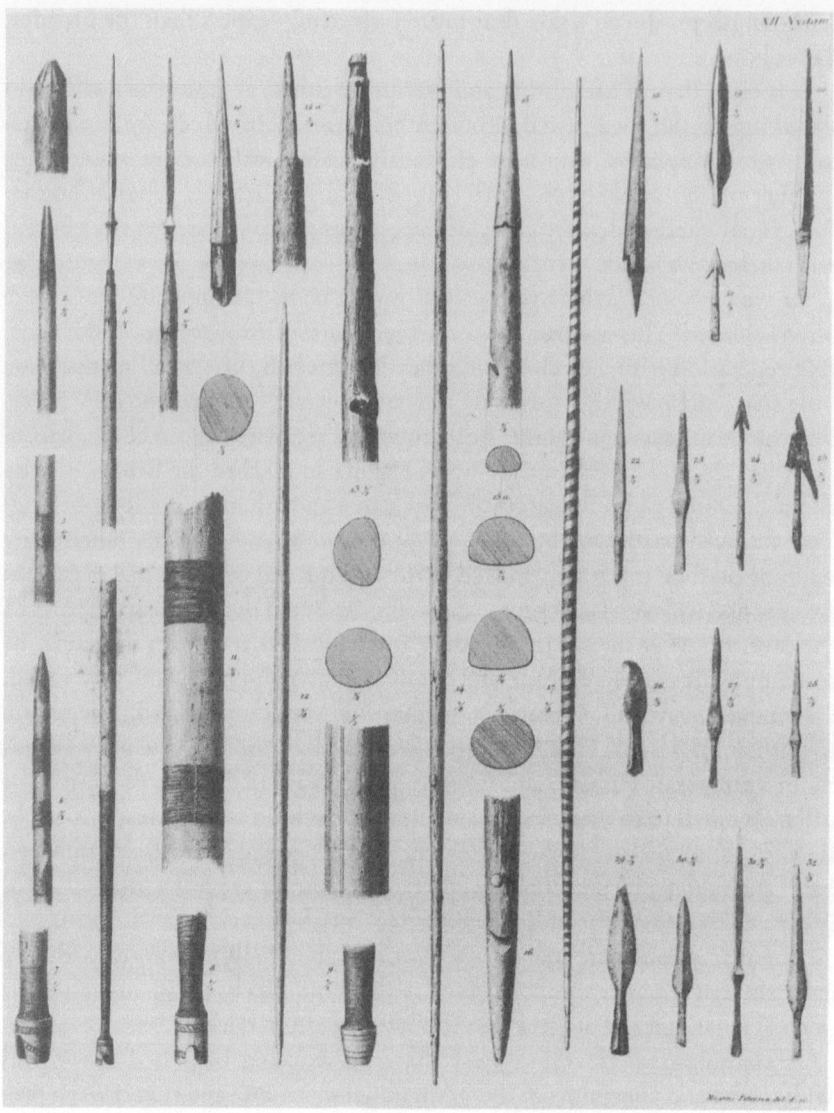

Details of bows and arrowheads from the Nydam finds (Copenhagen, 1865).

have contributed greatly to this book and whom we shall meet more formally later on, tells the story:

'In 1935 I found myself studying peat production in a big peat area in the centre of northern Jutland, in the county of Viborg. By sheer luck a day or two before my arrival the peat workers had dug up a long stick, pointed at both ends, unbroken and lying flat at the bottom of the peat layer some three metres down, at the level of what had been the clay bottom of a big shallow lake. The workmen were quite used to finding things in digging, and they would look after them, hoping to sell them to local museums. In this case the museum curators had little interest in their stick, and only offered a few Kroner, so it was preserved in wet moss in the hope of a better offer. I made a better offer and the bow was mine.

'I had it packed in wet moss, uncleaned, and sent off at once to the National Museum in Copenhagen for research and preservation. First research was done in 1936 by pollen analysis, and then it was examined again in 1961 by our greatest authority Dr Troels-Smith, using the much more reliable Carbon 14 method, which put the bow at the peak and towards the end of the Stone Age. The early Stone Age hunters had disappeared and others had invaded Denmark introducing agriculture and a new pattern of life, including trade in many parts of Europe, from Ireland in the west and possibly as far as Asia to the east. They may well have brought the longbow with them. At least we seem to know them from the primitive Bronze Age rock engravings in south Sweden. Longbows are there handled by naked men.

'Was it a "missing link" that I happened on that day in Jutland, between the Stone Age flat-bows and the true longbow? And – is it a bow? I think I can vouch for the latter. Everything lines up perfectly: length, 160.5 cm; largest diameter, at the exact middle, 27 mm with a circumference of 7 cm; a fine tapering of both limbs, apparently done by scraping away superfluous wood only from the inside or belly of the bow, which still shows a quite clear bend toward the drawcurve of the bow; in other words it has "followed the string", which most bows do after considerable use.

'One thing more than any other to my mind proves it a bow. In the whole length of the stave, which was probably cut from a young straight-grown tree, there is only one knot, some 58 cm from one tip, that is towards the middle of the bow and on the belly side where it would cause no trouble. Just below it is a typical "pinch" resulting from compression in use. Had the knot been on the back, I would have ruled out the possibility of the stick being a bow at all.'

From his great experience both as forester and longbowman, Count Ahlefeldt reckons that with a well-seasoned replica he could have tackled anything up to the size of a deer, or indeed a man. 'Rather intriguing is the fact that it is made of oak wood, which is rare but not quite unknown. There is the oak flat bow from Denny in Scotland, for instance, dating from 1,300 BC. Perhaps the choice of wood was due to lack of a better timber at hand. I feel sure the bow would be rather unpleasant, stiff and apt to kick in the hand.'

A general scarcity of bows after 1,500 BC is matched by the small number of arrowheads, either of flint or bronze, and where there are rock carvings of other weapons there are few representations of bows, and those that do exist, in Scandinavia, seem to be pictures of composite, not wooden bows. Perhaps it was because, with the development of bronze weapons such as swords, spears and axes, the fashion of fighting changed, and men, having found the wonderful new metal, preferred to pit their skill in the making of these weapons, and in their use against each other, hand to hand. The same thing seems to have happened to the composite bow in Greece. By the beginning of the Mycenean period the bow had become only a hunting weapon. The warriors fought with lance and sword, protected by metal armour, and the bow did not come into use again as a war weapon until it had returned with the armour piercing arrow in the late Geometric period, about 750 BC.

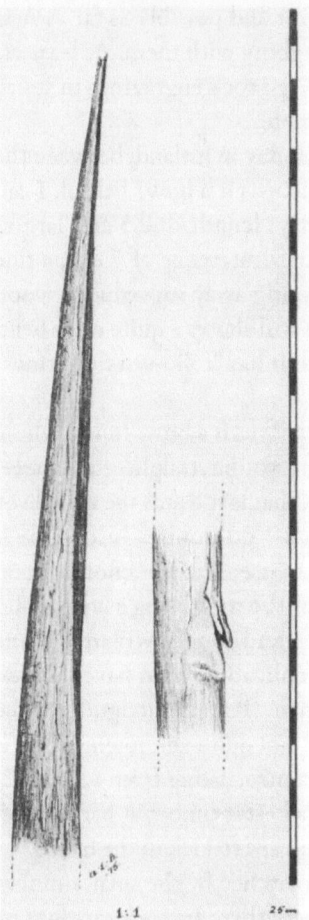

Below and opposite *The Viborg longbow in the possession of Count Ahlefeldt-Laurvig-Bille, drawn by him, clearly shows the knot, and the 'pinch' in the belly.*

Occasionally, in the seas of time the bow seems to disappear, though never for very long without showing its dolphin back in the sunlight again. By the beginning of the Christian era in northern Europe, it was very much in the ascendant once more and, alongside the growing use of horsemen in war, the use of the bow and arrow grew to counter the advantage of cavalry, and redress the balance.

The Romans on the continent of Europe equipped their cavalry with bows, the short composite weapons borrowed from the East, ideal for shooting from horseback. It is extremely difficult to cope with a longbow and manage a horse at the same time, but, as the mounted archers from all over eastern Europe, from Asia Minor and Asia, as well as north Africa proved, bands of archer cavalry that dash in and shoot and retire out of range as quickly as they came in can be devastating.

The Germanic tribes had forgotten the techniques of composite bowmaking by the end of the late Bronze Age, but their longbows gave the Romans considerable trouble. Once, in AD 354 they were prevented from crossing the Rhine by showers of arrows from the Alamans, and 34 years later the Roman attack on Neuss was repulsed by a hail of arrows 'falling as thick as if thrown by arcubalistae'. Shades of Crécy.

A very rich series of finds in Denmark and Schleswig-Holstein give some idea of the bows of northern Europe during the Roman Iron Age. Some were dug from bogland at Vimose and Kragehul, others from ship-burials at Nydam. Most of them were among spoils of war of every description which had been sacrificed to the gods after battle, and thrown into sacred mosses or lakes. They all date from between AD 100 and 350; they are all made either of yew or fir wood and they are recognisable longbows of deep stacked, generous 'D' section, cut from logs of wood, the outer surface of the logs forming the backs of the bows. The Vimose bows range from 66 in (167.6 cm) to 77½ in (196.9 cm). Two are 66.7 in (169.5 cm) and 70.3 in (178.5 cm). Of the 36 bows found in the Nydam ship, there is one of 72 in (182.9 cm), another 75 in (190.5 cm). Three are measured at 70.1 in (178 cm), 71.9 in (182.5 cm) and 77.8 in (197.5 cm). So in general they are longer than the height of the men who used them. They are not, on the whole, quite as sturdy as the *Mary Rose* bows recovered from a ship of Henry VIII's navy, wrecked in 1545, but they are very similar to look at; no Tudor archer would have found these bows anything but familiar.

One Nydam bow is 70.1 in (178 cm) long, 1 in (2.6 cm) deep (from back to belly) and 1.1 in (2.8 cm) wide, so the ratio of depth to width is 1 to 1.1. One *Mary Rose* bow is 73.6 in (187 cm) long, 1.3 in (3.2 cm) deep and 1.4 in (3.5 cm) wide, also a ratio of 1 to 1.1. A longbow maker in Britain today would probably aim at a depth, above and below the handle, of 1¼ in (3.2 cm) to a width of 1⅛ in (2.9 cm) or 1 1/16 in (2.7 cm). The kinship is obvious.

Some of these Iron Age bows were self-nocked, that is having slots for the string cut into the timber at each limb-end, others had horn nocks as we have today, or iron, some of them sharp enough to use as a weapon at close quarters. From the measured depths

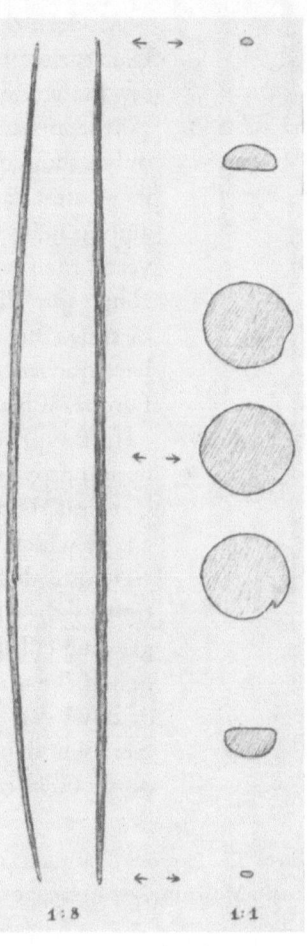

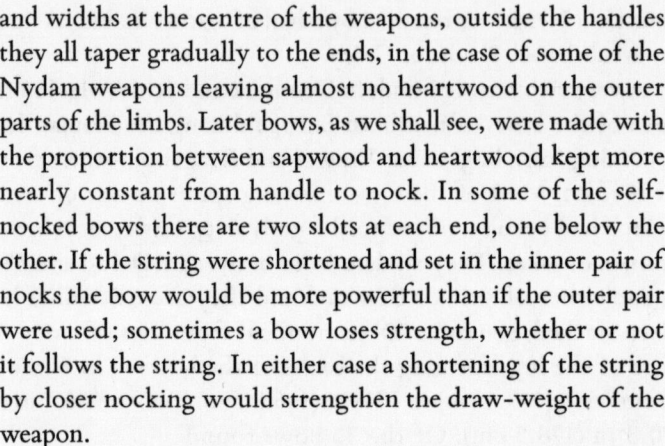

and widths at the centre of the weapons, outside the handles they all taper gradually to the ends, in the case of some of the Nydam weapons leaving almost no heartwood on the outer parts of the limbs. Later bows, as we shall see, were made with the proportion between sapwood and heartwood kept more nearly constant from handle to nock. In some of the self-nocked bows there are two slots at each end, one below the other. If the string were shortened and set in the inner pair of nocks the bow would be more powerful than if the outer pair were used; sometimes a bow loses strength, whether or not it follows the string. In either case a shortening of the string by closer nocking would strengthen the draw-weight of the weapon.

Modern though these bows seem, some of them inherit from earlier weapons the simple or criss-cross binding that we saw in the Meare Heath bow, more than 2,000 years earlier.

It seems as if the longbow has emerged, now that we are only a thousand years short of the medieval bow, to last until its greatest days; but though we can see this emergence from among many other types and through many thousands of years, the family tree, like many family trees, exhibits puzzling gaps. The longbow still keeps slipping below the surface of time. But from the total darkness of the distant past we have reached at least some certainties in northern and western Europe. What about the rest of the world's bowmen?

In Egypt some 8,000 years BC flint arrowheads of fine design prove that the bow was well established. Some heads were leaf-shaped and some were flat-headed and chisel-shaped, a type which lasted for thousands of years. Different points were introduced and the barb was developed well before the time of the Pharaohs. All this development indicates a progressing culture well capable of evolving excellent bows to match the arrows at a much earlier date than actual traces of bows can be found. In the fourth millennium BC there were wooden bows, but within 350 years there were composite bows of oryx-horn and wood, many not of Egyptian

Above *Two Egyptian bows of acacia wood, dating from 1900 BC (left) and 1400 BC (right)* (British Museum, Department of Egyptian Antiquities).

manufacture but imported from Syria, probably, and from Asia Minor. It was the composite that lasted as the more sophisticated weapon of war and of the hunt, right through to the dynasties of the Pharaohs, although the wooden bow survived as well and seems mainly to have been used by the peoples of the south. The same types of wooden bow are still used in the Sudan and by some African tribes, both simple segment longbows and doubly convex wooden bows, curving back towards the belly at the outer ends, bent by soaking in oil and forming with heat. In Egypt these bows were principally used by the mercenary Nubian troops, though their use spread among some elements of the native Egyptian armies.

There are four Egyptian longbows in the British Museum and a varied collection of arrows. The bows date from between 2,300 BC and 1,400 BC; three of them appear to be made from acacia wood, the other, which is lighter in weight and colour, from some other wood. As the illustrations show they are all doubly convex, probably shot with the outer limb curving back towards the archer, and so made, presumably, because the wood of their manufacture was not capable of the degree of tension and compression necessary to accommodate the long Egyptian arrows unless such concessions were made. Without being certain that the wood is acacia, and without knowing a good deal about the tension and compression properties of the component wood, it is impossible to guess whether the Egyptians ever used such bows with the outer curves facing away from the shooter, which, if the wood could accommodate the necessary strains, would make for a stronger and a faster bow. Representations of roughly contemporary date suggest that this was not done, but it remains an engaging possibility.

The bows vary in length from 68 in (172.7 cm) to 62⅞ in (159.7 cm), their weight in hand from 15.2 oz (431.5 gm) to 12.7 oz (360 gm). They are all leaf-shaped in section, two of them with a pronounced lateral ridge. Any estimate of their draw weight is guesswork, but they appear unlikely to have drawn at over 50 lb (23 kg).

The most remarkable thing about the six arrows examined is their extraordinary lightness. They weighed from 0.5 oz (14.5 gm) to 0.4 oz (10 gm) the heaviest being 37¼ in (94.6 cm) long, the lightest 34 in (86.4 cm). They are exquisitely made, the tiny flint heads set on to the hard foreshafts, some 8 in (20.3 cm) long, with a binding of vellum strips, the foreshafts set into the reed main shafts, bound with very fine linen thread (approximately a thousandth of an inch), gummed and painted. One has a pointed ebony head, head and foreshaft being one, and set into the shaftment in the same way as the flint-tipped arrows. The arrow nock in each case is cut very deep and always at a joint in the reed where the natural strength of the material is sufficient to take the

 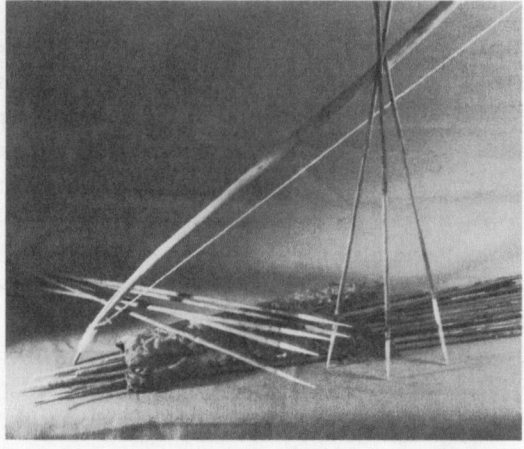

Above, left *Egyptian archers from a model in the Cairo Museum. They are carrying doubly convex longbows of the type shown in the British Museum illustrations.*

Above, right *Bow, arrows and quiver from Khozam* (British Museum, Department of Egyptian Antiquities).

thrust of the released bowstring. The fletchings are feather, about 3 in (7.6 cm) long, set on to the shaft with some sort of shellac, 0.7 in to 0.8 in (18 mm to 20 mm) from the depression of the nock.

The extraordinary effectiveness of Egyptian military archery of this period is grimly shown in the bodies of dead soldiers, killed at Neb-Hepet-Re, probably during the 11th Dynasty, in about 1,900 BC. Hardwood arrowheads had penetrated rib cages, lungs and hearts, some to a depth of 8 in (20 cm) and more, and reed shafts remained within bodies they had transfixed, both tip and shaft-end broken off outside the bodies.

We have already seen that in the Sahara area there existed simple short bows and longbows, and doubly convex bows. They lasted from about 8,000 BC through the time of the arrival of camels in the area, 2,000 years ago, and later than that. They were usually of acacia wood, and circular in section, so the idea of using sapwood for its elasticity and heartwood for its compressibility appears not to have occurred in the African continent, nor indeed elsewhere until it entered the Scandinavian and northern European scene during the first few centuries AD. Elsewhere to the east the composite bow, of many different designs, ruled, though simple bows did exist in some areas at the same time, and where there was bendable timber.

The simple bow has survived until the present day in parts of Africa south of Egypt and in south America, and in many other parts of the world almost

The Beginnings

to the present. An examination of Brazilian bows in use today shows that, quite independently of any European developments, almost every known section shape to be found in European bows can be seen in the weapons of the Brazilian Indians. They are flat, convex, concave, low-stacked and high-stacked in every degree and form. The major work in this field has been done by E. G. Heath, whose writings cover every kind of archery and whose books I recommend to anyone in search of the long history of the bow and arrow.

We can, happily, still put flesh and blood on the dry, skeletal remains of the past by looking at the weapons and methods of several groups of archers whose ancient practices seem to have survived into the world of today, none more vividly than the Liangulu and Kamba tribes of Kenya. It is impossible to prove that they shot with bows and arrows 10,000 years ago, but the links of time and usage, and the patterns of civilisation suggest that those who until a few years ago drew their longbows and killed elephant with their arrows, albeit against the law, are the true descendants of longbowmen using the same weapons, living at the time of the earliest archery finds in the north of their continent.

In *The Elephant People*, Dennis Holman has described the weapons and hunting methods of the Liangulu and the Kamba. The book was published in 1967, and by that time the forces of law and order had taken the Liangulu's bows away from them, and were busy resettling these bush people in more

Veddahs with longbows, Colombo, 1867 (Radio Times Hulton Picture Library).

respectable habits of life. Until they were overtaken by civilisation and the game wardens, they killed elephant by the hundred. They originated, it is thought, from the bush country on either side of the Galana river and spread among other tribes, so taking on the identity of their hosts as to be almost unnoticed by the census-takers and ethnologists. They moved about in small groups killing the elephant and living off its meat, until the prices commanded by ivory in the illicit export market turned them into an organised and highly skilled race of poachers.

The Kamba also killed elephant, but not with the exclusive ruthlessness of the Liangulu. The Kamba weapon was typical of the short, simple-segment wooden bows of the central African natives, about 3 ft (91 cm) long, of some 50 lb (23 kg) draw weight, and easily pulled by Game Warden Frank Woodley, in the late 'forties, when he was able to try out the bows of both tribes of poachers. Woodley was a big, powerful man, used to hunting-weapons of all kinds, but the first 5 ft 10 in (178 cm) Liangulu bow that he tried defeated him completely. When he handed it back to a Liangulu archer, the man held it with raised arms, the wrist-thick handle in his left hand, the string held in three fingers of his right hand, just as the longbow string has always been held

Above, left *Bill Woodley, Warden of the Mountain National Parks of Kenya, illustrating the difference between Kamba bow and arrows in his right hand, and the Liangulu weapons* (Dennis Holman).

Above, right *Liangulu hunter drawing the big elephant bow* (Dennis Holman, photograph by Bill Woodley).

in Europe. He brought his arms down 'with a powerful pulling and pushing action till his chest was framed between the string and the bow. Then he loosed with a clipped twang of the taut string, followed by the slamming impact of the arrow' in a tree.

That is the picture of the longbow shooter of all ages, and, apart from the raised arms, is a perfect description of how the medieval archers of Britain drew their bows. They did their pushing and pulling from a lowered arm position, raising them to the position of full draw, until they were framed by bow and bowstring, the chest forward, and almost within the drawn stave and string. That is why the action was, and is called 'shooting *in* a longbow'.

The Liangulu bows were made from any of five species of timber which afforded springy staves. They were usually slightly recurved in the upper limb, that is inclined forward in a slight curve, not manufactured to that shape, but chosen for it from the growing wood. The lower limb was usually straight, the section throughout varying from high-stacked convex with a flat back to low-stacked quasi-flat-bow shaping. The drawing weight was on average over 100 lb (45 kg). The strongest that was known to any white hunter weighed

Above, left *Liangulu hunters. The detachable foreshaft behind the barbed arrowhead can be clearly seen, and the spare bowstring on the upper limbs of the immensely powerful bows* (Kenya Information Service).

Above, right *A Brazilian bowman of the Erigbaasta tribe shooting fish* (Mme Schultz).

131 lb (59.4 kg). If we assume that the medieval war bows weighed from about 80 lb to 120 lb (36 kg to 54 kg) and over, we shall not be exaggerating.

Liangulu arrows, made from various shrub woods, were constructed in three parts, the head, foreshaft, and the main shafting. Their bowstrings, Count Ahlefeldt tells me, were made of giraffe sinew, so that for every well-strung Liangulu bow a fine giraffe was sacrificed. By the time Woodley came across them they were using metal heads adapted from nails and steel needles bought at the trading post. The Liangulu always used poison from the Acokanthera tree, smeared on the foreshaft, both it and the head being bound with protective ribbons of hide until needed for use. The poison ensured a kill, almost wherever the animal was hit, but the big bows could drive an arrow easily through the 1½ inches (3.8 cm) of an elephant's hide. The tang of the head fitted into the foreshaft, the foreshaft into the main shaft, bound with strips of sinew, applied wet, and thickly enough to form a lump or stop which prevented the main shaft driving into the animal. It would either bounce out on impact or be knocked off and recovered for later use. The length of the arrow was usually about 36 in (91.4 cm), so it deserved the name 'cloth-yard' of ancient, if uncertain fame, but arrows would be suited to the proper drawlength of each archer, fletched with vulture pinions, and individually 'crested' and marked by their owners.

The Kamba arrow, designed for a shorter, lighter bow and for fast shooting at swiftly moving targets, was a projectile of greater precision, made for brilliant marksmen, and matched in sets.

It is worth adding two descriptions. The first is from Woodley's diary, of September 1948:

'Today we found a terror trail of diarrhoea from a herd of about forty elephant. One had been shot with a poisoned arrow and we came across it after two hours. It was still alive but in terrible agony, and I ended its suffering with a bullet in the brain. This poison is horrible stuff, and the whole dirty business of poaching must be stopped.'

The other is a romantic archer's story. A Kamba poacher, Wambua, had been caught. He was a hero among the people for the legends of his marksmanship, and the warden who caught him wanted to lower his prestige a little. Hero poachers were not to his purpose. He organised an archery contest, pitting some forty other Kamba archers, whom he knew to be expert, against Wambua. The range was 50 paces, using headless arrows at a circle of five inches (12.7 cm) diameter cut in a tree. An arrow without its head is much less stable in flight than a complete arrow. The warden's excuse for this quirk was safety. Wambua was to shoot last. After all the others had shot, some very near the target but none actually within the circle, Wambua walked up to the shooting line.

'If he missed he would never be able to show his face again. He turned to the target, and a hush settled on the crowd. Then he quickly drew and loosed, and the arrow was flying through the air, and he was actually walking away, not even bothering to see where it went. There was the tight smack of the arrow striking, a sound that is characteristic, unforgettable in its sense of urgency, and for the thrilling jump it never fails to give the heart. A sudden, ecstatic cheer broke from the crowd.

'It was not just a hit, it was in the "pinhole", the exact centre of the target.'

In 1958 Gordon Grimley wrote *The Book of the Bow*. In it he said, 'The shadow of a million bows lie across the vast plains of inhabited time that preceded even the rudiments of material civilisation.' The sound of that 'tight smack', as a well shot arrow hits its mark true, echoes through time since first the sapling was bent to man's will.

CHAPTER 2

THE LONGBOW INTO BRITAIN

We know that various types of longbows were in use in many parts of the world in prehistory. We have seen that the efficient, high-stacked longbow which has survived until today in Britain exploiting the adjacent properties of sapwood and heartwood was foreshadowed, not absolutely for the first time, but for the first ascertainable time in consistent numbers, in Scandinavia, during the first three centuries AD. From there we have to trace its survival through centuries of very ill-recorded time.

The Romans were repulsed on the Rhine in AD 354 by German archery. That is all we know, but we can assume the archery to have been well organised and comparatively powerful. It seems unlikely that a Roman force would fail in the face of haphazard shooting, or of weakly shot arrows. Can we assume powerful bows? It seems so. Can we assume longbows? It seems, at the least, likely. The weapon existed not far away, and probably much more widely than actual discoveries are able to prove. Until there are more finds of later bows, dating from after the fourth and up to the tenth century AD, we can never say with absolute certainty that there was no break in the history of the longbow. The probabilities are that it did exist here and there, and side by side with weaker bows, more easily made from easier timber than the yew, but that among the very nations who used it most, in northern Europe, their fighting methods tended to push the bow down the scale of military value.

Viking laws codified in the middle of the tenth century AD, but possibly existing for a very long period before that, required that warriors be armed with axe or sword, spear and shield. The Vikings were born to fight; it was the only occupation thought worthy of the real man, and the quality of his weapons represented both his prowess and his wealth. The kind of men who in their cups would decapitate each other to see who could stand up longest after losing his head might be thought too obsessed with personal strength and bravery to think much of weapons that slew and maimed at a distance. But devoted though they were to hand-to-hand combat the bow was often among their weapons. At sea, and the Vikings were seafaring peoples, it was not always possible to come within axe or spear's length, though that was the aim of a sea battle, to grapple boat-to-boat and fight it out as on land. But in the Viking ships each man must have a bow and arrows, and some of the Sagas tell stories of their use.

At the same time as the setting down of those laws of Viking arms, other laws were recorded that regulated the number of bows and arrows to be provided by the peasants, not for their own use but for the warriors in the ships, and fines were fixed for equipment that lacked or was damaged. Later Norwegian laws, towards the end of the Viking period, regulated the weapons of free men called to arms: bow and arrows, spear, sword or battle axe, and shield. A Swedish law speaks of bows and arrows, and even uses the word 'bow' to mean a fighting man.

How long the Vikings had used their bows we cannot tell, but if, as was claimed, by the tenth century every Viking was an expert with this weapon, that it was a mighty weapon, and that from Viking hands no arrow ever missed its target, then we must admit these fierce and individual fighters did not disdain the bow much more than later archers who in close combat had recourse to dagger and maul.

Olaf Trygvason's Saga tells how Olaf, after successfully fighting Sven Tveskaeg and Olaf the Swede in a battle of longships, then had to contend with the Norwegian Erik Jarl. A terrible battle began, all manner of weapons flying and crashing among the longships. In the stern of one of Trygvason's ships, *The Long Serpent*, stood Einar Tambeskjelve with his bow, shooting harder than anyone else. He shot one arrow into the tiller by Erik Jarl's head and another struck the thwart on which the Norwegian warrior sat. In Erik's ship was an excellent archer, a Finn. 'Shoot me that giant,' cried the Norwegian. The Finn shot an arrow into the centre of Einar's bow even as he drew it, so that it shattered asunder with a mighty crack. King Olaf, hearing the noise, said 'What broke then?' Einar replied, 'Norway, King, from your grasp.' 'Not yet,' said Olaf, 'take my bow.' Einar took it and drew it, crying, 'Too weak, too weak, is our prince's bow.'

Here is a tradition of strong shooting and individual prowess with the longbow allowed high among the martial arts.

Harold Bluetooth had once blockaded the ships of Svend in a bay. In the early morning he went ashore determined to destroy Svend and all his crews. While he warmed himself at a fire on the shore, one of Svend's men, Palnatoke, longbow in hand, crept up behind, and as Harold bent over the fire to warm his chest, shot an arrow that struck Harold in the rear end, went right up his body and protruded from his mouth. So perished the Bluetooth.

The Vikings had settled in Normandy by about AD 800. Danish kings in the north of Britain ruled in York from 876 to 954, and the Isle of Man was a Norse kingdom from 820 until 1266. England and Denmark formed a single kingdom under Danish rule from 1013 to 1035. Norse customs of war and attitudes to the bow must have had considerable effect throughout these periods of cultural imposition and exchange. Scotland, the Isles and Ireland also felt the impact, and, though probably the longbow was not encouraged among subjected enemies, what is seen to be effective can be copied. Today, the use of high-velocity automatic rifles is not encouraged among guerrilla movements by the governments they seek to overthrow, but the guerrillas use such weapons nevertheless.

There were longbows in north Europe in the fourth century and longbows in Norse hands in all the territories to which their extraordinary energies drove them, during the ninth century. What about the Saxon invaders of England? When Roger Ascham, whom we shall meet again, tutor to Edward VI and to Queen Elizabeth I, wrote his treatise on archery, *Toxophilus*, he made diligent enquiry to find out when bows and arrows first came into England. He could not know about the Meare and Ashcott Heath bows, but he learned from Sir Thomas Elyot; 'That when the Saxons came first into this realm in King Vortiger's [sic] days, when they had been here a while, and at last began to fall out with the Britons, they troubled and subdued the Britons with nothing so much as with their bow and shaft, which weapon being strange, and not seen here before, was wonderful terrible unto them.'

King Vortigern was fighting in Britain in 449, and this date may well have seen the reintroduction of the bow and arrow. So that within a short period of its known use by Germanic armies in Europe it was also probably used in the British Isles by the Saxon invaders. Is it true that Roman and Celtic Britain did not know of the bow? It seems almost incredible that an island teeming with game, abounding in various timbers suitable for bowmaking, should for hundreds of years have done without bows and arrows. They must have been about, sometimes in hiding because of laws against their use, but always there, the weapons of the chase, the weapons of the common man, even when proscription made them illegal or military fashion banished them from warlike use.

An eighth-century Saxon archer defends his house with bow and arrow. Whalebone casket in the British Museum (British Museum).

About AD 490 Clovis, king of the Franks, put the Salic Laws in writing. In them there is a mention of a fine of 2,500 dinars for anyone shooting a poisoned arrow at a man, whether he hit or miss, and a fine of 54 solidi for cutting off the fingers of the hand a man would use to draw the bowstring.

In the midst of these dark, and so little recorded times, there is suddenly evidence of the utmost importance. In the eighth century in Germany longbows of yew wood were in use – eight were found at Lupfen in Wurtemburg – shaped so that the tensile sapwood was used on the back. It had been discovered again, or perhaps it was a knowledge never quite lost, that within a yew-log, rightly cut from the tree, are the natural components of a 'self-composite' weapon, the perfect natural material to resist tension, the sapwood, lying next to the perfect natural material to resist compression, the heartwood.

Not long after the Romans had gone from Britain, but before the Danes had really begun their main incursions into the country, in 633, Offrid, the son of Edwin king of Northumbria, was killed by an arrow in battle with the Welsh and the Mercians. We cannot be certain that the arrow flew from a Welsh bow, but since the inland races in Britain do not seem to have been addicts of military archery this may be the first reference in recorded history to the bow in the hands of the people who for centuries have been credited with introducing the longbow to the British islands and with being the first exponents of its general military use, which, taken up by the English and their Norman rulers some hundreds of years after Offrid's death, led to the great archer armies of the 14th and 15th centuries.

If the Welsh did not have bows yet, it cannot have been long before they adopted the weapon from the raiding Danes, but it is more than likely that, however they first came by it, the Welsh, among all the tribes in the British

Isles, either retained the use of the bow from much earlier times, or invented it for themselves long before there could have been any chance for them to have learned of its use from the Scandinavians.

By 870 the Danes were in many parts of Britain. In battle against the Anglian King Edmund, they defeated him, captured him, tied him to a tree and shot him to death. In the early 1900s the stump of the tree, which tradition said was the one to which Edmund had been tied, was torn and twisted in a violent storm. Deep in the wrenched trunk were found ancient arrowheads. Sadly, there is no report of any of them surviving today, or of their being examined for type or date.

Archery for hunting and for war was so general throughout northern Europe by the 11th century that it must be assumed to have been in considerable use in Britain throughout the Saxon period. One fact which may help to explain the lack of reference to bows in Saxon armies is that in Anglo-Saxon a single word is used to describe both arrow and throwing spear, but the song of the battle of Maldon, fought in 991, tells us 'bogan waeron bysige', bows were busy.

There is reliable evidence of Welsh archery 11 years before Hastings in the account of Ralph, Earl of Hereford, and an expedition he led into Wales. When

Martyrdom of St Edmund by Scandinavian raiders. 13th century illustration. The bow is plainly a longbow (British Museum).

the Saxon horsemen had ridden into the Welsh mountains they were ambushed by archers who shot so accurately and strongly that, according to the Abingdon Chronicle, 'the English people fled, before ever a spear had been thrown, because they were on horseback'. There was the lesson: cavalry was helpless against well-ordered archers. But this time the lesson was not learned. The Anglo-Saxon Chronicle explains that in England at that time archery was used for the killing of game, but was not much practised in battle. Had the English employed archers in their army that drew up on the ridge near Hastings, Duke William's many charges or horse and foot might never have won that battle.

In spite of the assertions in the Chronicle, it is likely the Saxon armies which gathered under King Harold's banner, 'The Fighting Man', contained some archers. The whole world knows the story of 1066. All the summer the levies had stood to arms expecting the invasion from Normandy, until men began leaving the armies to harvest the crops, and the feeling grew that nothing would happen that year. Then suddenly in the north-east, King Harald Sigurdson, and Harold of England's own brother Tostig, the traitor, invaded. On September 20 they met and defeated the Earls Edwin and Morcar at Fulford. Harold with his housecarls, and a very large force of the levies that he had managed to keep together, marched against them and met them on September 25 at Stamford Bridge. The battle was probably fought mainly between men on foot. On the English side there were archers, and horses may well have been used both for advance and pursuit. Once the Norse shieldwall had been broken, according to the *Heimskringla* (the Lives of the Norse Kings) 'the English rode upon them from all sides and threw spears and shot at them'. The Norse king fell, with an arrow through his throat.

If there were archers at Stamford Bridge, why were they so ineffective at Hastings, shooting down from the hill at William's horsemen and infantry, and William's archers below?

There are answers, which together explain the tragic lack. First, taking the Anglo-Saxon Chronicle's assertion that archery existed but was *not much* used in battle, together with the knowledge that archery helped in the victory in the north, it is almost certain that archers did not form part of the English professional army. When the levies, the 'fyrd', were called to the Standard, those that were skilled with the bow, as hunters, brought their bows with them and used them in battle, no doubt more or less well organised into battle groups and deployed to best advantage. So the archers at Stamford were levies, not only the northern levies, but from all over central and southern England. But is seems likely that the northern levies may have supplied proportionally more archers than those from further south; being in the old Danelaw areas they may well have had a stronger tradition of archery than the southerners.

Nine archers in the lower border of the Bayeux Tapestry. Proportionately, the weapons are clearly longbows.

It is incontestable that Harold marched with the major strength of the realm behind him, knowing that he must destroy the northern invasion utterly if he was not to find himself fighting both in the north and south, if, as he feared, William did invade, and there was every reason to suppose he would. Harold risked the northern throw and it succeeded. Then he needed a little time. He must have gambled on that time, hoping against hope that the early storms might deter Duke William. He lost that throw.

King Harold, in York, with an exhausted army licking its wounds, heard on October 1 or 2 of William's landing at Pevensey on September 28. He was back in London between October 7 and 11, and he drew up his battle line on Caldbec Hill on October 13. The march from York to Battle Hill was more than 250 miles. Every man who had a mount, horse or pony, followed Harold south. Few of the levies could have afforded ponies, fewer still of the archer levies, because they are likely to have been largely composed of those who, skills apart, could not afford the costlier weapons and armour. There are thirty odd archers shown in the Bayeux Tapestry, which was embroidered about 11 years after the Conquest and, of them all, only one is horsed and only one wears any defensive armour. This suggests that, even in William's invasion force consisting not only of Normans but mercenaries from all over north-western Europe, the archers were drawn from among those of humble position. Peasant bowmen, whose livestock was probably limited to oxen and donkeys, could not generally be expected to provide horses for war. So they did not come south with their king and their sorely needed help was not to be had on the day of the battle. It is one thing for mounted troops to accomplish that astonishing journey south, a feat that has been wondered at ever since. For men on foot it was another matter. More than 20 miles a day for 12 days was an impossibility.

It seems certain that on October 14 the English army stood to defend itself and the realm against the invader with very few archers. The degree to which Norman archery was decisive in their victory that day has been argued ever since. Certainly it did not win the battle alone, but bow-and-arrow men today

can be forgiven if they put a greater importance on the influence of archers at Hastings than was perhaps the case. Two questions remain. Did the Normans use the longbow as we know it? And was Harold killed by an arrow?

If the pictorial evidence of the Bayeux Tapestry is to be taken as truthful in any sense, then of the archers shown there some had longer bows than others and the longest bows were, in proportion to their users, longbows. In general the tapestry makers have taken great care to be accurate about the proportions of men and weapons, and men and horses. All the archers shown in the main central band of the work are using short bows and drawing to the chest. One of them, shown at the beginning of the battle, is a small Saxon, dwarfed by the housecarls in their ringmail byrnies. He is the only English archer, which suggests a proportion of about 30 to 1 in favour of Norman bowmen. There are 23 more archers in the lower border of that part of the tapestry representing the battle. There is no pretence of these small figures being in proportion to the main figures above them, but within the border bands elsewhere the human and animal and mythological creatures retain their own proportions accurately enough (even the lustful naked man with a maiden is proportionately possible, if unusual).

Why should the artists so explicitly have given the impression that of those 23 archers 12 seem to be using bows of their own height or thereabouts? The archers in the main tapestry look altogether smarter than the border men below them. Could it be that they were the horse archers, dismounted for battle, one of whom, on horseback again, pursues the Saxons at the end of the battle? If so, that would account for short bows. Do the border figures represent both longbowmen and shortbowmen of the Norman levies? The question must remain open. The tapestry does not give us the whole story; for instance it is certain that there were some crossbowmen among the Norman-French forces. It is possible that the bows were as mixed a bag as the bowmen, that the weapon was not at this time at all standardised, and that there were those on the Norman side who drew some kind of longbow.

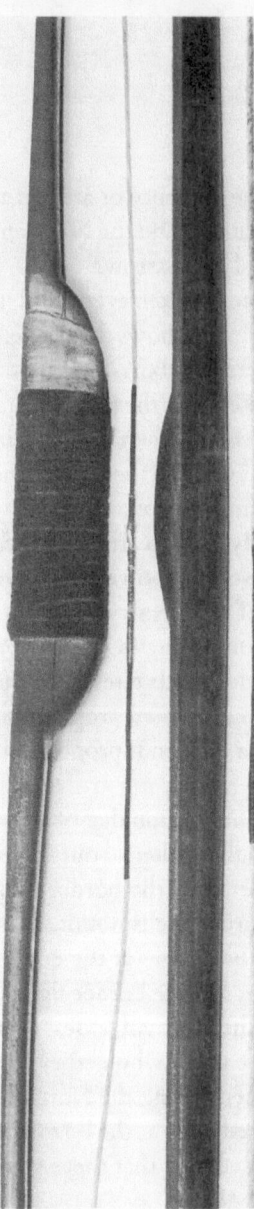

Below and opposite
Shows side and belly views of the handle section of a flat bow and a longbow (Photograph by Graham Payne).

As to the question of the archers' effective contribution, the tapestry shows Saxon shields riddled with arrows, both in the early part of the battle, and in the part depicting the death of Harold. Archer figures in the border greatly outnumber other living human figures, and are only outnumbered by the dead, the hacked and the maimed. Is that an intentional pointer to the importance of the archers' contribution?

Now, when we look at the Bayeux Tapestry we see King Harold apparently grasping the shaft of an arrow that has pierced his right eye, but experts suggest that the arrow is a later addition. Others point out that the falling figure to the right, a Norman horseman's sword slashing at his thighs, is Harold. Both figures have the moustache that Harold wears throughout the strip cartoon, and in the style of strip cartoons it seems perfectly possible that both figures represent the king, in position one, as it were, and position two. In any case, why add that arrow, at a later date? The answer can only be that it was done, if it *was* done, to fulfil a tradition. Why the tradition? Because the use of bows and arrows was always thought, whether in the actual slaying of the king or no, to have turned the battle in the Normans' favour.

The final answer must always lie in the realms of the imagination, and if anyone wants a truly imaginative understanding of that awesome day, let him read the rending description in Hope Muntz's *Golden Warrior*. There, at the battle's end, after the last arrow storm, men look to the king and see him bowed together, staying himself upon his shield. Though he fought on, and the Saxons fought on until Duke William with his last charge burst through to the Standard of the 'Fighting Man', and hacked the king to death, England fell when the king was struck by that one Norman arrow. That is the tradition. It must represent if not the fact, the basic truth of what happened.

Saxon England fell, not so much from a failure in her rule or her organisation, or from any decadence inherent in or produced by her kind of society, but from a desperate series of ill fortunes that befell her. It is permissible to think that among those ill chances was her inability to get archers from the north, back south in time. So after the defeat began the longer battle, not for the last time, for the soul of the nation.

Militarily, a curious result of the victory at Hastings was that the Conqueror seems not to have developed the archer arm, but instead, while retaining the use of cavalry, to have paid great attention to the Saxon infantry organisation which so nearly held him at bay. Many of the bowmen, both among conquerors and conquered, must have set about the business of the soil again, but it is inconceivable that a military commander whose archers had so successfully justified the training that he gave them, and who himself was an archer – 'none but Duke William could draw Duke William's bow' – should have allowed the arm to disappear from his forces altogether.

King William retained the Saxon obligations of the 'select fyrd' and the 'great fyrd' to be called on at need. The housecarls, that elite body of professionals introduced by Canute, seem never to have recovered from their near annihilation in 1066. However, the 'great fyrd', consisting of all those who by right of freedom in the old days had also the right and responsibility of defending their homelands, and the 'select fyrd', or those who, by obligation of owning more than five hides of land, could be called out when the royal armies were gathered, continued in the service of their new masters, though no doubt under very changed and very much harder conditions. A sort of compromise between the Anglo-Saxon system of customary law and free obligation, and the Norman feudal system of contracted service owed to the crown by separate charters of

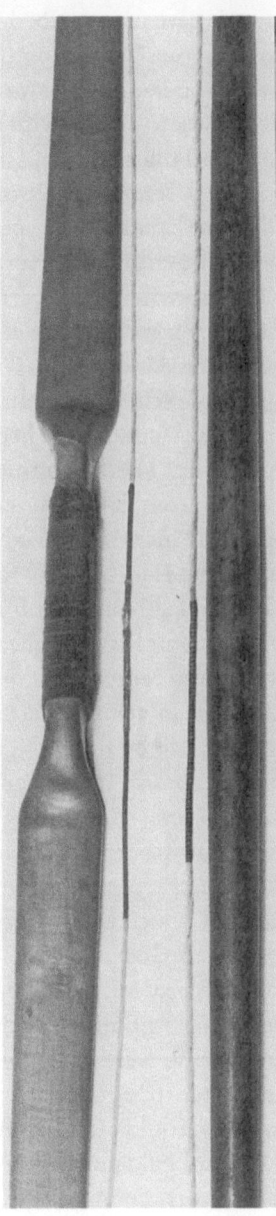

enfeoffment, began to emerge, which by the time of Edward III had developed into an extremely efficient and well-rewarded system of recruitment, owing more to the older, than the newly imposed duties of men to fight for their king and country.

Any attempt to fix the position of the bow, long or short, in Europe at the time we have now reached would be incomplete unless it paid some regard to the crossbow. On the continent of Europe the crossbow, of a simple type with a stiff wooden bow, or wood and horn bow, mounted on a wooden stock, drawn back to a catch with both hands, released by a simple trigger and shooting a short thick bolt, was superseding the simple hand bow, long or short, flat or stacked. It was capable of a good range and a flat trajectory and it required, unlike the handbow, no special skills – only the strength to get the string back upon its catch and the wit to aim.

No one will ever be certain whether William Rufus was shot dead by the quarrel of a crossbow or an arrow from a simple bow, though the evidence accumulated by Duncan Grinnell-Milne in his *Killing of William Rufus* seems to point conclusively to an arrow from some sort of longbow. Certainly the king's death could have been caused by a crossbow; there were plenty in use at the time. The Byzantine princess Anna Comnena, writing between 1118 and 1148, complained 'not only can a crossbow bolt penetrate a buckler, but a man and his armour, right through'. The weapon which was first pictured during the Han Dynasty in China in the second century BC, by which date it had probably existed for a long time (the pictures being of a developed and mechanically complicated type of crossbow), which was in Roman use by the third century AD, which according to William of Poitiers was among the Norman arms at Hastings, which Anna Comnena called 'the French weapon', had by the 12th century almost superseded the simple bows among which technical improvements seem to have ceased at the beginning of the Middle Ages.

By 1139 the crossbow had gained so much popularity as a military weapon of destructive power that it incurred the wrath of the Pope who through the Lateran Council issued an edict in that year forbidding the use of the weapon 'hateful to God', against. Christians and Catholics. Pagans were omitted from this protective ban and, among the archers of the Crusades, crossbows probably outnumbered other bows, not least because the timbers of which simple bows were made tended to weaken in the heat of the Holy Land and the greater strength of the short wooden crossbow, still more the composite crossbow, could offer a comparatively efficient shooting weapon.

Richard I did much to promote the crossbow during the Crusades in which he took part, finding the loophole of omission in the Vatican's anathema, but his death in 1199 by a crossbow bolt was taken among many to be God's vengeance on him for his wicked use of the evil instrument.

As many gamekeepers will tell you, and as the flourishing societies of the crossbow in Belgium and Italy attest, the weapon has survived until today, and more numerously than the longbow, but that is another story. We shall meet the crossbow again, but the importance of it in 12th-century Europe is that it bid fair to oust the simple bow altogether. That it did not has been thought largely due to Welshmen living 'between the upper waters of the Wye and the Bristol Channel', G. M. Trevelyan's phrase to describe the homelands of those he believed responsible for such continued and effective use of the long war bow that within a short time the crossbow was no longer at threat to the men who shot against it with the longbow, nor was it any longer regarded as an important weapon of war.

But the Welsh cannot have all the credit. The Normans continued the use of wooden bows (possibly longbows among them, as we have seen) and continued the employment of Saxon archer levies. King Stephen defeated the Scots near Northallerton in Yorkshire in 1138, using a large body of bowmen who routed the enemy just as the Scots swordsmen seemed on the point of breaking up the king's men-at-arms. Henry I issued an edict that any man who, while practising with bow and arrow, accidentally killed another man should not be indicted for murder or manslaughter. During the reign of Henry II Ireland was conquered by Anglo-Norman forces, using many archers, and led by Richard de Clare, Earl of Pembroke, the famous 'Strongbow', who was so called because he was reputed to draw the strongest bow in the kingdom. The Welsh might argue he found his skill and strength among the natives of his Earldom. That the Welsh did excel in the use of the longbow is proved by vivid contemporary accounts that have come down to us.

In 1147 Gerald de Barri, known as Giraldus Cambrensis, or Gerald the Welshman, was born at Manorbier Castle, near Tenby. He wrote: 'I am sprung from the Princes of Wales, and from the barons of the Marches, and when I see injustice in either race, I hate it.' He had set his heart on becoming Archbishop of an independent Welsh Church, an ambition he never achieved, but in 1188 he accompanied Archbishop Baldwin, who was preaching the Third Crusade, on a journey through Wales, and he wrote an account of the experience, *Itinerarium Kambriae*, noting many curiosities on his travels:

'It is worthy of note that the people of Venta (or Monmouth) are more used to war, more famous for their courage and more skilled in archery than any other Welshmen.' He goes on to describe an incident that happened in his lifetime at the siege of Abergavenny Castle in 1182.

'Two soldiers ran over a bridge to take refuge in one of the Castle towers. Welsh archers, shooting from behind them, drove their arrows into the oak door of the tower with such force that the arrowheads penetrated the wood of the door which was nearly a hand thick; and the arrows were preserved in that door as a memento.' Giraldus saw them six years after they had been so effectively shot.

Now 'a hand' is four inches, or about 100 mm. Giraldus' phrase is *'palmalis fere spissitudinis transpenetrarunt'* – 'they penetrated nearly the space of a palm'. We may take him to have described timber not less than three inches or 76 mm thick. To penetrate three inches of seasoned oak from no more than 25 yards (23 metres) requires not only a longbow with a draw weight of something like 100 lb (approximately 46 kg), but an arrowhead of a shape and temper that seems not to have been in general use by English archers until a good deal later. Giraldus then tells a story related by William de Braose, a Norman lord who unwisely tried to persuade the Welsh to forswear the use of their bows 'or other unlawful weapons'.

'One of his men, in a fight against the Welsh, was wounded by an arrow that penetrated his thigh, the casing armour on both sides' – which would have been mail armour – 'the part of the saddle known as the alva, and mortally wounded the horse. Another soldier was pinned to his saddle by an arrow through his hip and the covering armour; and when he turned his horse round, he got another arrow in his other hip; that fixed him in his saddle on both sides! What more could you expect from a ballista? Yet the bows these Welshmen use are not made of horn, or ivory, or yew, but of wild elm, and not beautifully formed or polished, quite the opposite; they are rough and lumpy, but stout and strong nonetheless, not only able to shoot an arrow a long way, but also to inflict very severe wounds at close quarters.'

It is worth looking at Giraldus' own words again in two crucial passages because translations can be misleading, and mine may be so. One translation, perhaps his own, has led a Swedish authority on the history of archery to believe that Giraldus 'states unequivocally that these Welsh bows were flat bows'. Professor Gad Rausing, whose wise guidance should be sought by any wanderer in the dark labyrinths of the history of archery, uses this, I believe false, deduction to argue that the ancestors of the medieval longbow may have been Scandinavian or may have been Norman French, and so originally Scandinavian, but were not Welsh. Giraldus says the bows were *'non formosos, non politos, immo rudes prorsus et informes, rigidos tamen et fortes ...'*. It would take the incumbents of at least two university chairs of medieval Latin to persuade me that there is anything in that description to suggest the bows were flat. Although the Vikings helped the Welsh to repel one Norman invasion of

Anglesey, when Hugh de Montgomery was shot in the eye with an arrow, here is a later Viking invasion of Welsh rights.

For a long time some writers on archery and military history have perpetuated a mistaken translation of these next words of Giraldus, which follow immediately after the passage describing Welsh bows as stout and strong: '*non tantum ad eminus missilia mittenda, sed etiam ad graves cominus ictus percutiendo tolerandos*'. Sir Richard Hoare, in 1806, following Camden, took this to mean that the bows were not such as would shoot a great distance, but that they could inflict severe wounds at close quarters. In 1866 James Dimock reflected that sometimes Hoare seemed 'scarcely versed enough in medieval Latin to be able always to understand his author', but the mistake has persisted. '*Non tantum ... sed etiam*' means 'not only ... but also'. Technically Hoare was also wrong: a bow that is strong enough to inflict frightful damage at short range must, with the rightly matched arrow, impart enough initial velocity to that arrow to send it a good distance as well.

The most important thing about Giraldus' description of those old Welsh bows is the way in which he makes it clear that by the second half of the 12th

Above, left *A sketch, in the margin of a document at the Public Record Office, of a Welsh archer drawing a broadhead arrow in a very crude short bow. One guesses that the naked foot is for better purchase in mountainous country* (Public Record Office).

Above, right *Unfortunate incident in the Welsh wars* (BBC).

century composite bows were well known in Britain if not much used ('horn or ivory'), that yew was appreciated as the finest bow timber ('nor from yew,' he says, implying that one might expect such powerful weapons to be made from the best material), and that here in Wales were tough crude bows and the bowmen to shoot them. There were all the ingredients of a new breed of bow, both mares and stallions as it were, and from the mixture came the great yew longbow.

In 1252, during Henry III's reign, the Assize of Arms of 1181 was renewed and extended. Commissioners of Array were from henceforward to choose out men to serve the king in his armies, for an agreed amount of pay. The better-off yeomen were to provide themselves with a steel cap, a buff coat of leather, a lance and a sword. Those who owned more than 40 shillings, but less than 100 shillings worth of land, must have a sword, a bow and arrows, and a dagger, as must townsmen or citizens owning more than nine and less than 40 marks of property. Yeomen or townsmen owning less than 40 shillings of land, or nine marks in chattels, must have a bow and arrows.

There was a proviso that those in the country who possessed bows and arrows could only use sharp hunting heads or broadheads if they lived outside the vast areas of Royal Forest. Those within must have blunt arrowheads or piles. The Norman French makes it quite clear, '*Arkes et setes hors de foreste, et en foreste ark et piles*'. The reason was obvious. The king's deer were for the king's people, not for the commonalty.

The moment the idea of controls upon peasant archery enters the picture, a figure can be seen fitfully among the forest trees, a figure in Lincoln green, with his companions, their longbows in their hands: the greatest outlaw of all history, but the hardest to pin down to a reality just as he was the hardest to catch in those distant days.

Two characters in Charles Lamb's play *John Woodvil* discuss the legend: ' 'Tis said that Robin Hood, an outlaw bold, with a merry crowd of hunters here did haunt, not sparing the king's venison. May one believe this antique tale? ... I have read of the tax he levied on Baron or Knight whoever passed these woods.'

Robin Hood was said to have died on December 14 1247, aged 87. That is a great age, but for Robin Hood it was merely a childhood. He is far from dead yet. He is the spirit of British resistance to harsh authority. When that fails, then Robin Hood may truly have died. He is the spirit of archery, still evoked when children see modern bowmen intent on serious target shooting, in town or country. 'Look,' they cry, 'look, Robin bloody Hood!' There can be few films more frequently shown on television than the one that plants Robin Hood firmly in the time of Claude Rains' Bad Prince John, when Errol

Flynn, the gentle Robin, finally put paid to the cruelty and avarice of Basil Rathbone's steely Guy of Gisborne. The only criticism I can allow of the film is that the outlaws use mid-20th century long American flatbows instead of true longbows, and that is a pity. If Robin Hood did not draw six foot of good English yew, then he ought to have done.

The idea of a Robin Hood character, though not the man himself, outlaw and rebel, probably originates with the Conquest, as a figure of resistance will always emerge at times of national suppression, and a figure of national hope or patriotism at times of despair. He is a good Anglo-Saxon, like Hereward the Wake, resisting the new overlords. He is first named about 1377, by William Langland who wrote *The Dream of Piers Plowman*, in which this line appears: 'Ich can rymes of Robin Hood and Ranulf, Erl of Chestre.' Ranulf has been presumed to be the great Earl of Chester who besieged Nottingham Castle in 1194 for Richard I, and Langland would thus seem to be indicating a real person who lived, like Errol Flynn, in Richard's reign. Robin appears again in verse in about 1400: 'And many men speken of Robyn Hood and shotte never in his bowe', which probably means that there are plenty of people who talk big, but who do not match words with action, and suggests that Robin was by 1400 very much a figure of legend. By 1420 Robin is rather confusingly assigned to Inglewood in Cumberland, and Barnsdale in Yorkshire. In 1424 the Parliamentary Rolls refer to a Piers Venables from Derbyshire who rescued a man from the clutches of the law and then 'assembled unto him many misdoers ... and in manere of insurrection, wente into the wodes ... like it had be Robyn-hode and his meyne'. In 1445 he is spoken of as 'the most famous cut-throat ... whom the foolish multitude are so extravagantly fond of celebrating'.

In 1495 was published the *Lytell Geste of Robyn Hode*, the first collection of ballads about the great outlaw, which give him a secret retreat in Barnsdale, and tell all sorts of tales of robbery and kindness done by Robin and Will Scathelock, and Little John and Much the Miller's son. The 'hye sheryfe of Notyngame', together with 'bysshoppes' and 'archebysshoppes' come in for a good deal of rough-housing. 'King Edward', in disguise, is captured by the outlaws but, when he bares his right arm and knocks Robin to the ground, Robin at once realises with whom he has to deal, and kneels at his feet:

'My lorde the Kynge of Engelonde,
Now I know you well.'

He is forgiven, of course, and king and outlaw ride together to Nottingham. A great deal is made in the ballads of Robin's prowess with the longbow: he splits wands, other arrow shafts and picks off deer and enemies with unerring aim, which is not surprising in a folk hero, a yeoman to boot, portrayed in poems that date from the great days of archery among the common folk of England.

John Maior, a Scottish historian and philosopher, who lived from 1469 to 1549, wrote of Robertus Hoodus, and was the source of some material for Grafton's Chronicle of Breteyne:

'About this time [the beginning of Richard I's reign] as sayth John Maior, in his chronicle of Scotland, there were many robbers and outlawes in England, among which number he especially noteth Robert Hood, whome we now call Robyn Hood, and Little John were famous theves. They continued in woodes, mountaynes and forestes, spoyling and robbing, namely such as were rich. Murders commonly they did none, except it were by the provocation of such as resisted them in their rifelynges and spoyles. [There, at least, is a tradition which English burglars have followed almost to the present day.] ... The aforesaid Robyn Hood had at his rule and commandment an hundred tall yeomen, which were mighty men and exceeding good archers ... he would suffer no woman to be oppressed, violated or otherwise abused. [The fashion there seems to have changed a good deal.] The poorer sort of people he favoured and would in no wise suffer their goodes to be touched ... but relieved and ayded them with such goodes as he got from the riche. Outlaws have been about, consistently, throughout history and the public attitude to them has always been a mixture of fear and admiration. The Great Train Robbery of 1963 bids fair to take over in popular myth where Robin Hood left off.

But happily Robin never leaves off. Wherever he hides he is pursued by scholars, both as a figure of mythology and as an historical character, and by film and television makers: friends of mine, longbowmen, are at the time of writing busy providing the necessarily astonishing archery for a new film, and this time there will be no flatbows, nothing but the real thing – and children pursue him as a darling hero.

The most persuasive of the historical character sleuths, Valentine Harris, whose book *The Truth about Robin Hood* examines with devotion and determination all possible clues, concludes that the legendary hero existed, first as a robber in Barnsdale, gradually acquiring a reputation for good deeds and remarkable skill with the longbow; that in 1324 he was to be found as a valet or yeoman of the chamber to Edward II, who was lenient over the question of his poaching and, being attracted to him, took him into his service; and that he died at Kirklees, possibly murdered by the Prioress there and one Roger de Doncaster, to be buried under Robert Hode's Stone in Barnsdale, a monument which was already known about in 1422 and is thought to have been six miles north of Doncaster. This Robin could well have been the Robin of *Piers*

Plowman as long as we allow his connection, in one line of verse, with Ranulf of Chester to have been fortuitous, and no sure indication that they lived at the same time.

The folk hero is undiminished by the possibility of a less epic man who lived and died at such a time, in such a place. In Britain now thoughts may well turn to the legendary promise that King Arthur will return in the hour of the nation's greatest need. So, too, may Robin Hood come again when there is great need for a hero to stand for all who would resist too much meddling with their lives, too much suppression of their liberty by those who wield great power.

There is always hope. There are sometimes signs. The name of the secretary of the British Long Bow Society today is Barnsdale, and he is a mighty shooter.

CHAPTER 3

FROM EDWARD I IN WALES TO EDWARD III IN FRANCE – THE LONGBOW COMES OF AGE

If the Robin Hood who was taken into the household of Edward II shot in a longbow it would not be surprising, because, in spite of the failure of English archers to stem the defeat at Bannockburn, it was Edward II's father, Edward I, who can really be called the father of the military longbow. His experience and development of the archer arm in Wales and Scotland laid the foundations for the exploitation of longbow archery and the tactical organisation of longbowmen and heavily armed infantry which so shocked the French between the 1340s and the 1420s.

Giraldus Cambrensis explained the efficiency of the south and mid-Welsh with the bow, and at the same time pointed out that the Welsh of the mountainous north, who were a very different people separated from the south by the mountains of the centre, were not archers but spearmen. Broadly these were the two types of guerilla fighters that the Norman Marcher lords attempted to quell, partly on behalf of their Norman kings, partly for their own profit and aggrandisement and in satisfaction of their lust for conflict and the exercise of power. The monarchs in England found that this was both an efficient way to tackle the subjugation of Wales, and a useful emollient to the more dangerous spirits among the baronage who tended to cause trouble at home if their attentions were not profitably engaged elsewhere. For two centuries no king set out to subdue the Welsh in any more systematic way, until Edward I succeeded his father Henry III.

When Edward started his campaign in Wales the Marches had been more or less subdued for a long time, from Chester in the north to Nether Gwent on the Severn estuary in the south and, among the Welsh who could now be counted largely as allies, the use of the Welsh longbow had spread from the south northwards until already the county Palatine of Chester and to a slightly lesser extent that of Shrewsbury (both of which William I created) had achieved a reputation as providers of strong archers which was to last into the 16th century.

Through the years there had been considerable Norman colonisation of these border areas, with the introduction of rules and rights from Normandy, gradually spreading as far south and west as Bideford and as far north as Preston. Some of them date back to the time of Edward the Confessor when a Norman colony was set up in Hereford. Now privileges were allowed to Norman settlers or natives who proved whole-hearted adherents of the new regimes. The same sort of colonisation was attempted by Henry V in France much later, taking back to the Continent ideas that had originated in Norse-occupied Normandy many hundreds of years earlier. Happily for the future of British archery the men of Gwent, those fierce and penetrating adversaries of Lord de Braose, those devastating users of the longbow noticed by Giraldus, became Edward's most reliable allies. Already, when the Norman invasions of Ireland, first under Robert FitzStephen and then under Richard de Clare, crossed the sea, south Welsh archers had gone with them, and Giraldus said that the combination of mounted lances and bows must always prove effective.

Among the Norman families who profited from the friendly assistance of Welsh archers were the de Bohuns. Henry V's mother was a de Bohun and he was born in Monmouth Castle; so, partly by his inheritance and partly from the experience he gained in border warfare as Prince of Wales, he was a natural heir to the command of the longbowmen who fought for him at Agincourt.

Edward I turned his fierce attention to Wales after the alliance of great parts of that country under Prince Llewelyn, and Llewelyn's union with Simon de Montfort and the rebel barons in England. He knew that Welsh archers of the de Montfort faction had badly galled the royalist forces as they marched through the Weald of Kent, and the battles of Lewes and Evesham, fought without the important use of archers, must have suggested to him that one way to see an end to the baronial power of mailed horsemen in the field was to develop the use of archer infantry.

There were precedents. Richard I had been among the innovators of the Third Crusade who relied on foot crossbowmen in conjunction with cavalry, but the crossbow never altogether found favour in British eyes, partly because, as in the case of Richard's archers at Arsuf, the great exponents of the crossbow were mercenaries, from Genoa and Pisa mostly, and Anglo-Norman Britain did not like mercenaries. The Pope may have condemned the weapon;

Magna Carta, for quite other reasons, condemned both the weapon and the mercenary. Edward went on crusade and must have seen the effective use of crossbowmen, and of Saracen horse archers.

At home Edward would not need the help of mercenaries if only he could organise the skills of the south Welsh bowmen which he already knew at first hand. He also knew of the English archers of Nottinghamshire and Derbyshire who in his father's reign were used, with mounted and foot crossbowmen, for garrison work and for pursuit of defeated rebels after Evesham (one of whom, appearing in yet another reign, was supposed to be Robin Hood). Edward used archers from these counties in all his Welsh campaigns, as did his famous grandson, offering pardon to outlaws if they would put their bows to the king's service. But at this time the only purely archer battalions came either from Gwent and Crickhowell, supplying 800 longbowmen in 1277, or from Macclesfield, whence 100 picked men served in the royal host that same year for the remarkably high pay of 3d a day, when the rate for mounted lances was 1s and for infantrymen 2d a day. It was Edward's genius to develop the longbow in tactical use and in individual power, starting with these 'picked men' from his own lands, and his allies across the border. The Welsh allies and the English together under his guidance learned the techniques and disciplines, gained through long and arduous practice alone, of the great longbow, drawn to the ear.

Without the sort of ruthless and intelligent leadership that gave the Conqueror victory at Hastings, the Norman feudal cavalry could have easily become unwieldy, extremely ill-disciplined and ill-adapted to any manoeuvre more complicated than the straightforward charge. Personal bravery was seldom in question but personal pride and individual courage did not contribute to organisation or adaptability. In that respect the Norman knights in mail armour on their costly, prancing destriers, their war horses, were rather like the French knights of the time of Crécy, valiant and emulous and hot-headed but not often capable of imaginative cohesion, or of the organisation that could exploit advantages, or work together to counter reverses.

Some gave their services under the feudal system, some already for pay, either occasionally or whenever they went to war, but the proper and systematic use of a structure of pay throughout the army had not yet succeeded in welding largely amateur forces into the cohesive professional armies that went to France 50 years later. Various earls with their retinues, barons and lords Marchers with theirs would turn up at the king's command, quite unable or unwilling to work to any system of subordination of command. Even after Edward had been working for 20 years at creating an organised army, the ill-timed rush of knights on the Scottish schiltrons of spearmen at Falkirk very nearly cost him the battle.

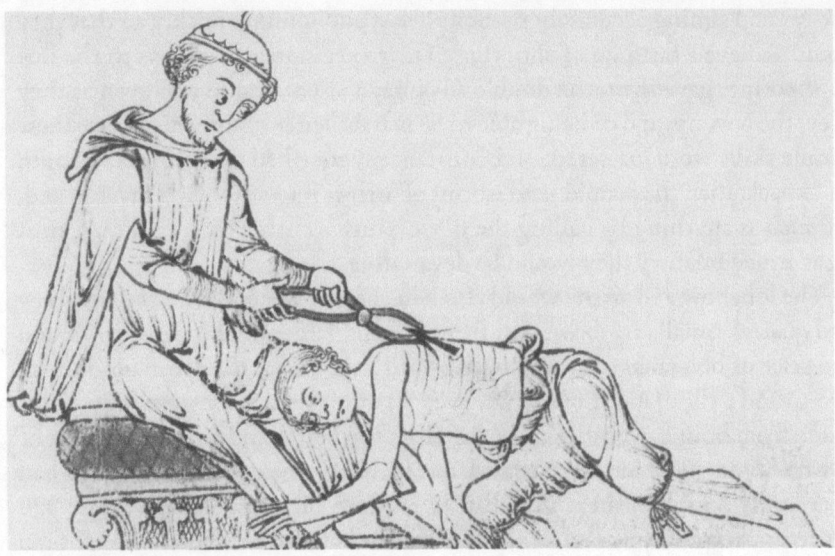

A surgeon withdraws a barbed broadhead (Trinity College, Cambrigde).

Infantry was raised where it could be had, and was seldom kept together through a whole campaign. Large numbers of infantrymen, from the better armed to the mere dagger men of the poorest groups, would be dismissed almost as soon as collected, because it was plain that such untrained and heterogeneous masses could do nothing but hinder the success of a campaign. In 1277, in one particular detail, four archers turned up with a bow and 25 arrows each, one with a bow and two arrows, one with a bow only, and one with a bow, but no bowstring.

Edward went a long way to rectify the inefficiencies of the armies he inherited and grew up with, evolving a strongly professional cavalry arm contributed by the owners of land who could afford the costly equipment and to whom horsemanship was second nature.

He decided not to try to make efficient crossbowmen of the peasants who provided the main part of his infantry, but instead, relying on the ability of the clever Welsh longbowmen to shoot arrows much more swiftly than crossbowmen with their cumbrous equipment, he determined to train his levies in the awkward but, once mastered, simple techniques of longbow shooting. He saw that the simple bow was a weapon they could take to, and that it was a weapon which could be greatly improved. As archer recruits acquired the strength, their bows could be made stronger, and the stronger they were the longer they must be made, so that they would be safe from breakage. The bowmen had to learn the knacks of fitting their arrows swiftly to the string, of drawing to the ear and aiming accurately without delay, and loosing quickly so that they could achieve a fast rate of shooting. Their very stance, sideways to the line of shooting, gave them the double advantage of coming to aim even as they drew the bow up, and of being able to be marshalled in close order. Once these simple skills were mastered, and constant practice had given them strength and knack, then they could send storms of arrows into advancing cavalry, and, as much as anything by galling the horses, turn a charge into a rout. Against light armed infantry they would be devastating.

The longbow and arrow would also be a cheaper weapon than the crossbow and quarrel. Small crossbows cost from 3s to 5s, larger ones from 5s to 7s, and the price of bolts ranged from 26s to 34s 4d a thousand, their iron heads from 14d to 16d a thousand. Longbows, even 50 or 60 years later were only 12d if made from bough-wood and 1s 6d if from the better and stronger timber of the tree bole. Two hundred years later, in 1470, when prices in general had more than doubled, there is record of a dozen branchwood bows and 120 arrows costing only 12s 4d, and in 1480 John Symson of London bought ten bows for 20s, 288 arrows (or twelve 'sheaves') for 34s 8d, a red leather quiver for 9d and a number of quiver belts for 2d each. At about the same time the price of longbows was limited by statute; the best were never to cost more than 3s 4d.

In March 1277, even for the comparatively small number of crossbows in use, at least 200,000 bolts are known to have been ordered for the king's campaign in south Wales and probably very many more were sent. Obviously it was going to prove far cheaper to equip with bows and arrows the growing number of longbowmen that the king encouraged, as well as far easier, than to train and equip a similarly increasing force of crossbowmen, especially as they were paid double the rate of longbowmen. Though the longbow still seems to have been outranged by the crossbow, the development of bows that would reverse the advantage must have begun by now for them to be in such general use 50 and 60 years later.

Another factor encouraging to the use of the longbow was that Edward in his Welsh and Scottish campaigns was able to count on calling to his standard the already skilled and powerful longbowmen of Gwent and Glamorgan, the Gower, Kidwelly, Llanstephan and Pembroke, and unfailingly they sent large numbers in response to his 'affectionate request'. It was these Welsh, almost unaided by the English, who defeated the partisans of Prince Llewelyn in the south, and in the campaigns of 1277 and 1282 they grew to outnumber the English infantry even in the north.

Welsh pay was generally on the same scale as the English. Companies were composed of 19 men at 2d a day per man with a 20th man, or 'vintenar' at

4d a day. These companies formed part of the main infantry unit of 100 men under a 'centenar' or 'constable', who was mounted and paid 1s a day or, in the case of some Welsh centenars who were considered poorly mounted, 6d or 7d a day. By 1282 some of these companies of 100 were further organised into thousands, and the whole picture of regular numbers of men and a regular chain of command, though it is by no means complete throughout the king's forces, represents an enormous advance from the largely disorganised rabble of the feudal levy, or of the fyrd that preceded it.

It is valuable to look at some representative wages and prices in England and Wales between the years 1287 and 1305, when, as we know, the crossbowman earned 4d a day, the archer 2d, the vintenar 4d, and, for instance, a 'sargeant' with a barded (armed) horse 12d.

Masons would earn from 9d to 2s a week; a repairer of crossbows 14d; an armourer 18d; a garrison kitchen boy 15d for 6 months; a plumber 3d a day, his mate 1½d a day. For a period of six months archers would be allowed 3s 3d per man for their clothes and footwear. At the same time wheat (not of the best quality) cost about ¾d a pound, oatmeal 2s for 28 pounds, and salt 16d for the same amount, though by 1300 it had risen to 3s. Wine, in 1300, was 2½d a gallon and bacons in 1305 were 9½d each. An ox carcase cost 3s 8d, a pig 2s 8d. Enough fresh and salt meat, cheese, eggs, butter, meal and salt, bread and ale for a garrison of 51 men for a year (1299) cost £24 5s 1½d, or roughly 10s per man. During that same year an archer in that garrison earned about 60s.

Between 1300 and 1305, arrows were two for ¼d, lances were 6d each, and helmets (bascinets) 2s 2½d; 20 ells of canvas (an ell being 45 inches) such as would be used for bowcases cost 6s 8d. Empty casks, like those used to store arrows, were 8d each. Sea coal was about 1d a bushel. As a footnote to this short survey, the London Carpenters' Guild made the following provision in 1333, which was representative of the efforts made by many guilds and companies of the period: 'If any brother or sister fall into poverty by God's hand or in sickness ... so that he may not keep himself, then he shall have of the brotherhood each week 14d ... after he hath lain sick a fortnight.'

The king demanded service from his Welsh allies without pay, so long as they were in their own county as a guarantee, perhaps, of good faith, but once over their county boundary they received money. Under certain special circumstances, such as a siege, they might receive pay even in their own counties after three days' statutory service without pay. Edward seems to have been the first to introduce a general identifying piece of 'uniform' into his armies, as distinct from the badges or liveries that local commanders occasionally provided hitherto. He ordered arm pieces of cloth for his infantry, bearing the

Longbowmen practising at the butts, from the Luttrell Psalter (British Museum).

cross of St George. There is no record of any Welsh reaction to this issue of an alien identification.

In creating an efficient army, one of Edward's chief difficulties was undoubtedly the unwarlike attitude of the average Englishman of the countryside. It is noticeable that the contributions of infantry actually used in the wars came mainly from the shires closest to the Welsh Marches and the scenes of fighting, and, as we saw, from the forest areas of Nottingham and Derby, where outlawry was as vigorous a training ground for war as the field of battle itself.

Edward plainly spent much time and trouble calling in the levies from elsewhere, sometimes in large numbers, hoping to find among them the raw material for training to the use of the bow and the disciplines of his new model army. He ordered 25,000 foot from the counties between Norfolk and Dorset to be at Winchelsea in 1295 to serve as bowmen and crossbowmen, knowing well enough that only a tiny proportion of them would have much aptitude for the weapons, and in that same year and the preceding one he offered pardons to poachers and to criminals, each crime and pardon being specified, in return for army service. He also ordered his captains and officers to treat the levies with courtesy and leniency, and though this tactic was not always successful – in 1299 and 1300 large numbers of men from Yorkshire, Durham, Derbyshire and Shropshire deserted after receiving pay and were punished for it – it marks the beginning of a more humane and more efficient system of recruitment which was to give the lie to the general opinion abroad that the Englishman was a poor soldier, the English army a rabble.

When Edward's grandson had unleashed the full power of his longbowmen on the astonished French at Crécy, the chronicler Jean Le Bel, looking back, was able to write:

'When the noble King Edward, still a youth, first reduced his kingdom to obedience, there was scant regard abroad for the English and no one, even after the wars in Scotland, spoke of their valour or their courage. They wore no plate-armour, no bascinets or beavers or gorgets, but heavy hauberks, and long surcoats emblazoned with their heraldry, gloves of padded cotton, and helmets of iron or hardened leather. But since the noble King Edward began to train them, the English have proved such apt pupils in the school of war as to be now the noblest, the ablest and most stylish warriors you can find.'

The grandfather of the victor of Crécy had struggled to create the groundwork of a national army and a national warlike spirit on which his grandson so notably and so successfully built.

Longbowmen in the armies of Edward I began to be counted in thousands not in hundreds, though the elite, like the bowmen of Gwent and Macclesfield with their high rates of pay, were still the small nucleus of the developing, learning archer corps. At the battle of Falkirk in 1298 the greater part of the archer corps was Welsh. By 1346 the proportion was reversed. It was Edward I who was largely responsible for the reversal, which marked the taking up by a nation in arms of what had been really no more than a local weapon among a small part of another, more warlike people.

By the standards of the time, the archers were remarkably well treated. They were paid regularly, every six days, sometimes every three days. Some were impressed; so were some of the workmen to clear forest or make roads, in cases where volunteers were not forthcoming, and there are instances of cavalry being detailed to guard against conscript desertions, but all were paid, whether conscripts or volunteers. There were bonus rewards too. There were extra payments, for instance, of 10s each to certain Shropshire centenars, an extra half day's pay to 450 Salopian archers, 40s to be divided among wounded archers from Morgannwg, an extra day's pay to 250 other bowmen, and so on. A detailed budget for the war of 1277/8 exists in the Pipe Rolls for the reign and can be studied today. From the total expenditure of £23, 149, crossbowmen, archers and spearmen, paid between July 18 and November 10, received £4,762, and other payments to archers of various categories, not including gifts and special rewards or payments to archers on garrison duties, amounted to at least another £400. When the king could not raise enough immediate money from taxes and customs, he borrowed, mostly from Italian bankers and merchants, but he always paid.

Although an absolutely clear picture of the financial arrangements of the armies of Edward I is difficult to achieve, because they were changing all the time, it is clear that Edward was moving towards an army that should be paid, from earl to spearman, and that his innovations were, in some cases, against the wishes of those lords who were determined to render their feudal duties without pay in order to preserve their independence, which they felt they would forfeit by becoming, as it were, mercenaries of the crown. But there were enough lords and barons ready to accept the king's money to make the gradual abandonment of the feudal idea inevitable.

By the time of the second Welsh war of 1282/3 archers are to be found acting as marines on shipboard, together with crossbowmen, and receiving the same pay of 3d a day. There were 350 of them distributed among 28 of the king's ships gathered at Rhuddlan in July, who, after their first three days' pay as soldiers, drew their wages from the navy.

Among the infantry at the king's northern headquarters, which varied in total during the months from 3,360 to over 8,000 men, the number of archers present was slowly rising. There were 1,000 encamped at Rhuddlan as well as 1,800 Welsh from the Marches, the greater part undoubtedly archers, and the proportion of crossbowmen was all the time dwindling. At Rhuddlan there were only 12. When Prince Llewelyn was surprised and killed at the battle of Orewin Bridge near Builth, the king's army was described as being a mixture of cavalry and archers.

During this campaign, Edward called to his service soldiers from Gascony, men whose grandsons would join in the campaigns of his grandson, and of his great-grandson the Black Prince half a century later. By January 1283 there were over 1,400 Gascons, including no archers, but 625 crossbowmen who brought with them 70,000 bolts, well over 100 per man, a figure worth remembering. The longbow was not a weapon of the French, but the Gascons saw to it that even the slow-shooting crossbow was very well supplied with ammunition.

The war in Wales once over and the country largely at peace, the Statute of Winchester was passed in 1285. This allowed for the regular inspection of the forces available in counties and the commissioning of arrayers to marshal the infantry. It emphasised the duty of every man between 15 and 60 years old 'to have in his house harness for to keep the peace'. The importance of the bow is, surprisingly, not much enhanced in the statute, which leads one to guess that Edward and his advisers were rather concentrating on an elite corps to use the big bow, a weapon newly come of age, beyond the handling of the ordinary levies, and therefore not specifically mentioned in the general setting out of men's duties.

When the war flared again in 1287, 11,000 infantry were on the march in August, of which only about 3,700 were English, and they came mostly from Chester, Shropshire, Herefordshire and Derbyshire. The rest were Welsh, 2,600 of whom came from Snowdonia and the cantreds of the north, and were probably not archers. There were a mere 105 crossbowmen, and the areas from which the large majority of both Welsh and English were drawn strongly suggest that

From Edward I in Wales to Edward III in France – the Longbow Comes of Age

archers represented a high proportion of the total. The battle at Conway, during the last Welsh rising of 1295, is one indication among many that the use of the weapon was spreading among the purely English forces. At Conway it seems there were no Welsh bowmen present on the English side, and the English force, being outnumbered, and failing in its opening push of cavalry, won the victory only when the mounted men were helped by the archers, who, 'intermingled with the horse', made havoc of the Welshmen's defensive stand.

In 1298 the battle took place which has always been regarded as the first classic victory for the longbow, though there have already been indications of its growing importance. Edward was campaigning in France when the Scots defeated an English force at Stirling in September 1297, and started to harry and devastate the border counties. By March 1298 he was home, and assembling an army to march against Wallace, numbering some 2,500 horse and 10,000 or 12,000 foot, which he led north in June. Wallace had the smaller

Assault on a castle. Well-armoured men using longbows (British Museum).

mounted force, his army being largely composed of infantry with long spears, about 500 cavalry and a company of archers from Ettrick. The Scots' flanks were protected by palisades of wood and their front by marshy ground. Here, near Falkirk, they stood, in four great 'schiltrons', hedgehogs of spearmen, with the Ettrick archers in between each group, and the cavalry force deployed ready to charge at any advantage.

Against them Edward had a large number of infantry including many Welsh and English bowmen, and probably 2,000 cavalry. Of the total infantry it is reckoned that five-sixths were Welsh, and, again judging by the areas they came from, it seems likely that more than a third of that Welsh contribution would have been bowmen. The English cavalry advanced in three bodies. Edward, leading the third, rode in some pain because he had been trodden on by his own or someone else's horse during the night, and had three broken ribs. Finding themselves bogged down in the marshy ground of Darnrig Moss, the first troop swung left towards the Scots right flank, but before they were on firm ground the Scots archers had them in range and caused great distress among them. The second English troop sought the firm ground to their right and charged the Scots left in sufficient strength to rout the greater part of the enemy cavalry, who streamed from the field. But the Scots infantry stood firm, and against the schiltrons of spearmen, the milling English horse could do nothing. Though they charged whenever they could regroup and though they achieved the breaking up of the Scots archers, most of whom were killed, wounded or driven from the field, they could make no impression on the masses of steady pointing spears.

Edward now came on, skirting the marsh, with the third troop of horse, the archers, and the rest of the infantry. The schiltrons now stood unprotected by their own bowmen, and not much protected by their scant armour, so that, when the Welsh and English archers were ordered right up to them and began massed shooting into their packed ranks, there was nothing the spearmen could do. They stood their ground until the arrows had so torn and thinned their formations that the English cavalry could plunge in among them. That was the end. Well-ordered bowmen with nothing to impede their shooting completely shattered a brave resistance that had held impetuous and repeated charges of cavalry at bay.

Earlier use of the archers on the English side, whose numbers were much greater than their opponents, and the range of whose bows was almost certainly greater too, would have opened up the Scots defence to cavalry attack after a few moments. But the lesson had not yet been learned that it was wise to restrain the hot courage of the mounted knights and allow men of the humbler sort to start the attack with their bows and their swarms of stinging arrows.

It seems probable, from all the figures that are available, that towards the end of his life, a life which, militarily, had accomplished the greater part of all he set out to achieve, the conquest of Wales and of Scotland and the restoration of his power in Gascony, the armies of King Edward I would go into the field with a proportion of five to one in favour of the infantry. The heavy cavalry of the past remained almost unchanged, but there was added to it now a force of light horsemen, called 'hobelars', which began to answer the need for greater manoeuvrability in the mounted arm. The proportions of archers to other infantry is harder to guess at, but that under Edward's direction it was rising cannot be in doubt.

Much that he achieved, his son lost after his death. The Scots' determination and the barons' insubordination outweighed Edward II's resolution and military skill, but the foundation of his father's reorganisation of the armies of Britain was well enough laid to last through many reverses, until Edward III, with as clear an eye and as hard a head as his grandfather, built on those foundations, true heir to all the improvements that had been laboriously achieved before him.

Robert Bruce in Scotland led his nation to overthrow the conquest of Edward I and the battle which set the seal on his steady successes in recapturing the castles taken by the English in the previous reign bears a name that has lingered in history among the most renowned, as a day of glory for Scotsmen and shame for Englishmen. Close to the scene of Falkirk, Stirling again being the point at issue, the armies of Edward II and Robert Bruce met near the Bannock Burn, two and a half miles south of Stirling. The Scots had mustered a large army, the English perhaps double the Scots number.

The position chosen by the Bruce was similar to Falkirk, his mustering place the very same that Wallace had chosen 16 years before. Marsh protected one flank (the burn ran across his front) and he had pot-holes dug, and camouflaged, beyond the burn, to upset any charge of English cavalry. He had fought on the English side at Falkirk and knew what charging cavalry could do if his schiltrons of spearmen were ever to be opened up by Welsh and English archers. During the afternoon of June 23, a Sunday, the English moved up to face him, their vanguard making contact with his. Sir Humphrey de Bohun leading the first English squadron of cavalry suddenly saw ahead the Scots king on a pony. He spurred straight for him, his lance point aimed at the Bruce's heart. Bruce galloped forward to meet him, swerving his pony just in time to avoid the lance and, standing in his stirrups as Sir Humphrey thundered past on his warhorse, split his head open with one blow of a massive battle axe. All the Scots king is reputed to have said is, 'Ach, I have broken a good battle axe!' The fighting became general, but disordered, and the English retired, much beaten about. Without archers, who had not yet come up, they could make no impression on the Scottish spearmen.

By Monday morning the English had moved round behind the Scots and were slowly getting into battle positions. The Scots faced about on this new front, advanced beyond the Bannock Burn and, as soon as they could see their enemy clearly, wisely started to attack before the English dispositions were firm. King Edward's infantry was still behind nine squadrons of cavalry, and with the infantry, naturally enough, were the archers. They had originally been intended as a screen in front of the cavalry, but so far none of this front line had reached their positions.

Four great masses, or as they were beginning to be called 'battles', of Scots spearmen moved against the unready English cavalry. The archers were hastily collected and brought forward as a right flank ahead of the main line of the English to shoot into the flanks of the schiltrons, whose advance into the English was in echelon, those further away from the archers engaging earliest. The bowmen first encountered Scottish archers against them but the English and Welsh outnumbered and outshot them and were soon able to concentrate on the masses of spearmen. If the bow work could have been well enough organised, the steady Scots advance might have been halted or confused, the English given time to put themselves in order, and Falkirk might have been repeated. But Bruce had learned more lessons than one with the English, and he now threw a cavalry reserve round the wet ground on his left. They swung right-handed into the flank of the archers and stove them in. Those that were not ridden down ran back among the infantry and could only shoot over their own army at an enemy they could not see and who by now were inextricably entangled with the English. They shot as many arrows into their own men as into the Scots. Meanwhile the Scottish archers had been re-formed and were able to add effectively to the slaughter of the English cavalry.

The defeat was complete. The lessons were three, and included the often repeated but never easily implemented certainty that organisation and discipline are all, and become progressively more difficult to enforce the larger the numbers concerned. The other two lessons were to do with army archers. Firstly, cavalry without archers were useless against well-controlled spearmen and secondly, archers without support, out on their own, succumbed to cavalry unless the bowmen were free to change front and face their attackers.

Bannockburn was the final necessary lesson to a nation that had begun to use the bow and arrow more importantly in warfare than any other since the Parthians defeated the Romans at Carrhae in 53 BC, bringing in their baggage train 1,000 camels carrying spare arrows.

It is not right to imagine that Edward II neglected the longbow. During his reign, and while England was largely on the defensive, the proportions of the forces called out was ten archers to one mounted man-at-arms, and two archers to each hobelar, or light horseman. Moreover it appears to be an innovation of his that counties should pay for archers' armour, slight though this may have been compared to that worn by the men-at-arms. Wages were announced publicly; pardons were offered to criminals (there were 215 granted in December 1324 before the Gascony campaign of the following year); and the recorded numbers of archers called up from counties were greater in proportion than in the previous reign. The attitudes and behaviour of men who made up at least part of the archer bands, and the rather slow and parsimonious methods of payment during this reign, are shown by the fact that, in face of large-scale desertions and looting in the countryside round the embarkation camp at Portsmouth in 1325, large numbers of infantry were sent on board ship and forbidden to land. No doubt the ships in question were put out beyond swimming distance.

Naturally, as the realm slipped into chaos and civil war, and the king showed himself less and less able to cope with the worsening situation, the organisation of the armies slackened, abuses abounded and the development of the archer corps suffered.

In 1327 the wretched, tortured Edward shrieked his last moments away under the repulsive inflictions of his executioners in Berkeley Castle and his 14-year-old son Edward, the third of the name, came to the throne.

Although two years later King Robert of Scotland died and some of the pressure went out of the Scottish cause, the English engaged in some quite fruitless campaigning in the north. But Edward soon began to look most carefully at military matters. Meanwhile, in 1332 a new invasion force against the Scots was mounted by the nobles, both Scots and English, who had been, they claimed, 'disinherited' from their lands north of the border by the success of Scottish arms. They got together 500 knights and men-at-arms and about 1,500 archers and, being forbidden by the 19-year-old Edward, loyal to the Treaty of Northampton which concluded the last hostilities, to cross the Tweed, took ship and landed in Scotland from the sea. They marched on Perth, but found a large army, perhaps 10,000 strong, under the Regent, the Earl of Mar, waiting for them on Dupplin Muir.

Mar was cautious, and the English held off attack until after dark when they raided the Scots camp and killed a number of infantry. In the early light the Scots main body advanced, all dismounted. The English drew back to the top of a slope, where they too dismounted and formed themselves into a single 'battle' of men-at-arms, flanked by long and irregular wings of archers, angled forward from the centre battle. The Scots attack was made *en masse* at the English centre, which was pushed back at the first shock. In the press there was little chance of cut or thrust and the conflict became a sort of vast heaving scrum, the English with the advantage of the slope gradually stemming the Scots advance and holding it. The flank archers closed in on the sides of the Scots masses and shot without pause until they gradually drove the Scots inwards on themselves. The result was a horrifying crush of men, helplessly trying to keep their feet. The arrows continued to pour into them; the crush became more frightful, very many were suffocated; and the pile of dead began to rise in a central mound until, the Lanercost Chronicle tells us, it was as high from the ground as the length of a spear.

Those in the rear of the Scots began to break and run and, as the rout became general, the English leapt to horse and pursued the fleeing. The result was a massacre. Only 14 Scottish knights escaped, and with the Regent and 76 knights died 1,200 men-at-arms and an uncounted mass of infantry. In the dreadful mound at the centre not a single man was found alive. The English lost 33 knights and men-at-arms and not one archer. No Scot had been able to reach them.

Edward III was not at that battle, but when he faced the Scots himself in July the following year he had with him Baliol, Umphraville, Beaumont and Atholl, the victors of Dupplin who had planned the tactics for the invincible use of dismounted men-at-arms and flanking archers. At Halidon Hill the Dupplin formation was developed, and made still more effective.

Edward, besieging Berwick, was attacked by a strong Scottish army. He took position on a hillside called Halidon Hill, his back to its wooded crest. Instead of a single central 'battle' of dismounted men-at-arms with flanking archers, he had three battles, each with flanking archers. If three such Dupplin formations are put together, side by side, the forward horns of archers on both flanks of the central, and each inward flank of the outer divisions, join together and form a sort of triangular wedge. Thus the whole formation presents the following appearance: left flank archers inclining forward of the left-hand division, whose right flank archers form half the triangle with the left flank archers of the central division. The right flank archers of that same central division form half the triangle with the left flank archers of the right-hand division. The flank archers of the right division incline forward, balancing those on the far left. Thus, each battle of men-at-arms forms, with its flanking archers, a re-entrant into which the attacking enemy must ride or run, exposed all the time to an increasingly close and murderous arrow storm, first at their front and then at their flanks. This was the English method of pitched battle that lasted for more than

From Edward I in Wales to Edward III in France – the Longbow Comes of Age

100 years, with devastating success. We have seen it developing. Now it has been achieved.

At Halidon Hill, as so often it would be in the future, the English force was smaller and handier than the army that attacked it. But for such tactics wholly to succeed the enemy must be made to attack. In this case the Scots *did* attack. They came on up the hill in four great masses. Few of them ever reached the English main 'battles'. They were simply shot down by the archers.

The splitting of one main battle into three meant that effectively the whole front could be covered by the bowmen, each central wedge of archers being able to shoot over a semi-circle with a radius as great as the range of their longbows. The outside flanking bowmen could cover more than a quarter of a circle. The importance of the two fights at Dupplin and Halidon cannot be overemphasised. Before them the weapon had been to hand, but usually in support of cavalry. Before them, often, the medieval commanders' preferred formation for conflict had been in three divisions or battles, usually one behind the other. The Halidon formation and its adaptation of the Dupplin flankers was entirely new. The situation at Dupplin, the need for a small force to stand and fight, and not to ride about the battlefield pursuing individual victories and personal glories, forced the use of a tactic which in a mere 12 months had been tripled in its effectiveness. We shall never know whether Edward said at Halidon to the victors of Dupplin, 'We will do today as you did last August, but we will do it in three instead of in one', or whether they said to him, 'Sire, what we did in August was effective but if you were to triple our formation we think it would work even better.' However, there is no doubt that with the names of Edward I and III, and Henry V, the historian of the longbow should greatly honour too the names of Edward Baliol, Gilbert Umphraville, Henry de Beaumont and David of Atholl. We shall see the method working in some detail, as it was used at Crécy, in the next chapter, but first we should look at the bow itself, and the bowman.

The longbow used by Edward III's archers was, at its best, made from local yew wood, at its very best, from imported yew. There were other woods available, but none to touch the yew. Brasell, or Brazil wood, was exported from India via Persia and so to Europe throughout the Middle Ages; thence eventually the name reached South America and the country of Brazil was called after the tree, not the other way around. There was elm, wych elm, which Cambrensis told us the Welsh used and which was probably the best of the secondary bow timbers, and ash.

Roger Ascham, writing in 1571, spoke with knowledge of the past great days of shooting when he said: 'As for Brasell, Elme, Wych and Ashe, experience doth prove them to be but mean for bows, and so to conclude, Ewe of all other things is that whereof perfect shooting would have a bow made.'

 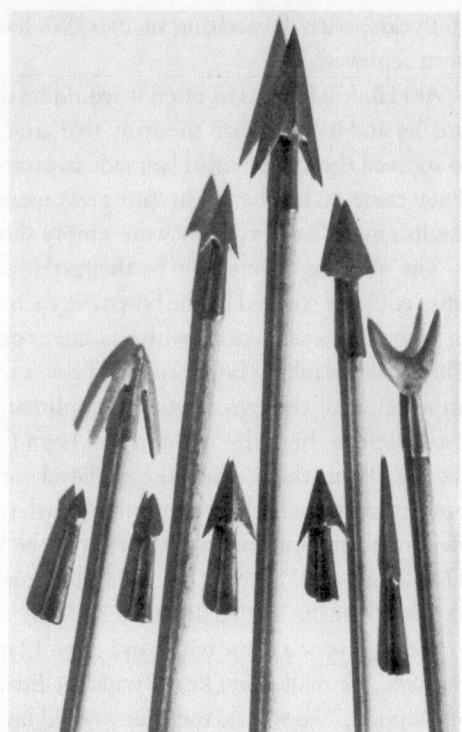

Above, left *John Gower, 15th-century poet, shooting a longbow* (British Museum).

Above, right *Modern reproductions of medieval arrowheads (the heads made by Ray Monnery, the shafts and fletchings by John Waller of the Medieval Society)* (Photograph by Graham Payne).

Of arrows, he said they were commonly made from 15 kinds of wood, including brazil, hardbeam (or hornbeam), birch, ash, oak, blackthorn and beech. Writing at a time when arrows were still made for the archers in the armies, as we shall see, he said: 'As concerning sheaf arrows for war, (as I suppose) it were better to make them of good Ash, and not of Asp (willow), as they be now a days. For of all other woods that ever I proved, Ash being big is swiftest, and again heavy to give a great stripe withal, which Asp shall not do. What heaviness doth in a stripe every man by experience can tell, therefore Ash being both swifter and heavier ... is much the best wood for shafts.' By a 'great stripe' he means a hard blow. The heavier the arrowshaft, consonant with good range, the greater the inertia when the arrow hits its target, and so the better the penetration.

There is one arrow remaining in England which is likely to be a medieval arrow. It is the property of the Dean and Chapter of Westminster, and was found in 1878, lodged on a turret of the Chantry Chapel of Henry V in Westminster Abbey. For years it has been thought to have been used at some period to scare pigeons in the Abbey roof, but it seems hardly the right missile for so vulnerable a target. It is shafted almost certainly of ash; its head is one of the commoner types of arrowheads surviving from the late medieval period (type 16 on the London Museum chart) being of a four-sided lozenge section with two very small barbs; it once had long fletchings of the type commonly in use at that period; and finally its 'spine' or bendability suggests, though it does not prove, that it would match a bow of enormous draw weight, in the 150 lb (68 kg) range. It suggests such a bow because according to spine and bow weight tables (see technical appendix) that is the weight of bow from which the arrow would fly best. It does not prove it because it *could* be shot from less heavy bows. But the medieval bow-and-arrow makers knew what they were about and I am convinced that, in the days of the great war bow, 150 lb draw weight was not out of the question. The great American longbow maker and hunter, Howard Hill, who died in 1974, habitually used bows of well over 100 lb draw weight, one weighing 172 lb (78 kg).

I believe we cannot be far wrong if we say that by the reign of Edward III his archers would have used longbows of 80 lb to 160 lb draw weight, achieving ranges up to 300 yards with the smaller-headed arrows, and progressively less in proportion to the increased size and weight of other types of arrowheads*. The use of such formidable bows could only be mastered by continuous practice, and we have ample evidence that the authorities were constantly encouraging such practice. The complaints often made, after the Hundred Years' War, of the decline of military archery, that 'our strong shooting is laid in bed', suggest that the ability to use such bows was only maintained with difficulty and easily lost. The complaints start at a time when war bows were still in constant use, but when evidently the great bows and the great archers were beginning to belong to the past.

So we must think of the bowmen of the early 14th century as having achieved a strength and skill in shooting the heavy longbow that very few men nowadays could easily attain. We know from the results of battle after battle that these heavy bows were superbly made, and the arrowheads adapted to pierce gradually improving designs and materials used by the armourers to resist them. By the end of the 13th century the big broadhead arrow was no longer used in war, though possibly the small barbed broadhead was still useful against cavalry, but in general these had been superseded by variants of the bodkin head, a three- or four-sided square or lozenge-shaped point, sometimes barbed but often quite plain, which tests have shown quite capable of piercing even the plate armour of the 14th and early 15th centuries. They varied in length from an unusual long bodkin of about 5 in (12.7 cm) weighing 1¾ oz (49.6 gm) to a more common 2 in (5.18 cm) head, weighing ½ oz (14.2 gm) or less. The heavier the head, within these limits, the shorter the range of the arrow.

Since no verifiable bows remain from the medieval period it is impossible to say with certainty that they were always made from a single stave of wood, and never jointed in the mid-section. But such bows as do exist, for instance the staves in the Tower of London, which were recovered from the sunk wreck of the *Mary Rose* and date from 1545, are simple staves. Why are there no survivors from the thousands and thousands that were made and used during the Hundred Years' War?

There are two main reasons. They were common and familiar things, which were not hung up as trophies on the tombs of the great, nor buried with the ordinary men who were so skilled in their use. Secondly, the useful life of a wooden bow is not very long. After some thousands of shots, the number of shots varying enormously with the quality of the timber and of the bowmaking, wooden bows will weaken or break. Also rough treatment, especially of yew wood, will produce bruises and cracks which develop into weak spots and points of breakage. Once a bow was no longer serviceable it would be thrown away or used as firewood; there could be no possible point in preserving it.

There is a bow belonging to the Spencer-Stanhope family, which at one time was on exhibition in the Wakefield City Museum. It is made of yew, with the proportion of sapwood along the back, and the high 'Roman arch' section, of the true longbow. It is 6 ft 7 in (2 metres) long, with a girth of 5¼ in (13.3 cm) at the handle, tapering to 1¾ in (4.4 cm) at each tip. The tips are broken off, and it apparently once had horn nocks. Since if it were drawn up it would certainly break, its weight can only be estimated, but it seems likely to have pulled a good deal more than 100 lb (45.4 kg) at 28 in (71.1 cm), and more again at the 30 in to 36 in draw that its enormous length would certainly have allowed. If it were possible to date this weapon by the Carbon 14 method, together with the Westminster arrow, and if they were found to be 14th or 15th century survivors, then several more definite conclusions could be reached.

At least we know how a medieval bow was made. The weaker staves would be cut from branches of yew, the heavier ones from the bole of the tree. The logs would be split and seasoned, the length of seasoning probably depending more on supply and demand than always on the wisest practice. Through the years of seasoning (three or four is ideal) a little work would be done towards reducing the log to a slim stave, the outer sapwood beneath the bark of the tree being carefully preserved. When the bowyer thought the wood seasoned enough he would shape it to a 'D' section, flatter or deeper according to his

From Edward I in Wales to Edward III in France – the Longbow Comes of Age

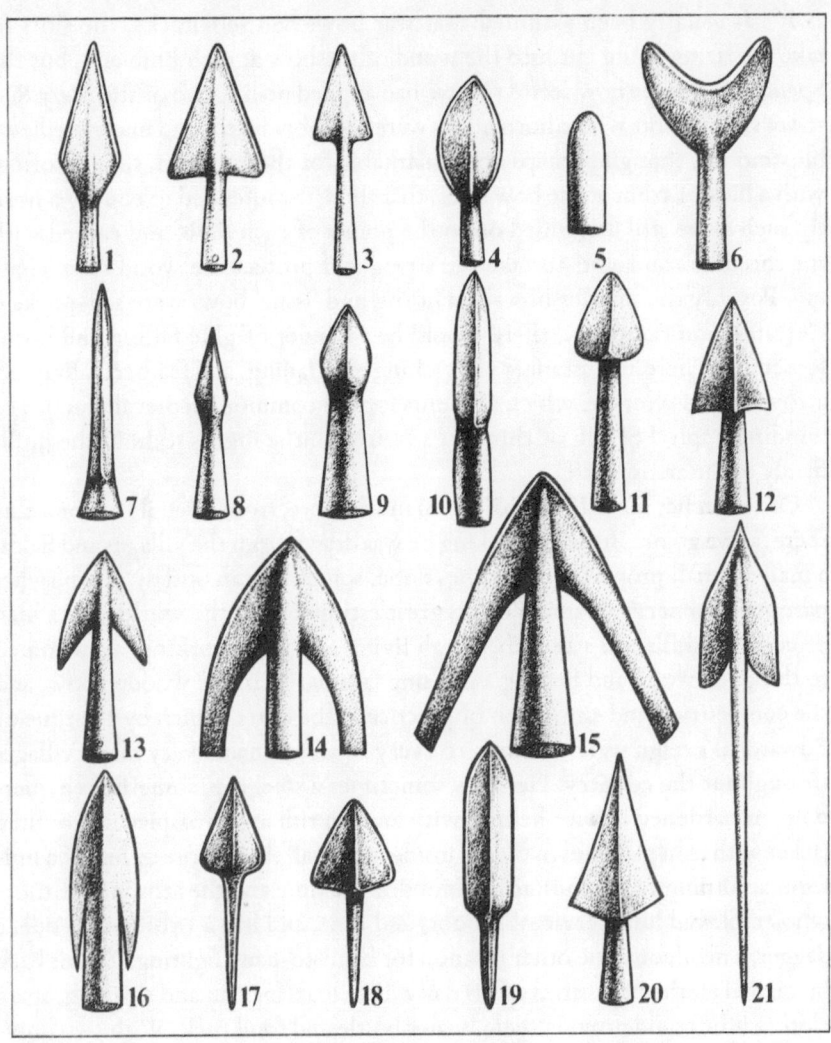

Medieval arrowheads, typed and numbered according to the London Museum catalogue. Numbers 7, 8, 9 and 10 represent bodkins; number 16 appears to be the most common medieval type, comparatively light and effective at long distance (Drawing by Clifford Anscombe).

inclination, and taper it, as regularly as the grain and the knots or pins in the timber allowed, from the handle to the tips, so that when drawn it would describe a semi-circle with a slightly flattened handle section. A bow that bends in the centre as well as throughout its length will jump in the handle on release and be hard to hold steady, so a foot or more is left thicker at the grip, allowing limb movement only above and below the handle.

It has usually been assumed that war bows had self-nocks, the slots to take the string being cut into the wood of the bow at each limb end, but the Spencer-Stanhope bow seems to have had applied nocks; two of the *Mary Rose* staves show marks where horn nocks were probably fitted; and many medieval illustrations, though perhaps not remarkable for their realism, suggest often, with a flick of paint at the bow ends, that the artist intended to convey a horn tip such as we still use, glued on to the points of each limb, and carved with the shoulders and slots to take the string and protect the wood from chafing. Possibly the smarter bows had horns, and 'issue' bows were self-nocked. Certainly, on campaign, there would be a danger of glue failing and horns loosening. There are instances of fletching glue failing, and feathers falling off arrows in bad weather, which accounts for the common medieval practice of winding a spiral of silk or thin thread through the flights to hold the quills firmly to the arrow shaft.

Of the archer himself we shall learn much more, from a detail here or a clue there, as we go on. Broadly speaking he was drawn from the villages and fields, a man of small property, sometimes none, sometimes an outlaw or a poacher pardoned for service; a man of no great estimation in the world, but a man of country skills and strength, rough living and hard working, accustomed to things of wood and finding a pleasing familiarity in the wooden bow, and the competition and emulation of practice at the butts, which by the time of Edward III's reign were common to every small town and very many villages throughout the country. He wore sometimes a steel cap, sometimes a querbole, or hardened leather helmet with an iron rim and crosspieces, a quilted jacket with iron plates sewn on the inside, or a mail shirt, more or less of a uniform according to the military fashions of the time and the attitudes of those who employed him, serviceable boots and hose, and had a sword at his side, a dagger, a maul, or some other weapon for hand-to-hand fighting. On his back he carried all the necessities of his daily life, cloak for rain and sleeping, spare clothes if he could provide them, water bottle and food pack. With two quivers at his back, he shouldered his bow in its canvas bowcase and marched his way to much estimation in Britain, and after 1346 to find great esteem and to inspire great dread throughout Europe.

It had been hard to train him to his best; it proved impossible to keep him to it; but at his best there was no man in the world to beat him, no matter the odds against him; and his breed lasted long beyond the longbow; he used the musket and the rifle; he endured in 1915 the same, and worse, than his forefathers suffered in 1415. There has been a fashion lately to deride, not his kind, but his service to his nation as an exploitation by his rulers of his servitude and simplicity. Neither he nor his nation has ever taken kindly to servitude, and often his simplicity turns out to have been a reticence, which once dropped when overt action has to be taken, is found by his enemies to have concealed both dogged and dashing courage, subtlety together with intransigence, and a total refusal to yield to pressures from outside his nation or from within it that are not acceptable to his not quickly formed but formidably defended attitudes. He will never entirely perish because, for all the sloth and the cantankerous emulation that lie side by side in his nature, he shares with the best of mankind, courage, clear sight and honesty.

Chapter 4

The Archer at Sea, in a River, and on the Downs at Crécy-en-Ponthieu

Edward III's claim to the throne of France is not the subject of this book. It dated from Henry II's marriage with Eleanor of Aquitaine, and his title to Normandy, which meant that an English monarch virtually laid claim to western France from Picardy and the channel ports right down to the Pyrenées, while at the same time he was the feudal vassal of the French king. This was pleasing to neither monarch at any time, and the long wars between the two countries stem from the attempts of the English kings to establish complete souzerainty over the claimed territories and of the French kings to oust them. War under these conditions was almost unavoidable, and was made quite inevitable by the importance of the wool trade to the English and the Flemish, the disputed succession to the French crown (which left Edward, as nephew to the last three French kings, a closer contender for the throne than Philip, a cousin, who was chosen by the French), and the aggressive intentions of the Scots against England, reinforced by their alliance with France.

In the long delayed end it was, and it could only be, the French who won the argument though, when Henry V's young son was crowned King of France and England by 1431, 16 years after Agincourt, it looked as if the outcome would be quite the opposite. That the English nearly brought off their long and impossible gamble was in great part due to the longbow. Their final failure was partly due to the fact that in the natural course of evolution the bow gave way to the gun.

The war really started with Philip IV moving warships into the Norman ports, as if threatening invasion of England, at the moment when Edward, in

1336, was on the point of a successful completion of his Scottish campaign. In May the following year Philip 'confiscated' all the French territories of his cousin Edward, invaded Gascony and attacked first the Channel Islands, and then Portsmouth and other towns on the south coast.

Edward withdrew his homage to Philip, claiming himself to be the rightful king of France and his cousin a usurper, and continued to foster his alliance with Flanders. In November he sent an expedition to the island of Cadsand, the garrison of which had resisted Anglo-Flemish attempts to occupy it. According to Froissart, Edward's cousin 'the Earl of Derby, and Sir Walter de Manny, bold warriors of the Scottish battles ... with 500 men-at-arms and 2,000 archers, embarked on the Thames at London'. They reached Cadsand on the afternoon of Martinmas, November 12, and having wind and tide in their favour sailed into the attack, with archers in the bowcastles of every ship, trumpets blaring and banners flying. Froissart says the garrison numbered 5,000, and was drawn up to meet this armada on the dunes and dykes. 'To the sound of the battle cry the archers drew their longbows, and they shot with such effect that the defenders, many of them struck by the first flights of arrows, had no recourse but to give way.' This falling back allowed the English to leap from their ships and form up, the men-at-arms in the centre, and two great masses of archers at either wing. The garrison, in spite of their Genoese mercenary crossbowmen, were dreadfully mauled, 'especially from the continuous and effective shooting of the English archers ... and lost more than three thousand men killed'.

In 1340, Edward, with a view to the confirmation of his alliance with Flanders declared himself King of France and quartered the lilies of France with the leopards of England on his arms. He returned to England in serious financial difficulties, leaving his queen, Philippa, as surety for his debts in the Low Countries. There she gave birth to a son, John, known as John of Ghent, which was englished to Gaunt. In England, Edward turned his attention to taxes and loans and the strengthening of his fleet in the Channel, while Philip assembled 190 warships in the harbour of Sluys. On June 22 Edward took his fleet out of Orwell, near Ipswich, and sailed for Flanders with 4,000 men-at-arms and 12,000 archers, under his own command, and having with him as 'admirals', the Earls of Huntingdon and Northampton, the good Sir Walter de Manny, and Sir Robert Morley.

Edward wrote to his son: 'We sailed all day and through the night, and on Friday at the hour of noon we arrived upon the coast of Flanders ... where we had sight of the enemy fleet, crowded all together in the port of Sluys ... As the tide did not serve us we hove to that night. On Saturday, St John's Day, soon after the hour of noon, at high tide, in the name of God, and trusting in the justice of our quarrel, we entered the port.'

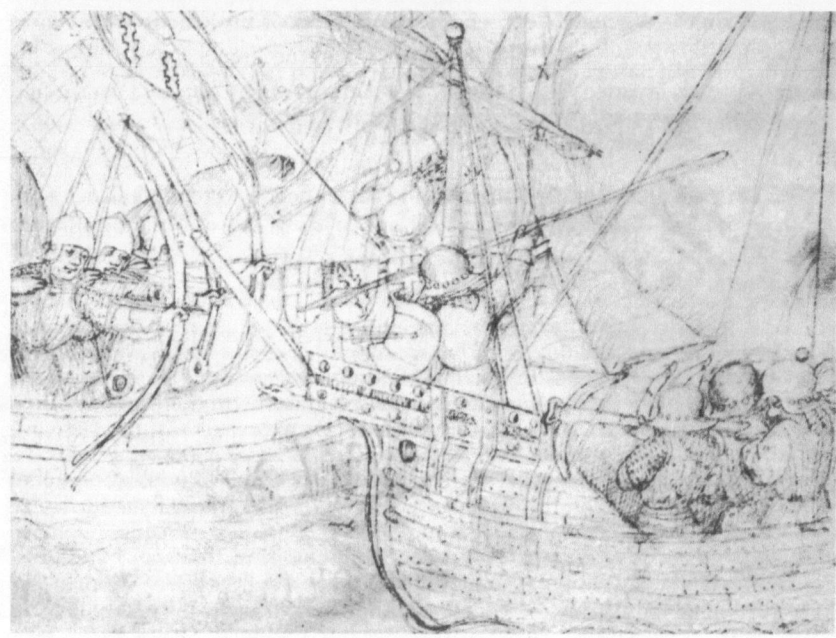

Longbowmen in a sea fight. Warwick Pageant (British Museum)

In the harbour at Sluys, which is now a great expanse of flat sand, the French and Castilian fleet was drawn up, like a land army, in four lines, the first three lines being chained and roped, ship to ship. The English fleet moved against them rather as the English armies now formed up on land, each ship of men-at-arms flanked on either side by a shipload of archers. In the front line of the French were two captured English ships, the *Christopher*, the previous year the English flagship, and the *Edward*. These two were quickly recaptured, their decks cleared with volleys of arrows, and, refilled with English and once again flying the Cross of St George, were put back into the battle.

The attacking ships grappled on to their adversaries, the longbowmen shot without pause, and as the French backed away over their own dead and wounded, from rank to rank of ships, so the Englishmen advanced from ship to ship, shooting, thrusting with lance, hacking with sword until their enemies were driven back right into the water. Only the fourth line of French ships escaped that night. All the rest were captured. King Philip's court jester asked him afterwards, 'Dost know, Sire, why the English be such cowards? I will tell thee, Sire; because they dare not, like the French, jump into the sea and swim.'

The years following Sluys saw much fighting in France by the English, in Gascony and in Brittany. They were campaigns of march and counter march, the ravaging of towns and countryside, as well as of fantastic deeds of rescue and attack, surprises and sieges in which no doubt the archers played their part. Froissart speaks frequently of the proportions of men-at-arms to archers, here 4,000 to 9,000, there 1,600 to 10,000, and though he may sometimes exaggerate the totals the balances are obvious; no English army would now expect to take the field with less than two archers to every armed man, and more usually with a much higher proportion of longbowmen.

In August 1342 a second invasion fleet with an army commanded by the Earls of Northampton, Derby and Oxford crossed the Channel. On September 30 this army faced the army of Charles de Blois on a ridge near Morlaix, its back against a wood, so that the flanks could not be easily turned and the wagons and horses could be conveniently and safely parked behind them among the trees. Just within bowshot they dug a camouflaged trench and waited for the attack, dismounted men-at-arms in the centre, archers on the flanks.

The French attacked with a first battle of infantry and Genoese crossbowmen who were shot to pieces before they reached the trench, and so, in retreating, could not warn the second battle of cavalry what lay in store for them. The horsemen charged headlong and full tilt into the trench, a plunging massed target for the archers. Each of the defeated French battles had been larger than the English force, but so was the third division now waiting to attack. In the lull, the archers, by now short of ammunition, started forward to collect the arrows that stuck out of the ground among the still and writhing remains of the first two attacks, but the third attack was on the move, which drove the English back upon their sheltering wood. By dark the two sides had had enough; the French withdrew, the English sensing the relief of pressure broke out of their wood, and the two forces separated.

Without archers the English would not have stood a chance. If the archers had carried more than the 36 arrows a man which they are said to have had at Morlaix, the French third line would have fared the same as its two predecessors. Ammunition was a problem that had to be solved.

While the war dragged on from hostilities to negotiations, from truces to hostilities again, renewed negotiations and renewed truces, in Flanders, in Brittany and in Gascony, Edward began to plan a three-pronged attack on France. One army should move from Bordeaux in the south-west, one from Brittany in the north-west, and his own new army of invasion from the north, towards the centre, splitting French power and French attention. Though there could be small hope of real liaison between the three forces, they would operate roughly at the same time, all aiming an arrow, as it were, at the heart of France.

Throughout the winter of 1345 and the early part of 1346 King Edward built up his English army. The cavalry were assembled in the same way that Edward I had raised them. Earls, barons and bannerets collected volunteers into their retinues and engaged them to serve the king for agreed wages. Anyone owning land had to either serve in person or contribute money, or find a substitute. Every man with land or rents to the value of £5 had to provide an archer to accompany him, of £10, a hobelar (or a mounted archer), of £10–£20, two mounted archers or hobelars, of £25, one man-at-arms, and so on, according to the quantity of their lands. In addition to these contracted forces, the Commissioners of Array and the sheriffs of the counties had to see to it that proper numbers of levies were raised, listed, and sent to the muster points.

Many of the archer levies were good men with their weapons but one is irresistibly reminded of Falstaff's recruiting drive, through Gloucestershire, on which he 'misused the king's press damnably'. He continued: 'I have got in exchange for a hundred and fifty soldiers three hundred and odd pounds. I press me none but good householders, yeomen's sons; inquire me out contracted bachelors, such as have been asked twice on the banns; such a commodity of warm slaves, as had as lief hear the devil as a drum ... I pressed me none but such toasts and butter, with hearts in their bellies no bigger than pins' heads, and they have bought out their services.' In their place he took the wretched who could not pay. 'A mad fellow met me on the way and told me I had unloaded all the gibbets and pressed the dead bodies. No eye hath seen such scarecrows. I'll not march through Coventry with them; that's flat.'

In the outcome, retinues raised by contract, though their proportions varied very much individually, would provide a roughly equal number of men-at-arms and archers and rather fewer 'armed men', or soldiers not wearing the full, expensive armour of the men-at-arms, but who would be mounted and who in battle would be ranged with the men-at-arms. Thus, in 1342, the Earls of Derby, Devon and Stafford had raised a total of 300 bannerets, knights and esquires, and 310 mounted archers. The total cavalry force for the king's expedition of 1346 was about 500 knights and 1,900 men-at-arms, and there were 2,500 'retinue' archers.

To this total were added the shire and town levies who were not part of a nobleman's retinue. From 28 counties south of the Trent – those to the north had to provide men to stand against Scotland, France's ally – came 3,780 archers, county contributions ranging from 280 coming out of Kent, 200 from Norfolk, 160 from Oxfordshire and 100 from Staffordshire, to 40 from Rutland. Chester provided an extra 100 special archers for the Prince of Wales' own guard. From the towns came more armed men than archers, as one might expect, totalling some 1,700. From Wales 3,350 men were to be raised, half archers, half spearmen; so, we may guess, half southerners, half northerners. From the Welsh Marches another similar force was to come, though in the event it seems no more than 4,500 actually joined the king's army. When all are counted, some omitted for defence at home, others added for the king's guard, a specialist company of crossbowmen and so on, it looks as though the royal army that sailed from Portsmouth mustered 2,400 cavalry and 12,000 infantry, probably more than half of whom were longbowmen.

A more detailed examination of the processes of recruitment, and the provision of bows and arrows, is made in the next chapter. It is worth looking at some known orders, which by no means represent the whole, but which were sent out to 35 counties. These required the supply in 1341 of 7,700 bows and 312,000 arrows, in 1342 and 1343 of 500 bowstaves and 12,000 arrows from each sheriff each year, which, if only 20 counties were involved, would total something like 20,000 staves and half a million arrows, and in 1346 of 2,380 bows and 133,200 arrows. The lower figures for the major preparations of 1346 suggest that the enormous orders in the intervening four years were sufficient to equip the archers who went to and from France. But the purpose of this chapter is to stay with the men who went to France in 1346.

As so often happened, the armada gathered at Portsmouth and Southampton and numbering 700 ships, great and small, was delayed by contrary winds through the whole of May and June. When they sailed, on July 5, they got no further than St Helens, where strong south-westerlies forced them to anchor for several days, and where the soldiery, crowded into the small ships for what should have been a few hours' voyage, suffered greatly as the fleet rolled and tossed at anchor. They sailed again on July 11, imagining, as they had been told, that they were bound for Gascony. In reality they were bound for Normandy, and for Paris. And there was another army bound for Paris from Ghent. They had no idea that Philip of France had enticed John of Hainault, Edward's brother-in-law, to fight on his side against the English. In spite of that a Flemish army with an English support group containing 600 archers did set out to join Edward's army, on the day that he reached Lisieux, August 2.

On July 12, after an easy crossing the English fleet made landfall at St Vaast la Hogue, 18 miles south-east of Cherbourg. From the first men ashore, it took until July 18 to complete the unloading of the whole war host, and on that day they started to march in a great column of horses, infantry, wagons and pack animals, keeping south-east, and parallel to the coast, until they reached St Lo, while the fleet kept slow pace with them a mile or two offshore.

The vaward, or vanguard, or first battle was under the command of the Black Prince, 16 years old, and knighted by his father within an hour or two

Edward III invades Scotland There are longbowmen on both sides (Bibliothèque Nationale, Paris).

of stepping ashore. The centre or main battle marched under the king's command, the rearguard or third battle under the Earl of Northampton. At St Lo, on the 22nd, the three battles spread out sideways and began their advance across country on a front perhaps two miles wide, probably with the pack animals behind each division and the wagon trains confined to such roads as then in Europe led from one small town to another. On the 26th Caen, the great walled town, was taken after a fight which routed the Normandy militia and a force of knights and men-at-arms commanded by the Constable of France, the military chief of the armies of France, who was taken prisoner. Before the English attacked they had formed into column again and marched ten miles, so the king ordered a halt and a meal, provided from the baggage train which was with the centre battle, and thus in the best position for a quick handout of rations.

While the army marched they seized all stores and food that they needed and burned the rest, as well as whole villages, and, as in the case of Caen, ran riot in some of the towns they took or passed through. Edward's orders for

An army on the march, with teamed horses and four-wheeled wagons. Chronique de Hainaut (Bibliothèque Royale de Belgique).

the march forbade pillage but, on this early part of the campaign at least, his orders were by no means always obeyed. The mass of the army was recruited with plunder in their minds, and they were in aggressive mood, after delays and seasickness, and thirsting for action. In any case, a medieval army would expect to do as much damage in enemy country as it could, and the tale of the Hundred Years' War is full of destruction and ravage far worse than the edicts of commanders might lead one to expect was their intention. We shall see in Henry V's orders to his troops a repetition of Edward III's, and they were not always enforceable even by that iron-willed and god-fearing young paladin.

After about 90 miles in nine days, one sizeable fight, and a great deal of destruction, the army stayed five days at Caen, and then moved towards Rouen. By August 2 they were at Lisieux, the Flemish army was setting out to join them, and Philip of France reached Rouen with his army, thus holding the main citadel of the Seine and standing between his two threatening enemies.

Edward continued due east, marching 16 miles on the 4th, 19 on the 5th, and resting on Sunday the 6th while a reconnaissance under Godfrey de Harcourt went forward towards Rouen, where his elder brother the Comte de Harcourt, loyal to Philip, was in command of the garrison. News was sent back to Edward that Rouen was heavily guarded and the Seine bridges were broken. On August 7 he made an 11-mile dash to the river at Elbeuf, where he found the bridge broken, and so led the army down the south bank of the river until they reached Poissy 60 miles further on the 13th, burning and sacking as they went, and paced so far all the way by a French army on the further bank. But at Poissy they discovered the French had moved on towards Paris, so they tried a crossing of the Seine at once. Archers went across the water in small boats and drove off the guard on the north bank, while carpenters and pioneers began to repair the broken bridge and even to construct others. On the 14th a French force attacked. Northampton led a counter attack, negotiating a 60 ft beam, only a foot wide, and beat the French. The same day, the Flemish army reached Béthune. If Edward had known this, his wisest course would have been to turn his back on Paris and try to reach his allies before Philip's main forces could catch him.

During the next two days, Prince Edward marched towards Paris, possibly to deceive the French, while the main part of the army crossed the Seine over the mended bridge on the 16th and set off due north. The Prince followed up behind, his diversion complete and apparently successful, because the French army did not start in pursuit until the 18th.

The English marched north as fast as they could, not pausing to destroy along their road until they reached Poix, where the troops disobeyed the king and attacked the town. The English, in spite of this delay and in spite of having to fight off attacks by the King of Bohemia's force, covered some 67 miles in six days, and reached Airaines on the 21st. But King Philip had done even better. Leaving Paris on the 18th, he was, incredibly, only nine miles from Amiens, more than 60 miles away by the evening of the 20th, though the bulk of his army was still stretched out behind him.

Edward was now within three days' march of his allies, but the Somme lay between them, and Philip with an enormous force was only a few miles from him. Worse than that, all the Somme bridges were broken and the French levies were out in great force on the north bank. Many horses had been lost, the food was running short, and Edward had news that the fleet had sailed for home, or the greater part of it, following the few ships that had definite orders to return to England, and in direct disobedience of the general order to be available to the army of invasion.

The English army, like a pent and savage animal, turned in its tracks and sacked a small town some eight miles to the west. King Philip crossed the Somme from his side and, almost before the English rearguard was gone, arrived at Airaines. The English had turned north and headed for the river again, the French on their heels. On the evening of the 23rd they reached Acheux. Edward had the prisoners taken in the neighbourhood brought to him. He knew or guessed there were possible fords near the beginning of the Somme estuary, but no one in his army knew exactly where. For any prisoner who would guide the army, he offered a large reward and freedom for the captive and two of his two friends. One man put out his hand and took the fee, a native of Mons-en-Vimeux, Gobin Agache, a name synonymous with traitor in French history. However, if among captured peasants standing before a warrior king, who for all they knew might be their lord with as much right as King Philip in distant Paris, one had not been found ready to help the enemy for 100 gold nobles and to avoid worse, it would have been astonishing. Agache said, 'By God, if you will keep your promise, I will take you to a place in the morning where you and all your army may get across before the hour of Tierce. There you can ford the river twelve abreast when the tide is low. It is the only place, but it is firm gravel and white marl.'

Before dawn next day, August 24, the Frenchman led the van of the English to Saigneville, where there was the ford that he promised, 2,000 yards or so across, and known as Blanchetaque, because at low tide the river was shallow enough to show a pale ridge in the riverbed, through the water, where a causeway had in the past been built. That it was a causeway rather than a natural ford is suggested by the fact that there is no marl formation locally.

They reached the river, six miles march, by dawn, and waited four hours before the ford was passable. That day the ebb started at dawn. While the van waited the rest of the army came up and crowded to the bank. The growing light showed them a strong force 2,000 yards away on the far bank. There were 500 men-at-arms and 3,000 infantry and Genoese crossbowmen there.

At about ten o'clock the water was getting shallower. Hugh Despenser with a vanguard of longbowmen, their quivers hoisted high on their shoulders, their bows strung and held over their heads, waded into the river. Slowly they pushed through the water that flowed at them from their right and swirled about their waists, those on the outside slipping into the deep water, scrabbling and being hauled back on to the causeway. Twelve abreast, they waded until – and how well they must have known when the moment would come – they were within crossbow range. As the bolts flew at them, feathering into the water, knocking men sideways, rolling them into the river, the longbowmen waded forward until they found foothold enough in the shallows to stand and draw their bows, and then they shot and drew and shot and drew and shot. The Genoese barrage thinned under the hail from the English bows, which increased as the ranks behind came up. The longbowmen opened a way to let the first of their cavalry ride foaming through them to meet French horsemen who came splashing into the water's edge against them.

Those French were driven back and the first of the English longbowmen and riders reached dry ground, the archers spilling on to the banks to either side, forming and shooting, running to their side, re-forming and shooting, until the French ran, chased by the yelling cavalry, and that small sodden band of bowmen who made the crossing possible threw themselves down in the grass and looked to their hurts.

By the time the last English troops and wagons were going down into the water, the French vanguard was on their tail. The rearguard of the English fought them off, but a few men and wagons were captured. The rest were across the river in the nick of time. The tide was making, and the water was already too deep for the French to attempt pursuit. Safely across, among almost the whole baggage train, were many carts full of thousands and thousands of arrows, and evidently at least one cart containing small newfangled cannon, made and supplied by the armouries of the Tower of London. Edward had two secret weapons. One was the curious gunpowder tube to fire iron balls. Their presence has been disputed, but the evidence in favour of their being in use on August 26 is strong, and now generally accepted. His other secret was, it is true, already guessed at, but no one on the Continent had yet a positive notion of the enormous power of his massed, highly trained and highly organised longbowmen. In the very campaign that saw the longbow reach its full and devastating development the seeds were sown that would in the end oust the longbow altogether from military use.

The English spread out on the north bank, as far as Le Crotoy on the flats to their left, where Despenser's men burned many boats, and towards Abbeville, where Northampton's division cleared the countryside of any lingering opposition. The same day, the 24th, the Flemish army, unaware of the closeness of their allies, broke up their camp before Béthune and began to retreat. Perhaps Edward did not know this, and perhaps the incredible success at Blanchetaque gave him and his men a feeling of invincibility. They slept that night near Forêt L'Abbaye in the countryside on the south borders of the Forest of Crécy, their leader determined, in spite of the fact that he might now join his allies, or retreat to Flanders, to chance his incredible good fortune and to find a place as soon as possible where he might offer battle to the French.

The French, having stared across the high tide waters of the Somme at the vanishing English, moved back to Abbeville where they spent the night of the 25th.

The English, on the 25th, almost certainly watched the river at low tide to make sure there was no French attempt to cross, then gathered their columns together and moved on, north-eastwards. Ahead of them the Forest of Crécy stretched, four or five miles deep and eight miles wide. The main army skirted round the south-east, reaching towards its northern edges by nightfall, but stopping short of the little river Maye, and the small town beyond it. That stream is full of watercress *(cresson)* now as it was then. While his army negotiated the forest, Edward rode straight through it, reconnoitred the region beyond, and chose his place of battle. It was ten miles from the ford at Blanchetaque.

Dawn, on Saturday August 26, revealed the open, shallow rolling country ahead of the English, the little town of Crécy-en-Ponthieu, and some tiny villages to their right front, Marcheville, Fontaine, Éstrées and, at the far end of the gentle ridge that rose above Crécy, Wadicourt, standing in its trees.

Along that ridge, facing south-east, the English army was deployed, early in the morning. The right flank was just above Crécy itself, and protected by a sharp fall of ground to the town. The little river Maye would be no obstacle to a flanking attack on that side, but the forest in 1346 was closer to Crécy than it is today, well within bowshot, and Edward felt his right secure. Two thousand yards away at the other end of the ridge were the buildings of Wadicourt, and enough woodland to protect his left. A few hundred yards behind the lane that ran from Crécy to Wadicourt was a large wood, le Bois de Crécy Grange, or Crécy Farm Wood, and there the baggage wagons were formed into a hollow square and the horses tied within them. The grooms and pages could run from there to the battle line with fresh supplies, and if necessary shelter there from attack.

It can all be clearly seen today, the changes being only that the woodland has now shrunk a little, and perhaps the contour in general slightly softened by long centuries of agriculture and the fact that almost every inch of the land is tilled, and presents an irregular chequerboard of cultivation, where there would have been more open downland 600 years ago.

The ridge is steeper on the Crecy side and its shape forms a re-entrant, with quite a steep terrace on its left which would be impassable to cavalry. The forward part of the terrace falls into the general slope of the ridge which runs the rest of its way to Wadicourt, north-east, becoming easier the further it goes. So the left flank would be the hardest part of the position to defend.

Almost in the centre of the slope is another terrace – they are called *raidillons* locally – which is steep enough and long enough to allow it to be lightly defended on its crest. There is one more such terrace, but a much lesser obstacle to attack, at the head of a slight declivity in the general contour of the ridge between the centre and Wadicourt.

The numbers of men concerned in medieval warfare are always hard to estimate, but it seems most likely, bearing in mind the 14,400 who reliably seem to have left Portsmouth, and allowing for some loss, principally at Blanchetaque, that Edward had with him that Saturday between 12,000 and 13,000 men to defend the Crécy-Wadicourt ridge. The right was held by the Prince of Wales with some 1,200 men-at-arms, between 2,000 and 3,000 archers, and 1,000 Welsh spearmen. Their central corps of men-at-arms was well down the slope, at extreme bowshot from the ridge bottom, known after the battle as the Vallée des Clercs, and the formation was several men deep. At their right were archers inclined forwards, above the drop to Crécy itself. At

Crécy field from the air. On the left is the town of Crécy, in the centre the Bois de Crécy Grange, on the right Wadicourt among its trees (Photograph by Graham Payne).

their left more archers, lining the left-hand side of the natural re-entrant of the ground and so matching the archers on the Crécy side. From the archers' left to the central terraces probably Welsh and Cornish spear and daggermen stood, as a hedge of weapons along the top of rising ground that becomes increasingly more difficult for cavalry as it approaches the steep terraces. No doubt there would be archers among them as far into the centre above the terraces as would allow them to shoot at an enemy advancing against the main body of the Prince's corps.

From the terraces, all along to Wadicourt where the ground fall softens, the Earls of Arundel and Northampton stood with their corps, probably more than 1,200 men-at-arms and spears and 3,000 archers or more, and on the terrace side some Welsh spearmen.

The king held the third corps a little back from the ridge top, 700 men-at-arms and 2,000 more longbowmen. They were, to begin with, rather more to the Prince's side than Northampton's. The king himself set up his command post in a windmill that stood a little to the right centre of the back of the Prince's battle. There is a mound there still, and in it now a waterpump, but about the windmill's position there is not the slightest doubt. A plaque and a flagstaff mark the spot, and from it one can see every part of the battlefield and the approaches to it.

Northampton's battle was a little further back than the Prince's, up the gentle end of the slope, to take full advantage of the ground.

The exact formation used by the archers has been argued a good deal. The problem of the men-at-arms is much easier. They stood in as many ranks as

The Archer at Sea, in a River, and on the Downs at Crécy-en-Ponthieu

The centre of the battlefield at Crécy, showing the Vallée des Clercs and the slope up to the central terrace with the Bois de Crécy Grange behind (Photograph by Graham Payne).

their numbers and the expanse of the ground they defended would allow; as far apart, man from man, as gave them ease to use their weapons; as separated, rank behind rank, as necessary to enable the second line to support the front, and so on, to the last rank, each rank stepping in to fill the gaps left by the fallen.

The bowmen were drawn up in 'wedges', 'herces', 'harrow-formation'. 'Herce' means a harrow, and the medieval harrow was usually in the shape of a blunted triangle, the cut off apex being to the front, widening to the back. We have seen the formation before, made up of the archer wings of two battles coming together to form the triangle. No doubt it was a formation adapted to suit the ground, both the shape and extent of it. But how many ranks could there be of men, shooting between those in front, or over their heads? Archers need to see what they are shooting at, so we can assume the ranks were staggered. One chronicler describes their formation at Agincourt as a coronet to the army. That tallies with the forward thrusting triangles, and even suggests that the main wedges were perhaps serrated, triangularly, like the teeth of a saw, which would allow, over a given space, many more men to stand and shoot. Looking straight forwards every man would stand so as to see clearly and, when the enemy to be shot at moved left or right, the archers would adjust their echelons by a pace or two to keep their aim open.

It has been suggested that the front rank knelt, like later riflemen, and it is possible they did. A longbow is an awkward six-foot thing to hold sideways and shoot well, and the more sideways it is held the further each archer must be from his neighbour, but it is quite possible for a man of average height to kneel and shoot, with his bow upright, and no doubt, where it was an advantage, this is what they did.

The whole extent of the battle line at Crécy was no more than 2,000 yards, probably a little less, and although it was indented, and in places re-entrant, there were a great many archers to deploy on ground not held by the men-at-arms. That they were in ranks is definite, and that being in ranks they could achieve a thicker barrage of arrows is certain. Beyond that we can only know that they were arrayed in such a manner as to be adaptable for changes of range and angle of shooting. Where enough slope gave them the chance, their ranks could stretch back, many deep. On flatter ground they had to open their order at the front to allow the staggered two, three, four or more ranks behind to see and shoot.

When each archer had shot off all his arrows, he could run back through the ranks for more, and his place would be filled by another, and no doubt too, as the archer corps grew more and more important and successful, efficient drills were developed to make such manoeuvres swift and easy. There must also have been efficient means evolved to allow for the use of light arrowheads to harass the enemy at long range, and the swift change to heavier, more deadly arrows as the range shortened. They had to get their arrows flying in great storms, which did not let up. They did so; and they found the best ways of doing it.

Each archer went into the battle line with two sheaves of arrows. That is not a great weight to carry. Forty eight arrows at an average weight of slightly under 2 oz (56.7 gm) means no more than 6 lb (2.7 kg) plus the weight of the quiver. They are said to have stuck some in their belts and some in the ground in front of them, which is quite likely in a pitched battle. Barbed heads could be drawn from the belt head foremost (the fletchings do not suffer), bodkin points either way, and arrow points put into the ground with fair care would not damage. The quality of steel used in their making has so astonished the metallurgists today that, as I write, it is being found extremely difficult to reproduce them for the planned penetration tests.

How fast did they shoot and how many arrows were available? It is perfectly possible to shoot, out of a bow of 70 lb (32 kg) or so, 15 reasonably aimed arrows in one minute. Ten is no great matter. If the archers of Crécy could manage their much heavier bows as well as we can cope with lesser ones, and I have no doubt at all they could, then we can confidently allow them to have shot off ten arrows a minute each during an enemy attack. Prince Louis Napoleon considered that 'a first rate English archer who, in a single minute, was unable to draw and discharge his bow 12 times with a range of 240 yards and who in these 12 shots once missed his man, was very lightly esteemed'. The Prince's 240 yards is a low estimate of the longbow's extreme range. If, at Crécy, it took big, trappered horses bearing armed men only 90 seconds to cover the 300-odd yards of uphill slope that faced them, and during which they would be within bowshot, 500 archers could shoot at least 7,500 arrows at them.

How many arrows were available? Some idea of the numbers ordered and supplied to the armies of the period has already been given. The next chapter looks at the question in more detail. If the basic allowance is taken at about 100 arrows a man, then there were half a million arrows, plus at least another 100,000, available at Crécy. Of course the number of archers engaged at any moment depended on the breadth of the attack, and on how many bowmen could see their targets. Mathematics of this kind are beguiling and dangerous but, if for any one period of two minutes, 3,000 archers could keep up the rate of 10 arrows a minute, the resulting storm of shafts would be 60,000. I do not suggest that such a number of arrows was actually shot off in any two minutes at Crécy or anywhere else, but I do suggest the total number of arrows shot off during the hours the battle at Crécy lasted must have been somewhere in the region of half a million. But let us look at the battle itself. The main contemporary sources for Crécy are four: Froissart, Le Bel, Le Baker and Villani. I quote from them or allude to them all.

The dispositions of the English force were made carefully, early in the morning. Edward knew that Philip had been at Abbeville. He might not have known that the French army had marched to Le Crotoy, drawn by the smoke of the burning the English left behind them and hoping to catch the invaders there, but he must have been sure it would not take the French long to discover his position at Crécy. So the preparations at Crécy were pushed forward with good discipline. Where the archers were not protected by sharp little falls in the ground, they dug potholes a foot square in front of them.

When all was done, and every man in place, the king, riding without spurs on a little grey palfrey, went from section to section of his army, the banner of England, quartering France, carried before him, and also the dragon banner of Wessex, the same that had flown above Harold at Hastings. He stopped frequently to talk to commanders and men, encouraging all, 'with a laugh,' says one chronicler, 'that the cowards became brave men'. He ensured himself that his orders were understood throughout the army. Above all he wanted it known that when the battle started no man, whatever happened, must leave his position to pursue the enemy if he fled. There must be no sign of a move forward until and unless a general order for pursuit was given. If the battle were won there would be time enough for booty or prisoners to be got, if lost there would be no need of either.

It was about ten o'clock. There was no sign of the French, so a meal that had been preparing was now ordered, and every man told to eat, drink a little wine and take his ease until the trumpet sounded. Each archer left his bow and his arrows on the dry ground, to mark his place, and, in due order, everyone streamed to the rear.

The French, finding the English gone from the Somme, turned from Le Crotoy to move up the road which runs from there to Hesdin, and spread about the countryside on either hand. This huge force of men and horses consisted of some thousands of Genoese crossbowmen in the van (there were said to have been 15,000 at the beginning of the campaign, but it is hard to believe they were all at Crécy), the great number of the king's retinue and those of his allies, Bohemia and his son Charles, king of the Romans, Jean de Hainault with the men of Luxemburg, James, the king of Majorca, the duke of Savoy, and many German mercenaries. Lastly came the *levées en masse*, thousands upon thousands of the levies of northern provincial France wherever the king's writ ran. The whole army numbered between 36,000 and 40,000 men, three times the English force, the cavalry probably matching it equally alone. As each mass of the army reached the high ground between Domvast and Noyelles-en-chaussée they could just have made out, past the edge of the forest, the English lines on the ridge three miles to their left.

The English finished their meal, packed away their pots, and moved back to their positions but, there being still no enemy in sight, they sat and lay about on the ground beside their helmets and their weapons. Four French horsemen were sighted, who rode about at a distance for a while and disappeared again.

At about four o'clock in the afternoon a cry from the lookout in the windmill turned every head within earshot to look towards the valley, and there they were, 'riding a soft pace', the vanguard of the French, with banners and pennons, all glittering in the sunlight, and the great red Oriflamme blowing slowly out, the sign of no quarter to the enemies of France. All down the English line men rustled to their feet, to stand and watch with awe.

King Philip, on hearing the report of his scouts, and with good advice, had decided to march to within sight of his enemy and halt for the night, because the day was well on and his commanders needed time to organise the vast incoherence of the French host. However, as the ranks of French came within view of the English in their three blocks, calmly lined up on the ridge, it was too much for the hot blood of the French nobility. There was much shouting and milling about and spurring and pushing from the rear. 'The king's orders were passed among his lords, but none would turn back, for each wished to be first in the field ... When the vaward saw the others behind pushing on, they would not be left behind, but without order or array pressed forward till they came in sight of the English.' The levies, for miles already, had been roaring,

'Kill, kill the English traitors!' and waving weapons, crowding in disorder from here and there behind the banners of the men-at-arms.

As the French gradually flooded forward and began to assume a front there flew, over both armies, a great flock of crows, heralds of a thunderstorm, and men looked up and spoke of portents.

King Philip, realising he could not control his host, and, it is said, enraged when he came to the front and saw the English, suddenly shouted to his marshals: 'Send the Genoese forward and let the battle begin, in the name of God and of St Denis.'

The Genoese commanders, Ottone Doria and Carlo Grimaldi, had quickly to organise their thousands to face in line of attack, and spread them out over the slope down to the little valley below the English. They had to halt them three times. The threatening black clouds towered above Crécy and exploded. There was a sudden downpour of torrential rain.

As the thunder cracked and rolled and the rain bucketed down, drenching the thousands that had gathered there to fight, dribbling down necks, under mail armour, soaking buff coats and hanging banners sodden against their poles, legend tells us the English and Welsh archers unstrung their bows, coiled the strings and popped them under their helmets to keep them dry. Science tells us that a linen string, such as the men would have had at Crécy, can be soaked for days in water without suffering any weakening, or stretching. But almost any archer, in heavy rain, would either shelter his bow, or do as legend says was done at Crécy, even if, scientifically, it were quite unnecessary.

'Then anon,' says Froissart, 'the air began to wax clear, and the sun to shine fair and bright, the which was right in the Frenchmen's eyes and on the Englishmen's backs.'

In the warmth of that August afternoon, while the French squinted against the sun and the English waited, there must have been quite a haze of steam in the air from drying clothes.

In the valley bottom the Genoese began a great shout, marched forward to within range, stood, raised their crossbows and shot, and, as the bolts flew towards the English lines, stooped to reload their cumbersome machines. It is possible that the thick twisted strings of their crossbows, being slack until drawn, absorbed moisture, which, followed by warm sun, could be enough to spoil their cast. Sir Ralph Payne-Gallwey once soaked in water a crossbow string of the type he supposed was used at Crécy. After an hour he found it drew a further inch on the stock than when it had been dry. If this had happened to the Genoese at the first volley, then their only recourse would be to remove the strings, twist them tighter and replace them before reloading.

Welsh longbowmen in their green and white, as shown in the BBC Chronicle film The Longbow, *written by the author of this study* (Photograph by Dave Watts of the Medieval Society).

Even as they bent to reload, or to restring, orders were shouted all along the lines above them on the slope, thousands of longbows were drawn back and the first flights of English arrows hissed, curving down at them, 'so thick that it seemed snow'.

The Genoese had never met such archery, nor at such range, thinking until they were struck that they could shoot further than any bow the English had. Pierced in their hundreds, heads, arms, legs, coats of mail, they began to recoil. Some are said to have cut their bowstrings. It is hard to see why. It is likely that the first English arrow storm caught some of them with their bowstrings off. Some threw down their bows and ran. There was confusion and pain, many wounded and killed. At the same time it is likely the first booms of cannon echoed over Crécy, and every horse pricked his ears and stared. French sources indicate there were three cannon, 'which much disturbed the Genoese'.

Seeing this fracas below and in front of them, the hot bloods among the stamping horsemen of the French nobility began to surge forward. The Comte d'Alençon cried out, 'Kill me this rabble, kill, kill!' and spurred forwards. They rode in heraldic splendour, trailing colour and gleaming in steel, thundering into the backs of the Genoese, yelling and deriding, and lancing, butt and point, as they galloped. Under this attack some of the Italians shot at the French who plunged among them. The chaos in the front of the French army must have been dire, and out of it, up the slope, came the first enraged charge against the Prince of Wales' position.

The next lines of French cavalry one after another but in no clear order wheeled to face the English and to spread out as far as the end of Northampton's battle on the English left. In groups and masses they spurred up the slope against the lines of English, and the archers shot and shot, and the arrows hissed, rattling and thumping as they struck. 'The bowmen let fly among them everywhere, and did not miss with a single shaft, but every arrow told on horse or man, piercing head or arm or leg among the riders and sending the horses mad. Some stood stock still, others rushed sideways, and many began bucking ... rearing ... tossing their heads ... and some threw themselves down ... the knights in the first French battle fell, slain or sore stricken, almost without seeing the men who slew them.'

Within a few minutes the ridge was strewn with fallen men and animals, but through them and over them rode the yelling French, horses skidding and plunging, twisting back unmanageable, stumbling and shrieking as the arrows pierced them, wagging the wounds open with frenzied movement, frantic with pain and fear.

The archers nocked and drew, closing their backs, opening their chests, pushing into their bows, anchoring for a second, holding the drawn arrows firm, thumbs of their drawing hands touching right ears or the points of jaws as they aimed for a heartbeat when the drawn shafts slid past their bow hands until the cold steel of the arrowheads touched the first knuckles; letting fly, right hands following the strings almost as swiftly as the shafts' flight past the brown bows, grabbing the next arrow from ground or belt or quiver, to nock and draw and anchor and loose, in deadly unrelenting repetition.

In the pressing forwards of so many men and horses, numbers got through, their sheer mass protecting those in the crowds' centres against the arrow storm. The English men-at-arms fought the men on horses, and the archers on the flanks poured their driving shafts into the clusters, piercing men right through, burying arrows in horses, grinding through the close iron rings of mail armour, punching into the thin plates of steel. The Prince's division was hard pressed; the boy was beaten to his knees. So his guardian Godfrey de Harcourt sent a messenger to the king for help, and himself ran to Lord Arundel on the near side of the left battle, begging him to attack the flank of the forward French. The messenger, Sir Thomas Norwich, ran to King Edward at the windmill and said: 'Sire, Lord Warwick and Lord Oxford and Lord Reginald Cobham, with the Prince your son, are hard pressed and beg you to send aid, for they truly think if the numbers of the French grow more, the Prince will be overcome.' Edward answered: 'Is my son dead, or fallen or too badly hurt to help himself?' 'No, Sire, I thank God, but he is sore pressed and needs your help.' 'Go back to him, Sir Thomas and to those who sent you and tell them not to seek help while my son lives. Say to them I would have the boy win his spurs, for the day shall be his, please God, and he shall have the honour and the glory of it, he and his men.'

The man ran back, astonished, to the fight. He lived, and told the story long after to a Frenchman, who told it to Froissart, who set it down, for our amazement now. The king did send some men-at-arms from his reserve, about 20, under command of the fighting Bishop of Durham.

Arundel's flank attack came to the Prince's rescue, moving behind the centre and making the archers of the Prince's left flank open up to let them through. The whole of Northampton's wing also seems to have moved down the slope until its front was level with the Prince's. That movement of troops across the king's eyeline must have prompted the answer he gave to the plea for help. When that messenger had pushed his way through to the Prince he found him and his men leaning on their weapons or sitting on the ground among the dead and dying, waiting for the next attack.

According to the king, there were 1,500 French knights lying dead in front of his son's battle.

There were 14 or 15, or, some say, 16 charges against one part or other of the English line, sometimes great waves of men against the whole reachable front, and wherever they came the archers braced and shot. Those in the reserve must have run behind the lines handing out new sheaves of arrows, and runners must all the time have been going to and from the wagons and the stacks of arrows to the men in the front who worked ceaselessly to drive their steel arrow points, cold chisel bodkins, through mail and through

The Archer at Sea, in a River, and on the Downs at Crécy-en-Ponthieu

plate, and into the wretched horses that were spurred again and again up the reeking slope.

In the lulls, while the French masses in the valley moved and milled before the next charge, archers ran forwards among the carnage on the slope and retrieved arrows where they could. These were the only movements away from the long lines on the ridge. The discipline was absolute.

Among the French, as captain after captain went down, and banners fell, the King of Bohemia, nearly blind, begged some of his knights to lead him forward so that he could strike a blow. They tied their reins together and galloped towards the arrows. Next morning they were found all together, men and horses dead together, their reins knotted together as they had ridden.

Still the French came on. They could not believe, and could not bear, that the mighty chivalry of France should fall before baseborn men. Here and there they broke through the archers, but always the gaps were filled. Those that charged through were dragged down and hacked to death.

Jean de Fusselles, carrying the banner of Hainault, was on a great black charger, who in the roaring panic of battle ran away with him, carting him straight through the English lines, and on behind, maddened now by an arrow lurching in its body. When he finally managed, hauling on one rein, to turn the animal, it stumbled and threw him, so that he lay winded in a ditch. Behind him, incredibly, had galloped his page, who pulled him up on to his own horse and, looking right and left to see the better way, rode for both their lives right round the English flank, and back towards the French host, until they were shot down.

The sun went down, and the stars came out. The moon rose, and by its pale light the French charges continued, but more and more desultory, until about midnight the fighting ceased. King Philip had left the field already, wounded in the face by an arrow, with guides and his own guard. Sir John de Hainault had taken his reins and made him ride away: 'Sire, today you have lost, but another day you will win. Come away. It is time.' So many of the other leaders were dead or dying on the Crécy slopes there was no one left to order the army. 'The flower of the chivalry of France lay dead upon the field.' The English losses were light, perhaps no worse than at Blanchetaque.

In the darkness some among the English made fires. The archers and men-at-arms lay and slept where they were, too exhausted to make camp or to eat, among the butchery, among the cries of the wounded. The noise, the frightful roaring of battle, was over; the weak cries persisted, and the stench.

Sunday dawned with a thick fog, so pursuit was out of the question, but the burial parties went out down the dreadful slope, and the monks, or clerks, of the abbey of Crécy made careful note of all the French of quality who lay dead. That is why the little valley where the Frenchmen massed their many attacks has been known ever since as the Vallée des Clercs, and the road along which the French army came to its destruction is still talked of as the Chemin de l'Armée.

When the fog lifted, a detachment, largely of archers, did go forward and made contact with a great number of confused French levies. They attacked and killed a large number and scattered the rest.

★ ★ ★

'That day the English archers brought great advantage to their side … it was by their shooting the day was won and the Genoese were turned back … the English held the field … they made great fires and lit torches, for the night was dark … King Edward came towards the Prince his son, and kissed him. "My fine son", he said, "you are indeed my son".' There were many fine sons who slept that night.

CHAPTER 5

THE MEN OF THE BOW

After Crécy, the great shocks to the French were that their chivalry could be shot off their horses and that crossbows were suddenly outranged by longbows. The crossbow of the time, with a wooden or composite bow, was probably not capable of much more than 200 yards. The Genoese could have had the 15,000 that some chroniclers attribute to them, and still the battle would have been no different.

When two such armies met again ten years later at Poitiers there were almost no crossbows on the French side. They were remembered as useless. Crossbows would remain in military use, and the invention of the steel bow would greatly improve their range and penetration, and turn them into a siege weapon or a marksman's bow of high value. In 1901 a 400-year-old crossbow was shot by Sir Ralph Payne-Gallwey across the Menai Straits, reaching nearly 450 yards. By the time of Agincourt there is no doubt that crossbows could reach something like 400 yards, and outrange longbows, but with their increase in power came complications in loading, and the hopeless inequality of shooting speed was increased.

There is one further footnote worth adding to the question of how many arrows went with the armies of the Middle Ages, in carts and on pack animals. By the time the rifle had reached a speed of fire that equalled the speed of shooting of longbowmen nearly 600 years earlier, most British riflemen carried 150 rounds each in their belts and pouches. There would be a further 100 rounds per man in the ammunition trains for each regiment, another 50 rounds each in Brigade ammunition columns, 50 more in Divisional columns, and a further 200 each in the base ammunition park and in reserve columns. The total for an infantry rifleman was thus 550 rounds. The long history of the handgun, which replaced the bow, shows that as the weapon became more efficient and capable of faster fire so the number of rounds provided for each soldier increased. I find it impossible to believe that commanders of the calibre of Edward III and Henry V, and the rest, wasted the enormous potential of their infantry weapon by failing to supply their archers with enough arrows. What those archers achieved proves that they had, apart from exceptional instances, enough arrows.

Less than two months after Crécy, at Neville's Cross, near Durham, a hastily gathered English army defeated the Scots invasion army of King David Bruce. On this occasion the English attacked and the archers were advanced in the same flanking formations as were used in their defensive role. They broke up the Scottish light troops and poured their barrages into the masses of spearmen, allowing the frontal assaults of the English men-at-arms on foot an inevitable success.

In the wake of these two victories, Edward III had no great difficulty in raising men and money for his new army to besiege Calais and at the same time hold off any French attempts to attack him or relieve the town. The role of his forces at Calais exists. In 1347 he had cavalry numbering 5,340 and foot totalling 26,963. 20,076 were archers, 4,025 of them mounted. That vast host of 32,303 men must have been the largest English army ever to take the field until that time, and the figures are trustworthy.

Within the year the Black Death had begun to ravage Europe and the British Isles, killing at least a third of the population before it died out, and it was a long time before such armies could be equalled again. But whatever the size of the armies, from now the proportion of longbowmen in them, though for various reasons not as great as it had been recently and would be again, was always as high as could be achieved.

Years ago in a documentary piece for BBC Television about the Agincourt campaign, called *The Picardy Affair*, I gave vent to a dramatic, fashionable, and I now know inaccurate cry: 'of the archers who won the battle, not one single one was named'. The Agincourt roll, which is incomplete, numbers but does not name archers. If we had the full roll we should find more archers, perhaps a separate named list. As it is, the combined Harleian Manuscript and the College of Arms roll name about 1,000 men-at-arms and list nearly 3,000 archers. That does not account for the whole army, which totalled between 5,700 and 6,000, 1,000 being men-at-arms and the rest archers. Two brief examples give the general style of the roll:

'Sir Henry Huse with his retenue': then follow 22 named men-at-arms, and: 'lances xxiij ... archers xxv'.

Or Sir Walter Hungerford, with 16 named men-at-arms 'lances xvij ... archers lv'.

Further search brings to light names that are shown nowhere in such 'government' lists. For instance there exists what appears to be a complete roll of the archers enlisted for the Agincourt campaign from the lordship of Brecknock. There are 159 named, and resounding some of their names are: Mereduth ap Trahan ap Jeuan Vachan (nowadays probably Meredith Trahern, or Meredith Vaughan) and Llewellyn ap Llewellyn ap Rosser. There is Jeuan Ferour cum equo cum Watkin Lloyd; did Jeuan come with his horse *and* with Watkin? Did Watkin come as groom with the horse? Or what? However he came, he is not lightly dismissed as 'i archer'. There are simpler names, less proud, or less certain of their ancestry, or perhaps belonging to those less able to explain to the roll-writers: Yanthlos, Res Weynh and David Coch.

From records of wage and other payments, letters of protection, and grants of pardon and reward, over 1,000 names can be listed of those who fought with the Black Prince at Poitiers in 1356. The list is mainly of those who served in the Prince's own company, and so represents only a proportion of a far larger total. A lifetime's search could possibly put together something approaching the full lists of those who fought with Edward III and his son, and for Henry V and his brothers, in their long attempt to win the crown of France.

By the time of Crécy the archers formed not only the limbs but part of the backbone of the invading armies, in a proportion with the men-at-arms often of three, four or even five to one, and it is quite wrong to suggest that they were a mass of nameless and exploited peasants. They were some of the finest, most highly trained and militarily efficient troops that any nation ever put into the field of battle.

In December 1355, there was an archer named William Jauderel serving in the Black Prince's army who was given a special pass to return home, signed by the Prince himself: 'Know all that we, the Prince of Wales, have given leave on the day of the date of this instrument, to William Jauderel, one of our archers, to go to England. In witness of this we have caused our seal to be placed on this bill. Given at Bordeaux 16th of December, in the year of grace 1355.' This William Jauderel, a descendant of Peter Jauderel who saw service with Edward I, was almost certainly a member of the Black Prince's elite guard of archers. He bore his own coat of arms; so do his descendants to this day, and they still possess the 'pass', which is written not on parchment but paper, a rare luxury at the time. The Jauderels, who were soon to settle for the spelling Jodrell, came originally from Yeardsley in Derbyshire. Jodrell Bank in Cheshire came into the family, taking its name from them, at a later date. William the archer went back to France a few months after the pass was issued, and having possibly fought at Poitiers, on his return to England in 1356 was rewarded with 'two oak trees' from the royal forest at Macclesfield to repair his house at Whaley Bridge in Derbyshire, where the name Jodrell is still to be found and the family blazon can be seen at the 'Jodrell Arms'.

It is plain that for a long time some archers had enjoyed a position of estimation above their fellow bowmen, which goes some way to explain the curious fact that, while the greater part of the archer force was simply and economically equipped, there survive many contemporary illustrations of archers in half-armour or even full armour. This suggests that they came from a wealthier and more privileged social position than the rustic bowmen in leather jacks and querbole helmets.

John Jauderel, William's brother, fought at Poitiers and, with other archers, was rewarded by the Prince after the battle for capturing a fine salt container in the shape of a silver ship which belonged to the French king.

The Prince also gave rights of pasture and turf-cutting as well as timber to his archers. A company of bowmen was raised at Llantrisant, in south Wales, from among the hereditary burgesses and inhabitants there, to go to France with Edward Prince of Wales as part of the archer corps of the royal army. Those who went became known as the Black Army of Llantrisant, and today their lineal descendants, wherever they may be, can apply to be enrolled as freemen. If their claim is justified they will be admitted to a company which embodies a true and living memorial to the longbowmen of the 14th and 15th centuries.

The claim of Vivian Thomas some few years ago was 'made as the son of an enrolled freeman, namely William John Thomas, who when admitted resided at Llwyn-crwn-isaf farm, Llantrisant, and whose roll number is 1525; William John Thomas being the son of Evan Thomas, the son of William Thomas, the son of Thomas Thomas, the son of Edward Evan Thomas, the son of Richard Thomas, the son of Rhys Thomas, the son of Morgan Thomas ...' and so on back to the Thomases who shot the longbow at Crécy, Poitiers and Agincourt.

There was promotion; there were pardons for civil offences, even pardons for murder or manslaughter, for archers who went into France and fought there.

In June 1338 French ships burned Portsmouth; in October Southampton was attacked; in March 1339 Harwich was burned; and throughout that summer the whole south coast came in for attacks of varying force and severity. Folkestone, Dover, Sandwich, Rye, Hastings, Portsmouth again and Plymouth suffered. The following year the Isle of Wight was attacked, in spite of Edward's victory at Sluys.

The first need was for the muster and array of men for the defence of the realm. In February 1339 the order of array for Hampshire demanded that the

The Men of the Bow

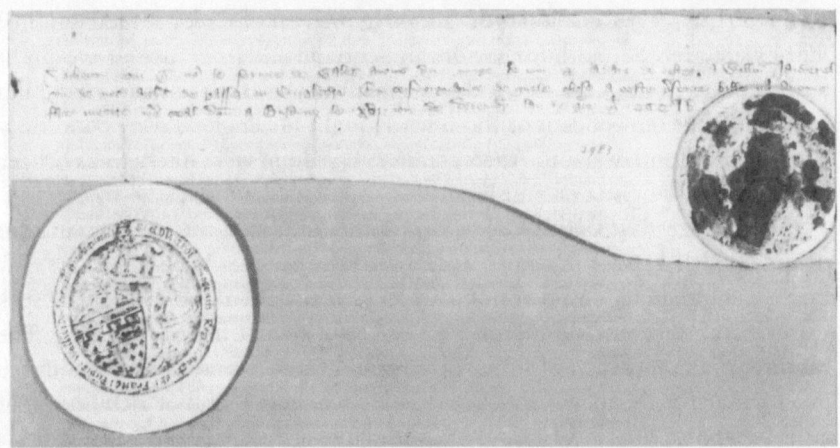

The Jodrell Pass (By permission of Mr. M. Mostyn Owen Jodrell).

county provide 30 men-at-arms (fully armoured, equipped and mounted men), 120 armed men (less heavily armed mounted men) and 120 archers. Sussex was to provide 50 men-at-arms, 20 armed men and 200 archers. Other counties had to offer similar forces. In general archers were not to be taken for service abroad from the maritime counties, which were being methodically put in a posture of defence, while the royal army was preparing to attack the enemy across the Channel.

The 14th century, in this situation of hostility between France and England, was witnessing the emergence of a kind of national consciousness in England, certainly an anti-French feeling in general, in which many Welshmen joined. It has been said that those who fought with the two Edwards at Crécy comprised an army that was part Norman, Saxon, Angevin and Celtic, but which in some way thought of itself as the army of England. Care and effort were taken to explain to people in general, often from the pulpit, often in the market place, what was happening, how the war was going, why the French were liable to invade, why England must go to war in France, and why there was need of men and weapons to make all safe at home and further the belligerent cause abroad.

The war allowed the king to take advantage of the desires of men outlawed for many sorts of crime to regain freedom within the law. So from that quarter came a few hundred willing volunteers to swell the numbers of those who offered to serve abroad for other reasons; because they wanted to turn national and anti-French feeling into action, because the pay was good, and the chances of loot enormous, and because going to war was honourable, manly and exciting. In 1339 and 1340, 850 charters of pardon were granted to men who had served as soldiers or sailors. In the year of Crécy several hundred more were granted, both for service in Scotland and in France. It would be foolish to overemphasise the outlaw element in the armies of the period but something like ten per cent of those who fought for England were conditionally pardoned outlaws, some three quarters of whom were likely to have been guilty of manslaughter or murder.

After the great victory of Crécy, the capture of Calais, the successes in arms of the English in Gascony and the defeat of the Scots at Neville's Cross, any English commander could count on a fervent rush of recruits ready to seek fortune and the comradeship of arms on a full tide of national feeling. The capture of a Frenchman of importance could carry a ransom that amounted to a fortune for the man responsible. There was every kind of rich finery, of arms, weapons, furs, gold and silver, and wine in abundance to be had after a battle or the taking of a town. The desire was there; the longing for chivalry in some cases, for personal renown in others, or for aggrandisement or material gain. The desire only had to be channelled.

The feudal organisation of obligatory service was giving way to new methods of paid service. By the 1340s recruitment was much more widely achieved than before with a form of contract for service, and a wage structure came into being which standardised the amounts to be paid to dukes, earls, knights, esquires, mounted archers, foot archers, indeed every kind and rank of those assembled for war. At home the wages of a highly skilled craftsman, or a master craftsman, were about 6d a day. Now, a mounted archer would be paid the same amount. This was a considerable sum at a time when a knight's chamberlain, if he did not 'eat in half, was paid a mere 2d a day. A mounted archer was not one who fought on horseback but who would travel on his horse, and fight dismounted with his fellows. There were 2,600 soldiers in the Black Prince's division raised for the 1355/56 campaigns; their pay records suggest that this total was made up of 1,000 men-at-arms (which is a likely high proportion in the commander-in-chief's 'battle'), 1,000 horse archers, 300 or 400 foot archers, and the rest, Welsh light infantry with spears. So, in the case of this long campaign at least, the proportion of mounted men over foot-sloggers, of sixpence-a-day men over the threepenny men, was nearly three to one.

The nobility and other magnates were in some sense the agents of this new recruitment, contracting with their superiors or with the king himself to provide specified numbers of men, men-at-arms, infantry and, vitally, archers, at specified wages for specified periods of time. These contracts were at first for 40 days, but could be extended at the proper rates of pay to last nine weeks, three months, six months, a year, or even 'as long as the king desires'. In 1341 the Earl of Northampton (who commanded the left flank at Crécy) undertook to raise seven bannerets, 84 knights, 199 men-at-arms, 200 armed men and 250 archers. A year later the Earl of Stafford contracted for three bannerets, 16 knights, 31 esquires and 50 archers, and in 1355 the Prince of Wales had a personal retinue of 433 men-at- arms, 400 mounted archers and 300 foot archers. As for wages, a duke received 13s 4d a day, an earl 6s 8d, a banneret 4s, a knight 2s, a man-at-arms 1s, a mounted archer 6d, as we have seen, and a foot archer 4d or 3d. Welsh lancemen (such as fought in the centre division at Crécy) earned only 2d.

There was also a form of conscription, by commissions of array, whose duty was to raise definite numbers of men for service from particular areas; in 1341, for instance, 160 archers from Northampton and Rutland. In this case Rutland was charged with raising 40 archers, but the county protested that Northamptonshire contained 26 Hundreds, 'whereof the smallest is larger than the whole of Rutland' and an adjustment was made. The number raised often fell short of the number projected, but the arrayers did their best to choose, test for the skills the archers should have acquired or maintained at the local butts, and array *les meillors et plus suffisantz, les plus forcibles et plus vigerous archers*, to clothe, equip, and where necessary mount them, to pay them, and send them with a leader to a collecting place, or hold them ready for departure. The feeble were to be avoided and the 'ailing and weakly' were sent home from the ports of embarkation, even from abroad. Men could buy out their services for 'reasonable fines', but, when the feeling for success was high, the proportion that did so was small. Exemption could be granted where there was need.

It was usual for writs to contain some mention of clothing, 'gowns', 'hoods', 'one suit' apiece and so on, which we can assume was uniform of a kind, though details of such dress are limited. Archers raised in Cheshire and Flint were to be provided with woollen 'short coats' and hats, half green and half white, green on the right, white on the left, and every group raised from those areas during Edward III's reign was so dressed. The uniforms were sometimes delivered at the points of array, sometimes in London.

Cheshire practice became gradually more widespread. In the following year recruits were given 16 days' wages in advance for the journey from Cheshire to Calais, and in 1355 they had 21 days' wages to reach Plymouth. From Chester they would march to Shrewsbury, following the Severn through Bridgnorth and Worcester, to Tewkesbury, Gloucester and Bristol, and so by Exeter to their destination. Obviously advance payment was essential if travelling troops were to eat properly on the journey without ruthlessly living off the land. Sometimes counties paid these travelling allowances to an agreed point,

Three sketches showing how armour was developed for greater protection in the face of the longbow. Soon after Crécy the left-hand figure took over from the right, and had been replaced by the late 15th century by the figure in the centre (Drawing by Clifford Anscombe).

after which they were 'at the king's charges'. The usual arrangement was for arrayed men to march to their county boundaries without pay, from there to the point of embarkation at the county's expense, and from then on to be paid by the king, even if there were delays before sailing.

In 1345, the year before Crécy, while the king was collecting his army of invasion, 125 Staffordshire archers assembled at Lichfield on May 25, and with six days' pay in their pockets (probably 18d) set off for Southampton and duly arrived in a week; 100 Shropshire men met at Bridgnorth on May 12, and set off for Sandwich with 1s 9d each, but 22 Buckinghamshire bowmen who gathered at Aylesbury on May 10 had to be content with 6d each to reach Sandwich. The organisation of archers and other troops on the march was by 'hundreds' and 'vintaines' of 20, each county contingent going in charge of a leader appointed by the king, who carried the wages and the nominal roll and was responsible for getting his charges safely to their port. Among the leaders of Cheshire men going south for the 1355 campaign were Sir John Hyde and Robert Legh with those from the Macclesfield hundred, Robert Brun from Eddisbury, and Hamo de Mascy and Hugh Golbourne from the Wirral and Broxton hundreds. Gronou ap Griffith led the men of north Wales for 1s a day, and a Welsh chaplain with him was paid 6d a day.

When possible the arrays assembled after the hay harvest, in the kindest time of the year for travel, and if the Cheshire contingent of 1355 is fair example, with their 21 days' pay in pocket to get to Plymouth, men were not expected to march more than 13 miles a day. Sometimes they billeted well at night and sometimes they slept rough, in barns or under the sky – no great hardship in a warm season. It could not always be a happy, comfortable spring march these recruits made, in their new uniforms with their bows slung and their first issue of one sheaf each of 24 arrows at their back. Archers moved about England in all weathers during Crécy year; August, October, December; and on January 2, 30 of them left Salisbury for Sandwich (130 miles), another 20 the next day from Cambridge (118 miles), 20 each from Cheshunt and Chelmsford the same day, and one sad group set out from Somerton, 30 of them, three days before Christmas with a march of 182 miles ahead and small prospect of Christmas cheer.

Whenever it was possible to beat the profiteers, contingent leaders, or the king's or magnates' agents, would buy up wine at the legal prices. Where overcharging was met, it was suppressed; where bad quality was found the stocks were destroyed, so that the new soldiery should drink good stuff at a fair price. Inns were encouraged to be reasonable in their prices, and at any time when many contingents were going through Kent into Sandwich all markets and fairs were shut down except for Sandwich itself, Canterbury and Dover.

Everything possible seems to have been done to make at least the first part of the march to war as pleasant as possible. Arrivals were spread over a period, and usually ordered as dates by which men had to be there: 'by three weeks from Midsummer', or 'by the first day of September'.

Attorneys were appointed for those with considerable possessions so that their interests could be watched while they were abroad, and letters of protection were issued; where law cases were pending, suspension was ordered until the men returned home. Once arrived, those for service overseas were billeted by agents of the Marshal of England. In 1346 Robert Houel had to lodge all the contingents, arriving in numbers far too large to be absorbed into the town of Portsmouth. There must have been encampments 'outside the verge of the household' (the royal household and retinue) to accommodate the main part of the army. Inevitably there were delays; ships did not arrive on time or the weather was adverse, so payment had to be made to the waiting men. One contingent at Plymouth in 1342, mostly Welsh and consisting of a chaplain, an interpreter, a standard-bearer, a crier, a physician, four vintainers and 100 archers, under the leadership of Kenwrick Dein, were given 21 days' service pay for waiting between November 5 and 25.

The men had boredom to overcome and new accents to comprehend; there was much more of a language barrier between Welsh and English, much more dissimilarity between the speech and usage of the northern and southern counties of England than today, and the language of the nobility was predominantly Anglo-Norman. Quarrels burst and were settled but, though little is said of it, it seems likely that waiting periods became training periods; archers practising both weapon and manoeuvre, groups and divisions learning their organisation, and the men getting used to their new military leaders. The best argument for the fact that the armies of the 14th and 15th centuries were on the whole well treated by their leaders lies in the knowledge that a great proportion of them returned after the campaigns, and a considerable number re-enlisted for further duty.

It was not always so. As time went on and the tide of victory in France ebbed, war weariness, discontent and depression, with all their attendant ills, took the place of the bright hopes and sun-glanced ambitions of the beginning of the long war. This evaporation of delight is marked in many ways, very much in the increasing severity of military discipline; but in these early days, all the youth of England was on fire, and in our imagination we have brought them in their thousands to Sandwich or Southampton or Portsmouth, where they wait, drilling, looking to their tackle, arguing, listening, working, lazing, as any other mass of troops encamped in any place, in any time; feeling homesick, boasting, whoring, laughing, sleeping, eating and drinking.

There were huge quantities of beef, pork and mutton, salted and fresh, more often salted because of the difficulties of storage, flitches of bacon, masses of cheese, oat and wheat cakes and loaves, peas and beans, and fish (usually dried 'stockfish' or herrings) which were caught in home waters, or imported, often from Gascony, an English dominion and ally in the French wars. Orders went out to the sheriffs of counties for the collection and transport of food and drink, and slowly it was drawn towards the army embarkation ports from all over England and Wales. From Cheshire, under the Prince of Wales' direct rule, came great herds of cattle, from north Wales cattle and sheep, and from Cornwall, again directly linked to the Prince, more cattle. The county of Lincoln, in Crécy year, sent to William de Kelleseye, the king's receiver of victuals at Boston and Hull, 552½ quarters of flour at 3s or 4s a quarter, packed in 87 tuns, 300 quarters of oats, 135 carcasses of salt pork, 213 carcasses of sheep, 32 sides of beef, 12 weys of cheese (312 stones) and 100 quarters of peas and beans. From Hereford went supplies to Bristol port, from Kent to Sandwich, and from Oxford and Berkshire over the Thames bridges and the ford at Bolney (oxen ford). But from Rutland went only half the demanded amounts; once again the tiny county complained she could not raise the total, and relief was granted. Where fresh meat could be had it was eaten in the camps, and the waiting ships were filled with the victuals meant to last. Though armies of the Middle Ages were expected to live off the lands through which they passed, as much supply as possible was carried, but transport was a grave problem. When an army sat down to a siege, supplies from home by sea and land became vital.

At Yarmouth in 1340, 30 ships were provided by the town for 40 days' service as troopers, and they were victualled at the town's expense. The bailiffs' accounts show that, apart from the foodstuffs, 60,400 gallons of ale, supplied by Johanna Hikkeson, Peter Grymbolp and John Gayter, were taken on board, at a cost to the bailiffs, hence the townsfolk, of £251 13s 4d. The allowance seems to have been one gallon a day to the men in the ships.

The mass of supplies moved slowly towards the armies, living beasts on the hoof, the rest in carts, in panniers and sacks on pack horses, by road and by river and sea, while the sheriffs and bailiffs and their staffs, the collectors about the counties and the king's receivers of victuals at the mustering points, sweated at their figures, at their droving, wheedling, threatening, bargaining and measuring. The wheels creaked, the millstones roared and rattled, the hoof falls pattered on the hard and whispered through the fields, splashing through shallow fords where the old droving roads crossed the rivers; the little broad-bellied boats plied the estuaries and inlets, until the men with the bows and arrows, and the men who fought in armour with lance, sword and maul, were fed full as their hearts were full with expectation.

From all over Wales and England came the weapons of war – swords, knives, spears and lances from the iron-producing areas, bought and provided by the counties, arrowheads of all the different designs and weights; there were 4,000 from Chester Castle in 1359, 52s 5d the lot; arrowheads had to be 'well brazed, and hardened at the points with steel'; shafts and feathers, mostly goose-wing feathers, bowstaves and made bows were constantly demanded. Later orders give an indication of earlier demands: in February 1417 six feathers from every goose in 20 southern counties were to be at the Tower by March 14. On December 1 1418, sheriffs were ordered to supply 1,190,000 goose feathers by Michaelmas. Year by year the orders went out to replace the stocks that were sent out to the archers.

Bows were divided into two kinds, 'white' which cost about 1s 6d each or 12 deniers, and 'painted' which were 2s or 18 deniers, a more expensive article. No one knows what these designations meant. Yew bows come into both categories, so the descriptions do not apply to yew and non-yew bows. My own belief is that the 'white' were fairly raw, not long-seasoned bows, 'green' staves, which in the case of yew would show a gleaming ivory colour on the sapwood back that fades with age. That would suggest the 'painted' bows were of thoroughly seasoned staves, treated with some sort of paint or varnish, as we treat them now, to inhibit the drying out of the last vestiges of liquid in the wood, that final ten per cent or so that stops a bow becoming too brittle.

Some consignments went direct to the seaports, and it is now impossible to establish the total numbers of bows and arrows supplied to our archers in any one year or for any one campaign, but, if my estimate of the possible number of arrows shot off at Crécy is within a shout of the truth, and I estimate a good half million, then the bowyers, fletchers, longbow-stringmakers and arrowsmiths of England and Wales worked miracles during the great years of the longbow.

As they were collected from local manufacturers, for instance at Bridgwater, Taunton and Sherborne and sent to Bristol (from six towns in Dorset and Somerset in 1346 about 11,000 arrows were delivered to Bristol), the bows were packed in canvas and the arrowsheaves corded together for stowage in wooden tuns. These were in turn stacked in carts and wagons which were hired for the purpose, covered with tarpaulins of horse-hide, pulled by teams of two to eight horses, and accompanied by clerks who kept the tallies, saw to the delivery and got receipts in exchange. The orders sent out to counties were not always filled in one consignment. Some would come in early and the rest would be promised to follow as soon as possible; and some orders were never completely honoured. A thousand sheaves of arrows ordered from Hereford on one occasion dwindled to 363 sheaves actually delivered. The orders were not regular yearly demands. In times of truce fewer orders were sent out, but when the war fire blazed the orders flew from the heat thick and fast. In 1360, in the two months of May and June, roughly 10,000 bows and half a million arrows were received at the Tower of London alone. In 1356 the Chamberlain of Chester learned that 'no arrows can be obtained from England, because the King has ... taken for his use all the arrows that can be found anywhere there'. He then had to get for the Prince of Wales, 1,000 bows, 2,000 sheaves of arrows and 400 gross of bowstrings, requisition all available immediately and make certain that production continued until the order was fulfilled.

Available figures show that in 1359, the year of a new royal expedition after the truce, the counties supplied over 850,000 arrows to the Tower, and about 20,000 bows and 50,000 bowstrings. That does not include already-existing stocks, nor the fact that the orders continued to pour in to the suppliers for more and more of everything, nor the fact that large quantities of arms went direct to the appropriate ports. If one accepts the idea of 6,000 archers shooting off half a million arrows in one of the rare major engagements, then the production of a million arrows in a year would seem too low a figure, but it should be remembered that, from a million arrows shot off, some proportion would be recovered. Every arrow that quit the string in battle was not a lost arrow.

In 1338 the king's 'artiller' was Nicholas Corand, who was ordered at one time to buy 1,000 bows and 4,000 sheaves of arrows, to make what he could not buy, and despatch them with all haste to John de Flete, keeper of the king's armour. The proportion of arrows to bows in this instance is 96 to 1; so here perhaps is an indication that each archer on campaign could count on about a hundred arrows, presumably replenished as often as possible. The cartage problem is a tricky one. A million arrows might weigh about 40 tons (40,648 kg), and that, in the wagonage of the day, and in relation to the poor roads and rough country to be crossed, would represent a large and cumbrous part of the baggage train; but the argument of the difficulty of carriage proving that carriage was not achieved is a poor one. If an army took with it only one sheaf per man more than each man carried himself, carts would have to be used. Armour had to be carried as well, tents and pavilions, food, spare bows, guns, sulphur and saltpetre, the whole equipment of the field kitchens and so forth and so on; the list is enormous. Forty tons of arrows, those vital components of English success in arms, may have been hard to transport, but it is quite certain that arrows were carried in great quantity. Otherwise England would not have had the successes she did.

By the time Henry VIII was campaigning against the French, taking a leaf out of Henry V's book, we can see many more details of transport. For example

An army on the march with wagon transport (Essenwein Medieval Handbook, Victoria & Albert Museum).

in 1513 his massive army of invasion marched in three 'wards' or 'battles' just as in Edward III's time, though the proportion of archers to other arms had dropped by then to one in three, or less. In one of these wards, of approximately 15,000 men, there were 90 vehicles allowed for spare weapons and equipment. Five thousand two hundred bows in parcels of 400 were carried in 13 wagons; 86,000 bowstrings in 20 barrels were in two wagons; and 240,000 arrows needed 26 wagons, a little under 10,000 arrows a wagon. Possibly in this case the artillery and its ammunition needed the big wagons, so that small carts were used for arrow-transport, and possibly in the past the bigger wagons with larger teams of oxen or horses were used for bows and arrows, which would have much increased each load and lessened the number of vehicles.

The great march from Harfleur to Agincourt is supposed to have been made without wheeled transport, for speed of movement. Therefore everything, including a good part of the crown jewels, which were rifled by the French raiding party in the rear of the English, and the appurtenances of the

royal chapel, had to be carried on sumpter animals. The English must have marched with a minimum of gear, though certainly with enough armour for 1,000 men-at-arms and 5,000 archers. But arrows? Could Henry V have risked cutting his bowmen short of ammunition? The possibility is there, and it is known that after the armies shocked together the archers threw down their bows and went in with sword and dagger, maul and anything else they could lay hand to. Was that because there were no more arrows? It could be. But I doubt it. Those commanders of genius who had armies in the proportion of five or six to one in favour of the archer arm would have been mad to risk that arm's paralysis. If the general contribution of archery to English victories in France has not been grossly exaggerated, we must grant that enough arrows were available however awkward the cartage involved.

In this context, 500 whips for 'great horses', 200 leather halters, 400 trammels and pasterns, 200 leather collars, 200 pairs of traces went across the Channel in 1339 from Yorkshire alone, and from Nottingham

10,000 horseshoes and 60,000 nails. For the same campaign, 80,000 horseshoes with the appropriate nails went from the southern counties via Yarmouth and Orwell. Again, for the retinues of 12 magnates in that campaign, which represent only a proportion of the whole, more than 6,300 horses went on inland from Calais, the number of those embarked in England being much increased by animals available in France.

It is not part of this study to examine the impressment of ships and the naval organisation necessary to take the armies assembled in the coastal towns over the sea to France and Flanders. In 1355 the little *Seint Spirit*, master Robert West, out of Plymouth, with a crew of 26, was paid for seven days, among many others of her hiring, at the rate of 52s 6d. There was the *Margaret*, from Weymouth, 80 tons, with a crew of 22; the *Welifare* of Shoreham, 50 tons and a crew of 16; the *Janette* of Winchelsea; the *Navdieu;* the *Cog John;* and the *Saintmarybote*. They came from all round the coasts, some late, some in good time to swell the number of the king's ships, to take on board the horses for whom loading ramps and wooden stallsides or hurdling separators for the voyage had meanwhile been built. They came to load all other supplies, as well as the men and their armour and lances, the archers, their bows and arrows, to ferry them over the sea at the wind's whim and land them for the start of their war expeditions on the Continent.

Repairs had to be made to some of the horse-transports on the 1355 voyage, although the crossing was calm, and the work was done by archers from the companies of Hamo de Mascy and Robert Brun, whom we have met before, and they earned extra pay for it, which they received when the ships reached the Garonne at Bordeaux on September 20.

In the case of a short crossing, as the 1359 expedition from Sandwich to Calais, the little ships could make more than one trip to and fro. Archers for that expedition were to be ready to start for Sandwich by July 15. Arrayed men were to be at the port by August 30, but it was not until the beginning of October that the first ships of the fleet got under sail, the king himself following on October 28. 'At last with a thousand ships and smaller vessels,' writes a chronicler of just such a departure, 'they began to sail wonderfully ... the masters of the ships did not know which way they were to steer, but followed the Admiral.' More than 1,000 ships, their gunwales crammed with country faces, about to fight an enemy nation of which few of them could have met one representative.

After the landfall, there were the immense complications of landing huge numbers from cramped little ships, and the encampment of men and animals long enough to allow for the organisation of the whole army for the march ahead and to give time, especially for the horses, to get back into condition after the crowding, lack of exercise and generally lowering effects of even the calmest voyage. That most of the ships used in such expeditions (if they were not sent immediately back into the supply-run) were supposed to stay at the port or beach-head of disembarkation, and that a number of them usually disobeyed such order, is well known, and possibly no great surprise to any who are used to the independent attitudes of those who sail the seas in the freedom of their own ships. It is hard for a commander-in-chief, even for a Black Prince surrounded by his mighty host on shore, to call back a hired merchantman which has stood out on a good breeze for home.

Imagine those archers, from the little country villages of England and Wales who on their journey to the coast had marched through bigger towns than most had ever seen before, arriving now in a warm September at Bordeaux, a great cosmopolitan city nearly as vast as London. Old campaigners ready to air their few words of French, some few who may have known Bordeaux a little before, and the younger recruits, wide-eyed at every new sight. There must have been many things to attract the men who served the Prince with bows and arrows in France in 1355, and they had 14 days this time at Bordeaux before marching out to find the enemy. The Gascon contingent with its *seigneurs* now joined the English army, eminent among them Jean de Grailly the Captal de Buch, Knight of the Garter. More language difficulties followed, and there was much strain on the marshals of the host to keep all together and ensure a maximum of comprehension between troops of different nations and speech.

Monday, October 5, the start: four miles' march to Villeneuve d'Ornon, where quantities of extra hay were bought. A slow beginning, the loading of the hay, some organisational difficulties, rough results from the good wine of Bordeaux, a contingent missing here, various groups in the wrong division, transport trouble, a line of wagons with no horses till well on in the morning, a company of archers separated from their barrels of arrows, the whole mass of men and animals and creaking wheels beginning their cumbrous movement in one enormous group, pushing ponderously through as yet friendly country, on a wide front, cavalry ahead, mounted archers, wagons, footmen, wagons, more archers, more cavalry, leaving a broad track across the land, but at this stage being careful of crops and beasts, and villages.

Tuesday 6th: The whips were out. Twenty-five miles (for a great part of the army at least) to Castets-en-Dorthe. Hay and oats were bought and a substantial sum paid for damage done by Welsh troops who had got at the wine, the combination of that and their fatigue after the march leading to trouble.

Wednesday 7th: After that very stern second day, to make up for all the awkwardness of Monday, and to allow some of the baggage to catch up, a rest day.

Thursday 8th: Ten miles to Bazas, where corn and wine were bought.

Friday 9th: Another rest day; there was a lot to be ironed out at this stage; a new army is not ready for campaign in a day or two.

Saturday 10th: Eleven and a half miles to Castelnau; meat was bought here.

Sunday 11th: A second gruelling 25 miles to Arouille. Those two heavy days' marching caused the loss of many horses, probably as a result of transport fever which only manifests itself after two or three weeks.

Enemy country being ahead, the host divided into its three war groups or 'battles'. This was done in open country near Arouille, and slowly the army manoeuvred itself into Vanguard, the Centreguard under the Prince himself, and Rearguard. The idea was now to move through the hostile territories, taking towns that stood out against the invader, and sacking, burning and looting in proportion as it was thought punishment was just and counter offensive likely to be weakened. All necessary corn and beef, mutton and wine would be seized, and the army, or, as in some cases of these warlike *chevauchées*, quite small marauding bands, were expected to live off the land.

Sometimes the seizure of villages, towns, food and forage, and livestock was conducted in a moderately orderly and disciplined fashion; sometimes, where there was resistance, there was inevitably horror and cruelty. Occasionally the troops would be impossible to govern, when control was broken down by siege, fight or march, by too much wine, or by the sheer lust of battle. Archers from the English shires could be kind or callous invaders, so could the Frenchmen marching with them. If the temperament of the commander favoured discipline, things tended to go better for the population; soldiers disobeying orders not to molest the priesthood, or the inhabitants of villages, or not heeding orders against looting, burning, rape and murder, might be hanged as an example to the rest, but these *chevauchées* were undertaken to cause havoc, not necessarily to bring the enemy hosts to pitched battle; to undermine the strength and prosperity of the invaded country, and bring it in that way to its knees. The Scots behaved in the same way, raiding into the northern counties of England, the French in a similar fashion when their marches led them through areas loyal to England and her allies. Until after the death of Henry V, the success of such methods in France seemed sometimes within English grasp.

Froissart, a contemporary observer of much that went on, describes the typical behaviour of a raiding force ahead of or on the flanks of the main columns of the host. 'When they found a fertile area, they would stay three or four days, until men and horses were rested and refreshed, then they would go on with their scouts out ahead, stripping the land of food and forage, often for ten leagues on either hand, stopping days together where there was good stock and store, collecting herds of cattle for the army, food and wine, more than they needed, and destroying what they did not take.' There were occasions, though, when the invaded would burn their stores and drive their cattle away in time ... 'sometimes the countryside was so wasted that they did not know where to go for forage'. The behaviour of armies does not greatly change from century to century, though now our potential for destruction is so infinitely greater.

On Tuesday November 3 1355, the Black Prince's army reached the great walled fortress-city of Carcassonne and the unprotected town lying at its feet. When the army appeared the town was full of refugees from the surrounding countryside, the fortress filled with great quantities of valuables that had been moved there. The army entered part of the town nearly unopposed and took up quarters. Another part was staunchly defended and in went archers and men-at-arms, leaping chains stretched across the streets, volleying arrows into the resisters until they were driven across the bridge over the river Aude, into the fortress that loomed 150 feet above, double walled, and tower upon tower, impregnable to anything but long siege. Siege was not part of the Prince's plan. Bargaining began, as so often, while the archers enjoyed the good stores and plentiful wine of the town, and a good deal else besides. Where girls and women could be coaxed or coerced without exciting the wrath of commanders, why not? Where coin was to be found, it could be slipped into bag, purse, satchel or quiver without anyone minding, except the French.

The town of Carcassonne (Author's collection).

Two hundred and fifty thousand *écus d'or* were offered to prevent the town being burned, and the French priests pleaded for leniency. The Prince saw his purpose as being both to punish those towns of France that stood against his father's claim to the crown, and to demonstrate the justice of that claim. So the offer was treated as irrelevant. Besides, he wanted to be on his way. The French armies were massing against him. Mobility must be retained. On Friday the town was burned, completely; it proved impossible to save the spread of fire to church property which the morning's orders had specifically excluded from destruction. Trèbes was sacked the same day, and the army quartered in and around Rustiques, leaving behind it, no doubt with regret, the good Roman road that must have made easy passage to Carcassonne for the baggage train.

A few days later Narbonne gave the invaders a very rough reception. When the army moved into the town, as at Carcassonne, the defenders in the fortress gave them a night of bombardment with catapult-thrown missiles, finally forcing the English to withdraw with their prisoners, of whom those who could raise ransom were released, the rest killed.

So the *chevauchée* went on, 300 miles from Bordeaux by this time and the army no doubt pretty well pleased with itself, as the order went to change direction for home. The shortening of the days, the beginning of bad weather, the sloth of the columns, weighed down with their booty, the growing hostility shown in broken bridges, burnt supplies, ruined water (the horses on November 11 had to be given wine to drink, which may have improved the pace, but probably not the direction of the mounted), and news of the main French army moving in their rear meant it was time to go towards winter quarters. On the 19th and again on the 28th, the army passed the night fully expecting a major battle in the morning. There was no battle; the French melted away; but rain, swollen rivers, mud and misery had to be endured, together with open bivouacs, dying horses, and the wounds and ill-health of the campaign. On they went, still firing towns, still pillaging, until Bordeaux was regained in early December; the Gascons were sent home with the promise of an even better, more profitable raid the following year, and rest and wound-licking in winter quarters was the pleasant prospect.

Such a campaign was only part of King Edward's whole strategy against France, but it helps to show the sort of life, rewarding and appalling, that the archer of the longbow's high days was likely to find. It could not be claimed that great military glory had been achieved that year, but that had not been the idea. France was being humiliated; and, as to strategy, a virus in France's blood was good English tactics. The Prince himself wrote: 'We took our way through the territories of Toulouse where were many goodly towns and strongholds that we burnt and destroyed, and the land was very

The Men of the Bow

Longbowmen, crossbowmen and guns at a siege (Bibliothèque Nationale, Paris).

rich and plenteous and there was scarce a day that did not see towns and castles and fortresses taken.' His steward said, 'there never was such loss nor destruction as hath been in this raid ... since this war began', while Froissart considered the expedition brought 'much profit ... our enemies are sore astonished'. That good news came to England in a ship, the *James*, out of Exmouth.

Military glory would come, perhaps next year. Meanwhile there was time to rest, gather new supplies, and make many visits to the accounts offices.

December 16: Nigel Loring got £5 17s 6d for grooms and shoeing, and wages of £9 13s 1d.

David ap Blethin Vaughan had £1 0s 6d for shoeing and £39 0s 6d wages.

January 2: The familiar Hamo de Mascy was paid £11 18s 6d for 63 Cheshire archers.

John Daniers for 18 archers staying in Bordeaux (so some presumably went home), £4 1s 0d.

Ralph Mobberly for 32 archers, £7 4s 0d.

Some of it was due, some was in advance, and payments in winter quarters were regular and fairly frequent. 'The Prince,' says Froissart, 'returned to Bordeaux before Christmas, and he told his lords and captains to quarter in towns about the district, and to rest and refresh themselves and their men through the winter. There was gaiety, noblesse, courtesy, goodness and largesse' for 'the Prince abode there ... the whole winter ... in great joy and solace.'

Much extra bread, wine, coal and firewood, candlewax, dried fish, pork, almonds, hay and corn, rice, honey and sugar had to be brought to the various places where the parts of the army were quartered, and paid for. It was at this time that archer Jodrell went back home, with six other archers, each with a gift from the Prince. They escaped at least the early start of campaigning in January 1356.

Five hundred fresh archers from Cheshire also missed those winter operations, and many more from other parts of England, who were requested in March. The Cheshire men, in their green and white, only reached Plymouth on April 17. Orders went for more supplies. Robert Pipot was sent to get another 1,000 bows, 48,000 arrows and 400 gross of bowstrings from the county Palatine; Little John of Berkhampstead was to arrange for transport of all such stock obtained; and the Chamberlain of Chester would pay the carriage expenses. Meanwhile Pipot dashed off to Lincolnshire on a similar errand and John de Palington went to London for more. Extra archers' mounts were needed, and more wagon horses, the 'strongest possible' each with a groom. Still more arrows were ordered in July.

By this time the Black Prince's second *chevauchée* was well under way. King John of France, with vast forces at his disposal, too vast to be as effective as their numbers suggest, had driven the English northern army, under the command of John of Gaunt, the Black Prince's brother, which was threatening Normandy, back into the Cotentin and had laid siege to and captured Breteuil. Now he could turn his attentions to the Black Prince and the threat in the south. He ordained September 1 at Chartres as the mustering point for his host.

France, the leading power in Europe, was still behind England in the development of a national spirit, a national army organisation, and that pragmatic recognition by one class of another, one specialist of another which allowed the English and the Welsh to forge a homogeneous military power out of a heterogeneous mass. The nobility of France, intensely proud, haughty, regretting the slow moves away from feudalism, mixed ill with the ordinary soldier, and with the newly evolving kinds of warfare brought to them from across the Channel. They despised 'all-in' war; they despised their own foot soldiers as paid men, which by now even in France they nearly all were; and they despised the mercenary crossbowmen hired from Italy, and the yeomen and peasant archers from England. Their idea of war was to arrive, with the enemy, at an agreed place and time, in marvellous panoply, and fight out the matter as a huge tourney, a massive joust to the death and the decision. We have seen and we shall see that, until such attitudes were relegated to heraldry alone, the way was constantly open for the English armies to destroy the high chivalric posture of those who led France in war, the French aristocracy.

By the time King John's forces were on the move south, the English and Gascon raiders, burning and seizing, punishing and harrying as they had the year before, were up into the heart of France in the valleys of the Cher and the Loire. Given the sizes of the opposing armies, the Prince in all wisdom should have scurried southwards, towards Bordeaux. But there was still a wide river between him and the French, over which the main bridge was broken, at Amboise; he had the advantage of time and mobility, and he wanted to join up with the forces of his brother John of Lancaster, who was prudently withdrawing before King John from his own *chevauchée* further north. Prince Edward moved westward towards Tours, hoping to effect the junction of the two forces and still be ahead of the French when he turned south for home. But, although the Prince waited for him at Tours, Lancaster was still too far away at Angers and the French were already near Amboise where they arrived on Monday September 12. The English burnt such parts of the city of Tours as they could reach, and on Sunday September 11 began to march south down the great valley of the Loire. So on Monday evening the French at Amboise and the English who had reached Montbazon were only 23 miles apart. Now began a grim race: the English covered 30 miles in the first two days' march, the French only 20 miles, but in the next two days the French managed 40 miles and only missed the English by hours. This was to be no agreed orderly tourney, but a swoop of revenge and rage.

The English reached Chatellerault in the evening of the 14th, and waited there all Thursday and Friday, apparently quite unsure, through bad intelligence and local hostility, where the enemy were. On Friday the enormous baggage train, loaded down with booty, was ordered to start crossing the river Clain and to be over and out of the army's way by the time it moved on Saturday morning, the 17th. The army was about 7,000 strong. Meanwhile King John had marched round the English and by Saturday was at Poitiers, though his army, at least 16,000 strong, trailed behind him over miles of roads and countryside. Elements of the two armies clashed, there was a brief fight and French prisoners were taken, but now each force knew the other's general position, and flight for the English was impossible.

At this point, as both sides drew up their forces for the inevitable battle, Cardinal Talleyrand, the Pope's emissary, who had already spent great effort to keep the warring princes apart, now made a dramatic intervention and managed to arrange a truce that lasted through Sunday. The two hosts stood still and the bargaining began. King John demanded the Prince's surrender with 100 knights. The Prince refused, and offered to surrender his prisoners, and not to bear arms against the French for seven years. Sunday passed, the English short of food, but with sufficient water in the streams about their position. The truce would end at sunrise on Monday the 19th. That was at about a quarter to six.

The English were drawn up in three divisions on high ground a little northwest of the big woods of Nouaillé, just below the fork of the road that ran south from Poitiers to Nouaillé and Gué de l'Homme. To their front ran a hedge across the slope. The left, under the command of the Earl of Warwick, rested its flank on low marshy ground; the right, under the Earl of Salisbury, a little further back from the hedge, had its flank protected by the southward turn of the hedge, and a barricade of weapon-wagons, and trenches that the 24-hour truce had enabled them to dig. Behind, in the centre, among rows of grapevines, was Prince Edward with the main 'battle'. The archers, some 2,500 of them, were formed up in the usual wedges on the flanks of the two forward divisions and between them, and probably all along the front for the first moments of an enemy attack, after which they would scatter sideways to their main wedges and leave the men-at-arms to cope. Some of their number were held in the Prince's reserve. The whole array was dismounted and the horses taken to the rear, except for a body of cavalry, again in reserve, with the Prince.

The French army was to their north-north-west, on a higher ridge than the English. Ahead a mounted vanguard, only some 500 strong; the rest of the army, learning from the fearsome result at Crécy ten years before, was on foot, their horses left behind in Poitiers four miles away. Their lances had been cut down to about five feet, and their riding sabotons (those long pointed foot armours of the medieval knight) removed. Behind the van were three massive divisions under first, the Duke of Normandy, behind that the Duke of Orleans, and then the King.

While the armies looked at each other and waited, the Black Prince rode round his host, repeating a short exhortation to group after group. Then he took pains that the archers, wherever they stood, should receive this special message: 'Your manhood hath bin alwaies known to me, in great dangers, which sheweth that you are not degenerate from true sonnes of English men, but to be descended from the blood of them which heretofore were under my father's dukedome and his predecessors, kings of England, unto whom no labor was painful, no place invincible, no ground unpassable, no hill (were it never so high) inaccessible, no tower unscaleable, no army impenetrable, no armed souldiour or whole hosts of men was formidable ... Honour also, and love of country and the desire of the rich spoyle of the Frenchmen, doth stirre you up to follow your fathers' steps. Wherefore follow your antientes and wholy be intentive to follow the commandment of your captaines, as well in minde as in body, that, if victorie come with life, we may still continue in firme friendship together, having alwayes one will and one minde: but if envious Fortune (which God forbid) should let us at this present to runne the race of all flesh, and that we ende both life and labour together, be you sure that your names shall not want eternall fame and heavenly joy, and we also, with these gentlemen our companions, will drinke of the same cuppe that you shall doe.'

Early in the morning the main English wagon train, with all the booty and supplies not needed for battle, began to roll south from the battle position. Obviously the idea was to get them away towards Bordeaux while the army stood and fought. The sight of this movement, together with a possible covering move on the Black Prince's part, urged the French mounted van, who thought it was the beginning of an English withdrawal, to charge down towards the hedge and the English front lines. As the French came sweeping down, the archers all along the line of the hedge began to shoot, their arrows striking or flying off breastplate and helmet, but doing enough damage among men and animals to break the steadiness of the charge.

The left flank archers pushed further out, leftwards and forwards, wading into the marshy ground and shooting from the side and then the rear of the plunging French cavalry. Froissart says, 'The hedges on both sides were lined with the archers who began shooting with such deadly aim from either side that the horses would not go forward into the hail of bearded arrows, but became unmanageable and threw their riders.' Some horsemen jumped the hedge into the lines of English but they were hacked down, and this first attack was routed. Riderless horses, many of them wounded, galloped and limped in front of the advancing columns of the slow French second foot divisions, which now came on, for the last 300 yards tramping into the arrow-storm that beat at them from the front and, as they neared, from flank as well; over their dead and wounded they came, many falling, clutching at embedded arrow shafts, on to the hedge, trampling it flat, and locking with the English and Gascon men-at-arms all along the line.

That fight was hard and bloody. It demanded nearly all the Prince's reserve division, and exhausted the better part of the longbowmen's arrows. But at last that French division drew back defeated. Some English followed forwards in pursuit; the wounded trailed or were helped to the rear; the rest leant and sat and got their breath. Water was brought up from the streams and scooped from the trampled marsh on the left; sheaves of arrows were rushed to the archers. Behind the northward ridge to their front, out of sight, were the two fresh divisions of the French.

On the French side the king ordered the Dauphin and his brothers off the field with a large detachment to guard them. There was a pause. The French third division did not come on. Seeing the defeated falling back towards them, watching the departure of the group with the young princes, aware of the awfulness of having to advance on slow foot, without the surging thrill of a charge of heavy cavalry, that division made off towards Chauvigny. They thought it no part of their duty to go on foot into the teeth of the arrow sleet. They thought it better to live, and fight another day.

That flight, and the fact that King John himself now hesitated, shows the effects of the mauling the first two French divisions had suffered. But in the royal 'battle' chivalry outweighed fearful or prudent caution and the king ordered his column forward: a long slow advance of heavily armoured men on foot, glinting plate and helm, bright banners and pennons fluttering, the rolling roars of defiance, 'Mont-joye! St Denis!', the great red Oriflamme floating high above all, down the slope towards the English lines who waited in the dust-filled air. It was often remarked that over a battlefield of armed men hung or blew a cloud of dust and steam from those who sweated between life and death below. The French column was fresh, and a larger mass than the English, who watched in great fear. An Englishman cried out in the Prince's hearing, 'Alas! we are beaten!' The Prince shouted, 'Thou liest, thou knave, if thou sayest we can be conquered as long as I live!'

At this moment the Prince showed his genius for battle. Instead of waiting with his frightened men for the cumbersome attack to break-in his front, he put some of his mounted reserve under the Captal de Buch, and ordered them down the slope to his right, past the north end of the Nouaillé woods, in dead ground to the enemy eye, and so to swing left again, still in the dip, towards the enemy left flank moving down the north ridge. When the Captal was enough up the ridge to see the Prince's position again he was to raise the banner of St George. At that moment the front division would be ordered to advance on the enemy directly ahead. The horses had been run up to the men-at-arms from the rear; the riders mounted (they had never taken their spurs off) and settled in the saddle, firm in their stirrups. The Captal's banner blew out on the distant ridge and the order was shouted down the English lines 'Avaunt! Avaunt!' The line surged forward with repeated yells 'St George for Guienne!', over the shattered hedge and down the slope below it to crash into the French mass at the bottom of the opposite slope, just as the Captal's cavalry burst on to the French left. The mounted archers, dropping their bows and leaping to horse as they could, charged in with the rest with sword and bludgeon, while the foot archers, deploying themselves with individual flair and best advantage, streamed to the French right and shot and shot into and along the French lines, continually closing on them, as they fought to hold their ground. The French column wavered and crumbled, and piece by piece disintegrated, leaving their king-commander and his 14-year-old son, Philip, prisoners. The king was found at the centre of resistance swinging a great battle axe, crying he'd surrender to no one but his cousin the Prince but, so hard pressed was he, he finally gave his right gauntlet to Denis de Morbèque, a French knight of Artois, who fought with the Prince.

The French lost 2,000 men-at-arms dead on the field, 2,000 more prisoners, and uncounted of their soldiery dead and captured. Some of the English, archers as well as knights, had five or six prisoners each, as they roamed the field of slaughter or scattered in pursuit of the fleeing. Froissart tells us: 'All who had accompanied the Prince of Wales were much enriched, in wealth as well as in glory, not only by the ransom money, but also from the gold and silver plate, rich jewels, whole trunks full ... and furred coats ... for the French had come richly and magnificently dressed, as to a victory ... In the evening the Prince gave a supper to the French King ... and himself served at the King's table.'

The King's salt cellar, as we know, picked up by a group of archers, was in the Prince's hands, the archers themselves rewarded. As Froissart said: 'The English archers were certainly of great service and caused great havoc. For their shooting was so accurate and so well concerted that the French did not know which way to turn, as the archers kept on advancing.'

The night fell, and cool air drifted over exhausted men, and over the dead, the wounded, the bruised, the broken bodies and broken bows. For a long time in the darkness raucous bursts of celebration were heard, but there must have been many among the archers whose aching shoulders and numb fingers made them think only of rest.

CHAPTER 6

THE LONG ROAD TO AGINCOURT

In 1360 the Treaty of Brétigny seemed to set the seal on Edward III's efforts throughout the long war with France. But King John of France, who had been a prisoner in England since Poitiers, except for three fruitless years in France trying to raise his monstrous ransom, returning to England in 1363, died the next year, still a prisoner, and Charles V came to the throne. Fighting in Brittany began again, and was ended quickly by Sir John Chandos' victory at Auray, assisted by the Captal de Buch, one of the victors of Poitiers. Archers were there, and did good work, but they were only 1,000 strong and were not the deciding factor in the battle. Except in times of major campaigns and national levy it was difficult to keep enough archers in the field to maintain the decisive proportion that had proved so effective in the past.

The Black Prince went with a large army to the aid of Pedro of Portugal and fought a victorious campaign in Spain ending with a battle near Najera, where the English arrows 'flew thicker than rain in winter time. They pierced through horse and man, and the Spaniards soon saw they could no longer endure.'

Meanwhile Charles V worked patiently and carefully towards a reversal of the situation allowed by the Treaty of Brétigny, and in 1369 the banked fires of the Hundred Years' War blazed again. But it was not to be at all as it had been. The French campaign was in the hands of a brilliant and evasive soldier, Bertrand du Guesclin, the first commoner to become Constable of France. Like the famous Roman Quintus Fabius Maximus who constantly refused a pitched battle to the invading Carthaginians, du Guesclin knew that

French armies stood little chance against massed archers, and decided to wear down the invader by guerilla tactics and the policy of 'scorched earth'. It was a long war of siege and counter siege, of *chevauchées* that achieved no decisive result, and of evasive action by the French. The English armies were progressively exhausted, the exchequer overburdened, and the spirit of adventure, of national pride, and the will for victory diminished.

The numbers of archers going out to France fell, foreign mercenaries increased, and there was growing discontent at home and abroad. The dreadful ravages of the Black Death were partly to blame, and as a result of the lowering of the population there was inflation of prices and wages. A ploughman in the first half of the 14th century might receive 6s a year. By the year after the Black Death he would earn 11s. Then in 1351 came the Statute of Labourers which reduced wages by law. The ploughman could only earn 7s from 1352 onwards, but he was still faced with growing prices. For instance, in one year the cost of a bushel of wheat rose from 10½d to 15d. Statutory limitations were applied to some staple goods, bread and ale among them, but the price limits were usually on a sliding scale related to the price of the raw materials and so subject to inflation. The miseries so well known today, of mounting costs and falling spirits, increasing curbs and lessening incentives, beset the country and its armies abroad.

In 1369 Edward III issued the following order to sheriffs: 'Cause public proclamation to be made that everyone of the said city [London] strong in body, at leisure times on holidays, use in their recreations bows and arrows ... and learn and exercise the art of shooting; forbidding all and singular on our behalf that they do not after any manner apply themselves to the throwing of stones, wood, iron, handball, football, bandyball, cambuck, or cock fighting, nor other such like vain plays, which have no profit in them ... under pain of imprisonment. Witness the King at Westminster, the 12th day of June.'

Nonetheless, it is plain that there was constant difficulty in providing enough archers, although by now almost every archer was mounted. The Duke of Lancaster contracted in 1369 to serve with 1,000 archers, plus 300 'lances and bowmen' (no doubt from Wales) and 499 men-at-arms. The totals for the campaign of that year were 1,343 men-at-arms, 3,858 longbowmen and 500 Welsh, but in the next year Sir Robert Knolleys' force had an equal number of men-at-arms and archers, which became a typical proportion in the armies of this period. In the army of 1373 archers were even slightly outnumbered. But, between 1369 and 1379 some 20,000 soldiers, many of them archers, were indentured for service at sea, and the demand was all the time greater than the supply.

Still, there are signs that it was no bad thing to be a soldier. Archers in the Dover Castle garrison in 1371 received five pints of wine each as part of their daily rations, and abroad the bowmen were often professionals, members of companies following professional captains, who could expect their share of plunder and ransom. There is a story told in the chronicle of Geoffrey le Baker, an Oxfordshire contemporary relating to the year 1352, but it illustrates well the sort of freebootery that began from then on to be more and more common.

'There was an archer named John Dancaster imprisoned in the castle of Guines, who having nothing to pay a ransom was released on condition he should stay and work among the French. Now this fellow used to lie with a laundrymaid who was his strumpet', and he learned from her that there was a causeway across the moat, just under the water, so built that he could escape from the castle, which he did, at the same time measuring the height of the castle wall with a thread. Then he made for Calais, and collected a band of 30 desperadoes from the English garrison, who came, in blackened armour and with ladders of sufficient length. They scaled the walls, killed the guard and threw them into the moat, and then fell on the unarmed 'ladies and knights that lay there asleep', and on others 'playing at chess and hazard'. They shut their prisoners into a strong chamber, released all the English prisoners that were there, and treated them to a feast.

In the morning they let all the ladies go, with horses and belongings, and sent for more help from their friends in Calais. The next thing was that two knights came from the Comte de Guines to find out who had taken the castle in a time of truce. The reply was that they had nothing to say until they had been in possession a little longer. Because of the truce, Monsieur de Guines sent to the English at Calais for an answer to his question. 'And therefore on St Maurice day (the King being busy in parliament)' the French enquired of others. An answer came back from the king that he did not know who had taken the castle and that it ought to be restored to its rightful owners. Back went the French to Guines, 'demanding in whose name they kept it'. The reply this time was 'in the name of John Dancaster'.

There followed a good deal of bargaining. The Comte offered thousands of pounds and 'a perpetual peace with the king of France' to the occupiers, provided they were liegemen of the English king. Dancaster replied they were indeed English, but that 'by their demerits they were banished from the peace of the king of England'. They would sell the castle to none other than the king of England 'to obtain their peace', but if he did not want to buy they would sell it to the king of France, or to the highest bidder. 'The king of England bought it in deede, and so had that place which he greatly desired.'

Why after the bitter lesson of Crécy and Poitiers and many smaller battles did the French not take to the longbow? There were, in fact, many efforts, now and later, to form an efficient longbow corps. Jean Juvenal des Ursins, writing shortly after Poitiers, says 'In a short time the French archers became so expert in the use of the bow, that they could shoot with a surer aim than the English. Indeed, if these archers had formed a close confederacy among themselves they might have become more powerful than the Princes and nobles of France. It was fear of just such an outcome that made the French king suppress the archer army.'

There was more serious unrest in France than in England, and the French king's attitude is easy to understand. In 1357 Froissart tells us, 'there were severe disorders in many parts of France, in Beauvais, Brie, Laon, Soissons and the Marne. Many people from the country towns assembled in mobs ... saying the nobility of France was betraying the kingdom.' They began a rampage of killing, burning, looting and raping: 'they went to a strong castle, seized the knight who lived there and bound him to a stake; then in turn they raped his wife and daughter in front of him, before killing both the wife, who was pregnant, and the daughter, and the other children, and finally they murdered the knight, with great cruelty ... they chose a leader from their number whom they knew as Jacques Bonhomme', and they called themselves the *'Jacquerie'*. They destroyed houses and castles in hundreds, and their numbers grew and grew. Froissart talks of 100,000.

When the fury of the persecuted turned against them and they began to be the victims themselves, killed in great numbers, hanged on trees by the roadside where they were taken, massacred wherever they were found, they were asked why they did such things: 'they answered that they did not know; that they only did as others did, and that they would kill all the nobles and gentlemen in the world'. Eventually that ally of the English, the Captal de Buch, joined his cousin the Comte de Foix and confronted an enormous mob that had invaded the town of Meaux, with a small but well-ordered force, attacked and killed them in great numbers, afterwards setting fire to the town and burning with it all those who had been on the side of the *Jacquerie*. 'After Meaux, the brigands never rallied in any great numbers' ... for some 400 years.

The unrest in England came to a head in the great revolt of 1381. The story of King Richard II's confrontation at Smithfield with the forces of Wat Tyler is too well known to need repeating. The interest from our point of view is that the rebels were drawn up in 'battles' in just the way they had learned from the wars, and we can be certain that a great many of those who watched their leader's death were longbowmen. But the rebel forces were argued away with promises, the hard core of the movement dealt with summarily, and there was never any question of proscribing or limiting the use of the longbow. Richard himself maintained a strong guard of longbowmen, said to number 4,000. The weapon and its use were so much a part of the national scene that any efforts to limit it would have met with small success.

It is not so very curious that after the long wars of Edward I, dispirited and embittered men turned against each other, nor that after Edward III's exhausting wars, bitter struggles broke out again, nor that after the final phases of the Hundred Years' War in France, the Englishman turned upon himself in his tired and wounded rage.

After the revolt of the peasants came the usurption of Bolingbroke, the murder of King Richard and the civil war between the new king, Henry IV, and his enemies, many of them his disenchanted allies of a short while ago. The Scots, under the Earl of Douglas, invaded England and reached Newcastle. They were brought to battle at Homildon Hill, near Wooler, and there a small force of English archers shot at the Scottish schiltrons with such precision, advancing and shooting, retiring when the Scots advanced, shooting, retiring again and shooting again, that the invading force was broken up and pursued to the Tweed, with enormous losses.

The Welsh, led by Owain Glyndwr, fought a long guerilla campaign, during much of which the young Prince of Wales learned the trade of war on the Welsh borders. He wrote models of precise and polite, though also urgent and sometimes hopeless requests to his father, for more troops and more money. Naturally in these Welsh wars the archery was not all on the Prince's side. There is a lyrical piece of toxophily that survives from the writings of Iolo Goch, bard to Owain Glyndwr, which translates roughly thus:

'Suppose I was in that wood yonder, and had in my hand a bow of the red yew, ready braced, with a strong taut string, a straight, round arrowshaft with a well shaped nock, with long low fletchings bound with green silk, and a sharp-edged steel head, heavy and strong, of green-blue temper, that would draw blood out of a weather cock; suppose I had my foot on a tussock, my back to an oak tree, the wind at my back, the sun to my side, and the girl I love best, close by, to watch me, I would shoot such a shot, so strong and deep drawn, so sharply loosed, so low in flight, that it would be no better for my enemy if he wore a breastplate and a Milan hauberk than a wisp of fern, a kiln rug, or a herring net.'

In 1403, the year after Homildon, the king and prince Henry, aged 16, met the rebel forces of Henry Percy, the great Hotspur, with the Douglas, Worcester and Vernon, at Shrewsbury. The battle was a dreadful foretaste of the future Wars of the Roses. The king's force was larger than Hotspur's, but the numbers engaged are very uncertain. The guesses range from 5,000 and

4,000 to 14,000 and 11,000. The smaller figures seem the more likely. Both sides came hastily to the confrontation, and the rebels were not joined either by Northumberland's or Glyndwr's forces which were expected. Both armies had a large proportion of longbowmen.

The royal force advanced among hedges, fields of peas and little ponds against the rebels on a small ridge, and so were badly galled by the archers in the early stages of the battle. The prince, who commanded the left wing, was wounded in the face by an arrow but refused to leave the field. The king, on the right wing, had taken the precaution of dressing several men in royal surcoats which became immediate targets for the rebel archers and men-at-arms. Like so many internecine battles, Shrewsbury was very bloody, and with bowmen against bowmen the toll of life was high. Sir Charles Oman in his *Art of War in the Middle Ages* suggests that 1,600 were killed out of a total of 9,000 involved. This was no fight for prisoners and ransoms; it was a grim struggle for the control of England, which the king and his son decisively retained. Hotspur was killed, the other rebel leaders executed, but the Douglas, caught two miles from the field with a broken kneecap, was pardoned. A legend has it that Harry Percy, in the thick of the fight and sweltering in the July heat, opened the visor of his helmet for a moment to gulp air, and was immediately killed by an arrow in the mouth. True or not, it was a likely enough way for any knight to go, who was rash enough to present any vulnerable part of himself to a quick-eyed longbowman.

Ten years later the king was dead and Prince Henry became Henry V, at 25 years of age.

The new king lost no time in setting about the reopening of war with France. He made unacceptable territorial demands, claiming as his right all the lands granted to his great-grandfather Edward 111 at the treaty of Brétigny, and reviving Edward's claim to the French throne. He believed it was his by right and by the particular wish of God. Henry was a pious man, his greatest ambition, as his father's had been, to throw the infidel out of Jerusalem. He was also an athlete who could bring down a buck 'without any manner of hounds, or without bow or other engine'. Churchill describes him as 'orthodox, chivalrous and just. He came to the throne at a moment when England was wearied of feuds and brawl and yearned for unity and fame.' He had a vision of a Christian Continent united under the crowns of England and France, both of which he himself would wear. It would be wrong to think of him as a tyrant who dragged an unwilling nation into foreign wars. Church and Council, Parliament and the City showed the greatest willingness to back his projects with encouragement and more money than was ever voted to his predecessors.

Longbowmen on both sides at the battle of Shrewsbury (British Museum)

France was still a much-divided kingdom, split by factions and parties, ruled in name only by Charles VI who was ageing, weak and frequently mad. Sometimes he thought he was made of glass and was terrified of breaking if anyone touched him. At other times he would become so violent that his attendants greased themselves to avoid his clutches. He spent long hours strapped helpless to a bed, while his Queen, the Dauphin, the princes of the blood and the great lords of France ruled his kingdom by division.

It was the Dauphin who replied to Henry's claim and challenge: 'Since you are a youngster, you are sent little balls to play with and soft cushions to lie on, until one day, perhaps you shall become a man.' With that message came the box of tennis balls, one of the most celebrated royal gifts in history.

The Dauphin had this reply from Henry: 'If God so wills, and my life lasts, I will within a few months play such a game of ball in the Frenchmen's streets, that they shall lose their jest and gain but grief of their game.'

It was the afternoon of August 11 1415 when the king boarded the *Trinity Royal*, of 500 tons, Master, Stephen Thomas. At three o'clock Henry ordered the mainsail hoisted to halfmast, as a signal to weigh anchor, and his armada's sails began to feel the light wind. Fifteen hundred ships, large and small, carrying a fighting force of 10,000 men with almost as many horses, swung their prows south and, accompanied for a time by a flotilla of swans, sailed slowly past the Isle of Wight. The swans were thought a particularly good omen as Henry's personal badge was a swan.

The army of invasion was made up of 2,000 knights and men-at-arms, some 8,000 longbowmen and 65 gunners. There was a mass of siege equipment; towers and scaling ladders, and guns from Bristol and the Tower of London (one was called 'London', one 'Messenger', another 'the King's Daughter'). There were sappers and pioneers, smiths, painters, armourers, tent-makers, bowyers and fletchers, masons, cordwainers, carpenters, turners, carters and farriers. There were dukes, at a daily rate of 13s 4d, earls at 6s 8d, barons at 4s, knights, esquires and men-at-arms at 2s, 1s 6d and 1s, and longbowmen by the thousand at their new rate of 6d a day.

The fleet entered the estuary of the Seine on the afternoon of Tuesday August 13 and anchored off the Quai or Clef de Caux, three miles from Harfleur, the main French port on the west coast. Landing was forbidden until the next day. The archers gazed at the looming landfall of chalk cliff, discovered its name, and with inalienable and unchanging Englishness christened it 'Kidcocks', no doubt making a good deal more of it than that.

The unloading of the fleet took until Saturday the 17th, and the advance on Harfleur began over marsh and mudflats in broiling weather. Waiting for news from his advance party, Henry halted his sweating troops and sent a message through the ranks: 'Fellows, be of good cheer. Breathe you and cool you and then come up at your ease, for with the love of God we shall have good tidings.'

Orders had already been issued for the control of the invasion army which like any other at that time assumed loot, rape and every form of warlike excess to be its prerogative. The inhabitants were to be encouraged to go about their lawful work. There was to be no wanton destruction. Churches and monastic buildings were not to be harmed, and no man must lay hands on a priest, a woman, or any servant of the church unless caught in armed violence. There were prohibitions against the taking of farm animals and implements, or the hindering of those who worked with them.

By Sunday night, August 18, Harfleur was encircled. Through the rest of the month and well into September the siege continued. The city was bravely defended and the army outside suffered dreadfully from war, heat and disease.

Dysentery became as formidable an enemy as Harfleur itself. The blazing summer days dragged by. Expectations of a French relieving force grew and waned, and grew again, but no force came, and at last unaided Harfleur yielded to the invaders on September 22. The siege was over but the dysentery continued, and the desertions, for which this opening phase of the campaign was remarkable.

The question now was, what next? Should Henry march directly on Paris before the French co-ordinated their strengths, and while they were separated by dissentions and factions, and by widespread floods that came with sudden and heavy rainfalls? Or should he go home? The season was late, the army was weakened by casualties, disease and desertion, and would be further depleted by the loss of a strong garrison for Harfleur. After the heat of the summer the unseasonably bad weather would now hamper English movement as much as French. The arguments for returning home, and abandoning any further activity till the spring, were strong, and they were firmly put by the king's war council. But, at some point before October 1, Henry's belief in his destiny made him decide to overrule advice and to march north to English-held Calais. He would make a *chevauchée*, like his ancestor's, carrying his banner of France and England through the country to show that the rightful heir of France could march free in his own dominions, but he would do it without destruction.

'We will go, if it please God, without harm or danger, and if they disturb us on our journey, we shall come off with victory, triumph and very great fame.' So said King Henry. He sent a challenge of single combat to the Dauphin, waited eight days for a reply which he knew would never come, and on October 8 marched out of the north gate of Harfleur.

The army had about 6,000 fighting men, of whom 1,000 were men-at-arms, the rest archers. All, or nearly all baggage wagons were left behind, and each man carried rations for the eight days it would take to reach Calais if all went well. Historians both then and now have disagreed as to whether or not the force was all mounted. No doubt as many were mounted as possible and, since an army's speed is that of its slowest members and eight days for 160 miles, the distance to Calais, would mean an appalling race for men on foot, it is likely the king took pains to mount the entire force, just as it is certain he must have had enough pack animals to carry arms and equipment. I guess some 20 tons of arrows were loaded on horseback; it is known that a crown, a sword of state, a gold cross set with jewels and a piece of the True Cross six inches long were included in the sumpter baggage of John Hargrove, sergeant of the royal pantry.

J. H. Wylie, in his extremely detailed biography of King Henry, described the army of October 1415 as 'a crumpled and ragged squad, with harness out of joint, and jacks all tattered'. Harfleur, in its estuary marsh, was below and behind them as they gained the open ground to the north, and Normandy opened broad and fair before them, when it was not raining. The wet season, which had started early, was becoming worse. They saw little villages and farms sheltered by trees, low forest on the skyline, the occasional grey loom of a castle, all calm and pleasant, except for the sodden weather.

By October 11, marching an average of 15 miles a day, they reached Arques, 60 miles on, 100 more to go. There, after a cannonade from the castle, a threat to burn the town at its feet, and a parley, food and wine were sent out to the English.

The same threats the next day at Eu produced similar results, but only after a sharp fight with the garrison that charged out of the gates to be met by a storm of arrows. From Eu, Henry made for the ford at Blanchetaque, hoping like his great-grandfather to force the Somme crossing, but when they reached the ford, the king probably having Edward III's map in his hands, they found it staked and heavily guarded by a large force of French on the other side.

There was only one thing to do. Henry turned right and began to feel his way along the south bank of the river for a crossing. The French were out. The plan was going wrong. Now, as they marched along the waterlogged valley of the Somme they must have felt less and less protected from watching eyes, as the thinning leaves revealed them more and more.

With the English army rode Thomas Elmham, chaplain to the king. He saw it all, and wrote it down: 'We had no other expectation but that we must go quite to the upper parts of France, and to the head of the river, which was said to be upwards of 60 miles away ... When it was reported that a multitude of the French were preparing to fight us ... we then expected that after having finished our eight days' provisions, they would proceed artfully before us, laying waste the country, and strike us with famine ... Without other hopes ... so very few, and wearied with much fatigue and weak from want of victuals ... the next day, Monday, we left the town of Amiens about a league on our left. On the following day, Tuesday, we came to a village by name Bowys.' This was Boves. Here below the castle on the hill, that stands in phantom ruins among the trees today, the English found great vats of wine 'to the great refreshment of the army'.

The wine caused some kind of a riot among the archers. Henry ordered that drinking must stop at once. 'What need?' said someone. 'The brave fellows are only filling their bottles.' 'Bottles!' said the king. 'Their bellies are their bottles. They are drunk!'

A French chronicler said of this campaign that, while the English did not rob or rape, their armies in the area consistently did both.

At Corbie prisoners were taken, who told the English, says the chaplain, 'that the French had appointed many companies of horsemen, in hundreds, on armed horses, to break through the battle and strength of our archers ... therefore the king gave orders that each archer should provide himself with a square or round pole or staff, six feet in length, of sufficient thickness, and sharpened at each end; directing that whenever the French should approach to battle, with troops of horse of that sort, each archer should fix his pole before him in front, and those who were behind other poles intermediately; one end being fixed in the ground before them, the other sloping towards the enemy, higher than a man's waist from the ground'. One more thing for the men to carry. They were living on dried meat and nuts, and many were suffering badly from the flux.

At Harbonnières, on the 17th, an archer 'stole from the church a pix of copper gilt, which he mistook for gold'. It was found in his sleeve when the priest complained of its loss. Henry had the culprit bound, led through the army, and hanged from a tree. On the 18th 'we were quartered in moderate sized farm houses near the walled town of Neel [Nesle] ... News was suddenly brought to the king that about a league off there was a convenient ford over the river.'

On the 19th they reached two fords, close together, one at Voyennes, one at Béthencourt. The approach causeways over the marshes had been partially destroyed. 'The king had the breaks filled up with wood, fascines and straw, until three could ride abreast; and he ordered the baggage of the army to be conveyed over one of the causeways, and his army across the other; where, stationing himself at one entrance [Voyennes], and some chosen men at the other [Béthencourt]' a disciplined crossing was made 'until great numbers soon collected beyond the river ... We began to cross about an hour after noonday, and it wanted an hour to night when we had entirely passed over.'

French horsemen had come up during the crossing, but held their distance and eventually disappeared. This same day, Saturday the 19th, a second French army moving from the south-east had arrived at Péronne. French and English slept seven miles apart that evening, with no river between them.

On Sunday the 20th Henry let the army rest on high ground above the river valley. He rode ahead with a reconnaissance party. They were met, to their great surprise, by heralds from the French, who told Henry: 'Before thou comest to Calais they will make thee to fight with them', and they asked what route he would take. The king, says Elmham, 'with a courageous, firm look, and without his face changing colour', replied that he would march 'straight to Calais; and if our adversaries attempt to disturb our journey it shall not be without their utmost peril. We seek them not, nor for fear of them shall we move slower or quicker.'

Armour was put on and Henry drew up his army on the defensive, between woods on the Athies-Péronne road. But there was no attack. Evidently both French and English were feeling their ways tentatively, uncertain of each other's strength and position. The heralds had had much trouble finding the English on Sunday.

On Monday the English moved on, prepared for battle at any time. At Péronne 'after we had passed the town about a mile, we found the roads strangely trodden by the French army, as if it had gone before us in many thousands'. That is one of the most vivid moments to survive from any account of medieval warfare. Elmham did not claim to speak for the leaders, he said, but 'we who were the rest of the people raised our hearts and eyes to heaven, crying for God to have compassion upon us, and to turn away from us the power of the French'.

A French chronicler (Laboureur) gives the other side of the picture: 'The French assembled all the troops that were dispersed and ordered them to follow Henry's route, and to keep in the fields without lodging in the villages, except at night. The king of France came to Rouen ... with an army capable of conquering the best disciplined forces ... The city of Paris offered 6,000 men ... the Duke of Berry spoke much in praise of this militia ... but Jean de Beaumont replied with contempt: "What do we want with the help of shop-keepers, since we are three times the number of the English?" ' He spoke of only one part of the assembling army.

The English marched on, north-west now, towards Albert and Acheux. For days they had passed through, and would for miles to come pass through, little towns and villages, a whole land known field by field, ditch by ditch and wood by wood to soldiers, British, French, Belgian and German who, 500 years later there, fought and died innumerably. The silent armies that stand now in white untroubled rows, corps, squadrons, 'battles' of the dead.

The English went on through the rain, to Frévent, hungry, in many cases ill, and tattered. Somehow Henry kept his men together. On October 24 they reached their last river, the little Ternoise, at Blangy. They drove off a small party that was starting to destroy the bridge and crossed over. Seven Lancashire archers were taken prisoner in this skirmish.

After the crossing there was a steep climb. In a moment of clear weather the advance guard reached the top of the ridge and looked right. They saw the amalgamated armies of France moving slowly up the open valley to their right, and already ahead of them. Three great masses of men and horses, gleaming with armour and lances, forested with bright heraldic banners, guns, carts, wagons, and countless glossy caparisoned cavalry. A breathless scout spurred to the Duke of York with the vaward, who in turn clattered down the hill to fetch the king.

By the time Thomas Elmham could get a view of the three French columns, they were 'about a mile from us. At length they were formed into battalions, companies and troops, in multitudes compared with us, and they halted little more than half a mile opposite us filling a very wide field, as with an innumerable host of locusts, a moderate sized valley being between us and them.'

The king drew up the English 'with great courteousness', expecting battle. All were confessed that could be – the chaplain complains of a lack of priests – and busy as he was Elmham heard, 'amongst other speeches', Sir Walter Hungerford regretting in the king's presence 'that we had not 10,000 more of the best English archers who would gladly have been there. Then the king said "thou speakest as a fool, for by the God of Heaven, on whose grace I lean I would not have one more, even if I could ... this people is God's people. He hath entrusted them to me this day, and he can bring down the pride of these Frenchmen, who so boast of their numbers and their strength."'

The French, 'having for a little while examined and considered our small force', moved on, and the heads of their columns began to disappear behind the woods to the right. The English waited. All through the afternoon the French, re-emerging from behind the woods, spilled ponderously across the plain to the north, across the little road to Calais, filling the ground between the woods of Tramecourt and Azincourt. 'When at length daylight closed, and darkness hid us from them, still we continued on the plain, and heard the enemy as they quartered their people, each one ... calling for his comrade or servant, or friend, who might be at a distance in so great a multitude. Our men began to do the same, but the king commanded silence throughout the whole army ... He turned immediately off in silence to the village nearby, where there were houses for some to rest in, but very scanty gardens and orchards, where we were exposed to much rain through nearly the whole night.'

The village was Maisoncelles. There the 6,000 sheltered as they could, while they listened to the confused roar of the vast French host settling less than a mile away.

Among the French, wine was drunk; tomorrow's prisoners were played for at dice; a cart was gaily painted to drag Henry captive to Rouen; and the lackeys ran to and fro fetching straw for their masters' beds, fodder for the animals, and walking the horses up and down, to stop them shivering and stiffening in the cold and wet, up and down, trampling and deepening the mud across what would be tomorrow the very front of their own position of battle.

The dawn came cold and wet, 'that is Friday, the feasts of Saints Crispin and Crispinian, the 25th of October. The French arrayed themselves, and ... took their position in terrific numbers before us in the said plain named Agincourt, through which lay our road towards Calais; and they placed many companies of horse, in hundreds, at each side of their vanguard, to break up the line and strength of our archers; the van being a line of infantry, all selected from the nobles, and choicest of them, forming a forest of lances, with a great multitude of helmets shining among them, and the horse in the flanks, making a number, by computation, thirty times greater than all ours.' By the end of the battle, when he knew the numbers better, he spoke of 60,000 against 6,000.

We know the numbers of the English. We know their condition. They had marched for 17 days with only one day's rest, the Sunday after the river crossing, and they had covered 260 miles, averaging a little over 15 miles a day. We know how they were drawn up on the morning of October 25. The Duke of York had the right wing, the king the centre, Lord Camoys the left, by Azincourt woods. Three small battles, made from a total of 990 men. On either flank, and between each battle, were the archers, in their wedge formations, numbering 5,000 on a front of about 1,000 yards. They were a little in front of Maisoncelles, a very gentle slope running down from them, and rising again towards the French, forming so slight a dip as to allow almost no dead ground between the armies. The king 'made them fix their poles before them, as had been before determined, to prevent them being broken through by the horse, and when the enemy learnt this ... they kept at a distance opposite us without approaching ... He ordered the baggage of the army to the rear of the battle ... together with the priests ... in the before mentioned village and closes ... for the French plunderers had already, on every side, their eyes upon it.'

Our eye witness was now well to the rear of the army. The morning was a long and awful wait. There were exchanges between the leaders, but neither side suggested anything that was acceptable to the other. Eventually at midmorning the French were seen to be sitting down about their banners. Fires were being lit, and apparently they were having a meal.

They were in three massive battles. In the first some 8,000 heavily armed nobles and knights, on foot, stretching from wood to wood. Armour had changed a good deal since Poitiers. Now the men-at-arms were encased in steel. The casque and cuirass alone of one French knight, Ferri de Lorraine, which has survived, weighs 90 lb (40.8 kg). Four thousand archers and crossbowmen, intended as a screen, had been elbowed aside by the nobles, and were now uselessly deployed behind the cavalry and the first infantry battle. In the second battle were 14,000 lesser nobles and mercenaries, ranged behind the

The field of Agincourt from the air. Maisoncelles is on the left; Azincourt in the centre of the picture; the old road to Calais is shown clearly running through the defile formed by the woods of Azincourt; and Tramecourt on the right (Photograph by Graham Payne).

first. On each flank were some 1,000 heavy cavalry. The third battle, half a mile or more behind the other two, was a corps of mounted men and foot some 18,000 strong, mostly raised from the communes on the march, ready to pursue the flying enemy. There were some guns on the French side and one English soldier is known to have been killed by a gunstone.

In the front ranks were 12 princes of the blood, and in the first two battles thousands of dukes, counts, barons, bannerets, knights and esquires, all proud of their birth, jealous of each other or each other's factions, vying for command and position. The Duke of Orleans, 23 years old; Philippe Comte de Nevers, 26; the Comte d'Eu, 20; three princes of the house of Bar, the eldest 22; the Comte de Vaudemont, 29; the Due d'Alençon, 30; the Due de Bourbon, 34; the Duc de Vendôme, 28; Arthur de Bretagne, 21 ... the banners were so thick that early in the morning many had to be furled and taken to the rear.

King Henry was 28. He rode up and down the army, his helmet on his head, his crown on his helmet, in which were set four great and four small rubies, 20 sapphires and 128 pearls. The largest ruby was given to the Black Prince by Pedro of Portugal and can still be seen in the Imperial State Crown.

When he addressed the army, earlier, 'in a loud, clear voice', telling them of the justice of their quarrel, reminding them of their families at home, he also told them of the French boast to cut off three fingers from the right hand of every archer taken, so that he would never shoot again at man or horse. The archers grew restless as the hours crawled by, and the king realised that he must attack, or stand and starve.

Some time before 11 o'clock a Welsh captain, Dafydd ap Llewellyn, known to the army as Davy Gam, came back to the king from reconnoitring the French. Henry asked him how many he thought there were out there. He replied, 'Sire, there be enough to be slain, enough to be taken, and enough to be chivvied away.' Henry took advantage of the spread of this report through the ranks, rode once more down the lines on his little horse, cheered by the men, rode back to the centre, dismounted, sent his horse to the rear, and knelt on the ground. The whole army knelt, made the sign of the cross, and kissed the ground in token of being ready to return to earth.

The old white-haired Sir Thomas Erpingham, the marshal of the army, shouted orders to the archers, who pulled up their stakes.

The king cried, 'In the name of Almighty God, in the name of Jesus and Mary, in the name of the Trinity, Avaunt banner in the best time of the year and Saint George be this day thine help.'

The marshal's baton was flung into the air, so that all could see it as it twisted and fell, and the whole line swung forward with a yell, which startled the French to their feet and echoed between the woods.

The advance was slow, in the heavy plough. Twice they halted for the heavily armed to get their breath, for the flank archers against the woods to steady their formations, and twice went on, each time with a yell.

'But I who write this, sitting on horse-back among the baggage in the rear of the battle, and the other priests who were there ... with fear and trembling ... cried unto heaven, beseeching God to have compassion upon us and the crown of England.'

When the archers were within long shot of the enemy, they drove their stakes into the ground and ran behind them, bows braced. An order was yelled all along the line, 'Now – stretch', or 'Knee – stretch', or 'Now – strike'. We shall never know which. The French said it sounded like 'Nestroque'. If it was 'knee – stretch' then we can see the front rank of archers going down on one knee, angling their bows a little but, however we try to listen through the silence of the years, we cannot quite hear that far call.

Oman says a shiver was seen to pass along the whole front of the French line; it was the lances coming down to the position for attack. Their flank cavalry started to charge but because of the chaos and crowding less than a quarter of the intended number rode forward, across the thick, poached mud.

'The horsemen of the French posted all along the flanks, began to attack our archers on both sides of the army ... but they were quickly compelled amidst showers of arrows to retreat and fly [ten arrows a minute from every man within range] ... with the exception of a few who ran between the archers and the woods, yet not without slaughter and wounds ... a great many, both horses and horsemen were arrested by the stakes and sharp arrows, so they could not escape far.'

Wylie wrote, in 1919: 'When the French launched across in answer to the English shout, they had not counted on clay furrows neshed with rain, and as they stumbled up the archers caught them like a butt, the foremost of them toppled over turve, got haunched on the stakes, the horses flounced and plunged and would not face the huzzing arrows, but turned tail and galloped back in plumps, tearing huge gaps in the vanguard ranks.'

The chaplain's account continues: 'The enemy's crossbowmen, behind the men-at-arms on the flanks, after the first too-hasty discharge in which they hurt but few, retreated from the fear of our bows.'

The French first battle had begun to advance, in a massive crowd through the mud-clogged stampede of the cavalry remnants.

'When the French nobility, who at first approached in full front, had nearly joined battle, either from fear of the arrows, which by their impetuosity pierced through the sides and beavers of their bascinets, or that they might more speedily penetrate our ranks to the banners, they divided themselves into three troops, charging our lines in the three places where the banners were.'

The chaplain, on the rise towards Maisoncelles and on horseback, must have had a good view of this separation of the front attack, as the crowded French pressed inwards upon themselves, and away from the arrow storm.

'Intermingling their spears closely, they assaulted our men with so ferocious an impetuosity, that they compelled them to retreat almost a spear's length. Then, we who were assigned to the warfare of the spirit, fell upon our faces ... crying out in bitterness of spirit for God still to remember us and the crown of England ... to deliver us from this iron furnace and dire death ... nor did God forget ... Our men regaining strength ... repulsed the enemy, until they recovered the lost ground.'

The English, in tight discipline, could with their shallow formation give, or push as they needed. The French in their mass, constantly pushed from behind, only jammed themselves more and more until they could hardly lift hand to wield sword or lance.

'Our archers pierced the flanks with their arrows ... and when the arrows were exhausted, seizing up axes, poles, swords and spears which lay about, prostrated, dispersed and stabbed the enemy ... Merciful God, who was pleased that England should, under our gracious King, his soldier, and that handful, continue invincible ... as soon as the armies were thus joined,

Massed soldiers in battle. Chronique de Hainaut (Bibliothèque Royale de Belgique).

increased our strength, which had before been debilitated and wasted for want of victuals, took away our terrors and gave us a fearless heart.'

The French second battle pushed up behind the first. When they were drawn up they had a front of over 1,200 yards. Where the press of the fighting was, in the neck of the space between the woods, they were forced into a space of 900 yards.

'They were seized with fear and panic. There were some, even of the more noble of them, as it was reported in the army, who on that day surrendered themselves more than ten times. But no one had leisure to make prisoners of them, and all without distinction of persons ... were put to death without intermission ... When some of them were killed and fell in the front, so great was the undisciplined violence and pressure of the multitude behind, that the living fell over the dead, and others, falling on the living, were slain; so that in places, where the host of our standards were, so great grew the heap of the slain ... that our people ascended the heaps, which had grown higher than a man, and butchered the adversaries below.'

Somewhere in the mass, the fat Duke of York was squashed to death. In the centre, Henry was beaten to his knees and the crown sheared through on his helmet, losing a fleuron. On the flanks, where the English archers could shoot from the edges of the woods, the French began a confused withdrawal. The English stood, unable to follow, standing among the crawling heaps, exhausted with butchery, blind with sweat.

'And when, in two or three hours, that front battle was perforated and broken up, our men began to pull down the heaps, and to separate the living from the dead ... but behold, immediately (in what wrath of God is not known) there arose a clamour, that the hinder part of the enemy's cavalry, in incomparable and fresh numbers, was repairing its ranks and array, to come upon us.' Men came running from the baggage train and told of guards killed, a crown and the sword of state and many horses carried away.

The Duc de Brabant, who had ridden all day from a family wedding in the north, with a handful of men and without his proper armour or heraldry, borrowed armour from his chamberlain, Goblet Vosken, took the cloth of arms from his trumpeter, cut a hole in it, pulled it over his head, and tried to rally the fugitives. Thus there was threat from the front and the rear, at the moment when the surrendered French outnumbered their exhausted captors.

Henry ordered every man to kill his prisoners. When there was stubborn refusal at the order he brought up 200 of his archer guard to do the work. The French stood about in crowds, their helmets off. The ground was thick with weapons. If there were now another French attack the prisoners could pick them up and use them. The English fell on the French, paunching them, and stabbing them in the face. Many were killed before it was seen the French attack was not developing. This massacre was never laid to Henry's blame by contemporary chroniclers. Twenty years before the French had cut the throats of 1,000 prisoners, before the battle at Nicopolis, for mere convenience. The fault, to them, lay with the three French nobles who had led the raid on the baggage park, and who were severely punished: Isambard d'Azincourt, Robinet de Bournonville and Rifflart de Clamasse.

The three-hour fight was over. The remains of the second French battle and the unscathed third fled on their fresh horses or were hunted off the field, and melted away on the roads to Ruisseauville and Canlers, and along the paths in the woods.

'We returned victorious through the heaps and piles of slain.'

Ten thousand French were dead, half the nobility of France, perhaps another 2,000 made prisoner. The English loss was incredibly, but certainly no more than a few hundred, probably less than a hundred.

'Nor could several refrain from grief and tears that so many soldiers of such distinction and power, should have sought their deaths ... to no purpose.'

Heralds from the French knelt before Henry, who asked to whom the victory belonged. 'To whom should it belong but to you?' He asked the name of the castle that rose above the trees to the left. 'Azincourt.' 'Then let this battle forever be called the battle of Agincourt,' he said. 'It is not we who have made this slaughter, but Almighty God.' Then there was sung 'Non Nobis' and 'Te Deum'. When Henry was 17 years old he had written to his father, after a small victory against the rebel Welsh, 'It is well seen that victory does not rest with the multitude of people, but in the strength of God.'

'After a long space of time during which the king kept the field, and when day declined towards evening ... he returned to lodge with his army in the same village as on the preceding night ... The night being spent, the king returned with his army through the middle of the plain where the battle took place, as the most direct way, and found all the bodies of the slain naked and stripped.

'There is not a heart of flesh or stone, if it had seen and contemplated the dreadful destruction and bitter wounds of so many Christians, but would have dissolved and melted into tears.'

★ ★ ★

Wylie wrote of the archers, 'Trained from their boyhood by constant practice at the bowmarks that were fixed near every parish church, these quick-eyed longbowmen could hit the prick of the oystershell in the centre of the butt with a nicety of a Thames fisherman garfangling an eel, while for nimble

The crucifix over the French grave pits at Azincourt.

readiness in the field they stood unequalled in the western world ... they wore no cumbering armour about the chest, but a loose-fitting jack belted at the waist, and nothing on their head but a wicker brain cap, stretched over with querbole, or pitched leather, strengthened on the crown with two crossbands of iron.'

The French crossbowmen at Agincourt we know about. They also had archers with longbows, and all we know of them is that after the battle, among the wagons that were left behind, were found many bows and bales of arrows not even unpacked.

Among the archers who fought for England on October 25 1415 at Agincourt were Jankyn Fustor, Lewis Hunte, William Bretoun, William Gladewyne, John ap Meredith, Morgaunt Filkyn, John Grafton, Richard Walsh, John Hartford, David Whitcherche, Richard Whityngton, David Taillor, William Glyn, Thomas ap Griffith Gogh, Matthew Bromfeld, John Hert, Yevan ap Griffith and Thomas Tudur.

The French dead lie buried in the centre of the battlefield. Over the pits, in a little spinney, is a cross set up by the family who own and farm the battlefield today, as they did in 1415. They lost a father and two sons in the battle. A few hundred yards away is a memorial to another father and two sons of the family, killed by the Germans in the Second World War.

In 1961 I visited the family to ask permission for cameras to film the battlefield. They were at first reluctant. I was told, 'It was a bad day, for you as well as for us, and many dreadful things were done by you to us, here.' I have a letter from them in which is written, 'We defended our fields in 1415 and in 1915, in 1939 again, and often in between.'

In the 19th century an English antiquarian who had met great courtesy from the family sent them his thanks, and a picture of Henry V. He received in return a picture of Joan of Arc.

The much-argued figure of Harold and the fatal arrow. Bayeux Tapestry.

Right *Norman archers with short bows among the Norman cavalry. One wears a ring mail byrnie. All carry quivers. Bayeux Tapestry.*

Opposite *Illuminated page from the Hours of Catherine of Cleves, showing longbows, crossbows, a centre-shot stone bow, arrows, bolts, quivers and quarrel bags, circa 1440 (Pierpont Morgan Library).*

Below *English armies attack a French town, longbowmen to the fore (Bibliothèque Nationale, Paris).*

post hoc exilium possim dei fili
um. contemplari cum beatis. ĩ
conspectu deitatis. Amen

De scis fabiano et sebastiano oro.

O quam mira refulsit
gracia Sebastiani
martir incliti. qui
milites portans insignia. sed

Above *Panorama of the south-east facing slope of the battlefield of Crécy, up which the French charged. The Vallée des Clercs is in the foreground, Crecy on the left, Wadicourt on the right. The windmill position can be seen as a small hump to the right of a standing tree on the left of the picture. Below that fought the Prince of Wales. The Bois de Crécy Grange looms on the horizon (Photograph by Graham Payne).*

Right *English longbowmen against cumbrous French crossbows. It is plain to see who has scored the most hits (Bibliothèque Nationale, Paris).*

Opposite, centre *Pillaging (British Museum).*

Opposite, bottom *Longbowmen in battle, their shafts in their belts and on the ground. One is re-stringing his bow (Bibliothèque Nationale, Paris).*

Above *English longbowmen at the battle of Poitiers. The effect on horses is horrifyingly shown, and arrows in this picture are unequivocally piercing plate armour. Note the soft quiver in the foreground (Bibliothèque Nationale, Paris).*

Right, centre *The field of Agincourt from above the village of Azincourt, looking across to the woods of Tramecourt. The French grave pits are marked by the rounded clump of trees to the left of the picture. It was between there and the east-west road running across the centre of the picture that the heaviest fighting took place (Photograph by Graham Payne).*

Right, bottom *Jeanne d'Arc with longbowmen on her side and against her (Bibliothèque Nationale, Paris).*

Left *Archer's equipment including bow, arrows, pocket quiver, tassel, bracer and tabs, and a reproduction of a Tudor quiver made by John Waller of the Medieval Society (Photograph by Graham Payne).*

Below *The charge of the French at Patay (Boutet de Monvel).*

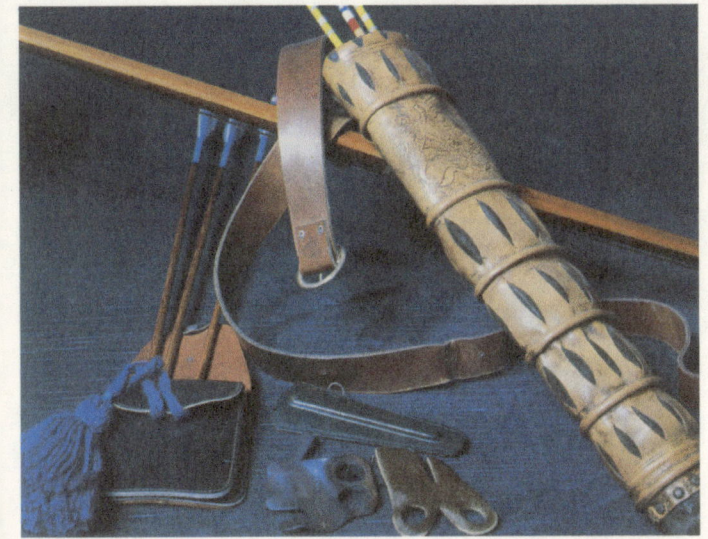

An American Indians with longbow. From John White's water colour, circa 1584-90 (British Museum).

Above, top *The Kings ship* Mary Rose *from the contemporary manuscript in the Pepys Library, Magdalene College, Cambridge.*

Above *Details from the same manuscript giving, in columns, the ship's list of ordnance, artillery, munitions, equipment etc. including (in modern English):*

Right *Dr Margaret Rule, CBE, Archaeological Director of the Mary Rose Trust, swimming at the wreck site with one of the longbows found during 1981 (Mary Rose Trust).*

Below, left *Jon Adams holding the first longbow (No. 80 A812) brought to the surface from the wreck site of the Mary Rose (Mary Rose Trust).*

Below, right *Some of the thousands of arrows being cleaned at the Mary Rose Trust. The blackened, unidentifiable heads are readily seen (Mary Rose Trust).*

Left *Dr. Margaret Rule, spray in hand, with the first box of longbows to be brought to the surface (Mary Rose Trust).*

Below, left *Modern reproduction and similar arrowhead of 14th/15th century which certainly represents one type of head commonly used in 16th century warfare (Author).*

Below, right *Detail from the Martyrdom of Saint Sebastian by Hans Memlinc (1440–1494). The bows of more than half a century earlier than the Mary Rose longbows might be portraits of the Mary Rose staves as we know them. Note the small horn nocks and no bound handle (Musée du Louvre, Paris).*

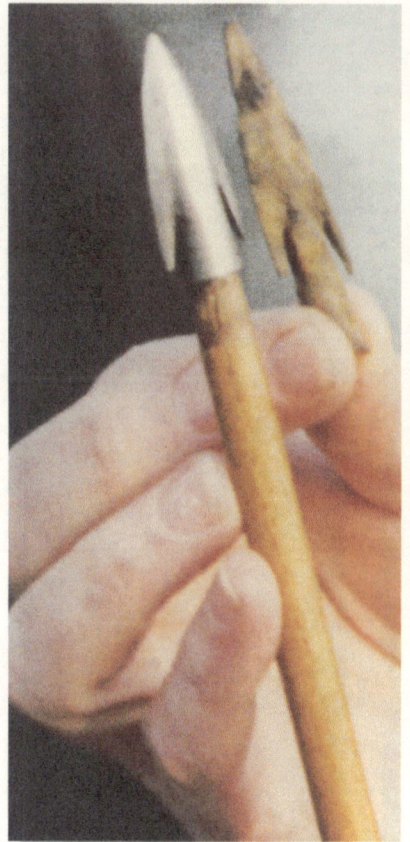
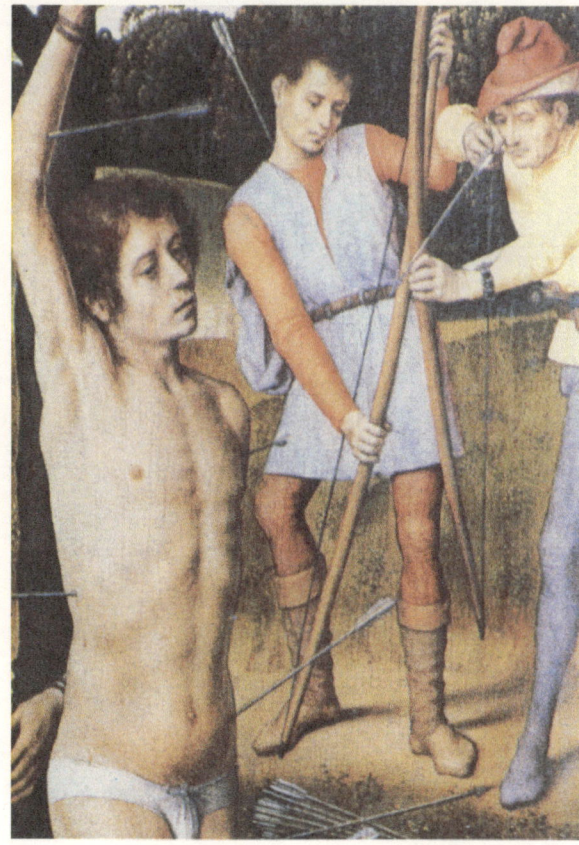

Four steps in the emergence of a Mary Rose *Approximation bow, made from fine Oregon yew.*

Left *She begins to bend under the hand of Roy King, bowmaker to the Mary Rose Trust (Roy King).*
Above *MRA 1 on the Tiller (Roy King).*
Below, left *MRA 1 being drawn up with a 30 inch (76.2 centimetre) bodkin arrow by Simon Stanley (Author).*
Below, right *MRA 2 drawn up by Simon Stanley (Author).*

CHAPTER 7

FROM JOAN OF ARC TO ROGER ASCHAM

From the day of Agincourt until he died of dysentery contracted at the siege of Meaux, Henry V's strategy was the unspectacular but careful, piecemeal, subjugation of Normandy. He died on August 31 1422 at the Château de Vincennes, and his body was carried, guarded by the archers that he used to call his 'yew hedge', to Abbeville, past the field of Crécy, to Hesdin, a little short of the field of Agincourt and so 'to Calais and to England then ...'

During all these seven successful years of slow conquest and consolidation the French avoided a pitched battle. Curiously the largest fight was an English defeat, at Baugé. Henry's brother Thomas was pursuing with cavalry a mounted force of French double the size of his own. He was without infantry, who had not come up yet, and the French suddenly turned in their tracks, charged, and cut the English and the prince to pieces. Had he waited for his archers, no doubt he would have won the fight, but few archers were well enough mounted for swift pursuit.

There were two other 'Agincourts', at Cravant in 1423, against a French and Scottish army, and a much greater engagement, when half another Franco-Scottish army was killed or captured at Verneuil the next year. In both these bloody battles the archers were used much as before, though not in such great proportion, not because their effectiveness was forgotten, but because there were large numbers of Burgundian infantry present and they were not long-bowmen. The Douglas, inveterate and unfortunate foe of England, fell at last to English arrows at Verneuil.

The future looked bright for England. Its baby monarch was, in name at least, king of France and England. English armies, though over-extended, were in fair order and on the whole well led. But France was at the point of a renaissance; about three years before Agincourt Jeanne d'Arc was born at Domrémy. By the time of her fame her surname must have had a curious ring for English bowmen. In 1429 the siege of Orléans was raised when the French, in vivid attack, charged behind Jeanne's banner and bright spirit. In 1435 Burgundy, England's ally, made peace with France. The decline had begun.

There are many signs that during the slow recoil from success the armies had to proliferate rules and regulations. There were stricter investigations of equipment and training, more frequent musters and returns of men, and pay was constantly reviewed. Soldiers were forbidden to leave one retinue for another, and captains were prohibited from holding on to men from other companies. Muster rolls were made in duplicate, and sometimes at the ends of lists empty parchment was carefully scrawled over to prevent the insertion of extra names. Failure to 'pass muster' because of absence, illness, or inadequate equipment meant loss of pay. In 1423 there came a remarkable advance towards a treasury system that we can recognise today. Hamon Belknap was made Treasurer General, and Pierre Surreau Receiver General; both were civilians. The old days were gone. The Civil Service, the professional overseeing and implementing of army accounts, and the red tape, had begun.

From now on, captains must provide lists of names under oath, and at each muster archers must show their bows and arrows, and demonstrate their shooting ability – '*monter les arcs et tirer tous ceulx qui seront passes pour archiers affin de veoir la souffissans*'. In 1424 commissioners had to swear that they had carried out such tests on every archer named: 'all the archers, one after another, were made to shoot at the butts', and the men had to show they possessed swords, bucklers, caps and cloaks, as well as proper uniforms, that they had sufficient arrows, and were properly mounted. Archers who could not bring a horse to the muster, or were detected borrowing horses to meet the requirements, were sometimes fined half their pay.

Great insistence was put on the proportions of archers to other soldiers. One lance or man-at-arms should always be balanced by three longbowmen. After a muster at Dover of 44 lances and 135 archers in 1424, when all was checked and declared in order, £616 8s 5½d was paid over and signed for. Payment could be balked to captains who failed to provide muster rolls at the proper time. The only way captains could avoid the obligation of muster and review was to contract at a flat rate to provide a force or a garrison of a definite number, or a number sufficient for the job; then it was their own responsibility and their rate was at risk in proportion to their failure in the contract.

Equally it was to the contractor's advantage. Such great captains as Talbot, who mustered large numbers of men, were entitled to one third of the 'gains of war' of their companies, if the captains were responsible for wages rather than the Treasury. In such cases the overall rate would be lower than the total of wages previously paid by the Treasury.

Both the government of Henry VI, '*par la Grace de Dieu Roy de France et d'Angletonne*', and his captains took the greatest pains not to pay for men who had gone absent or were away on little 'vacations', though there are often examples of fair-minded treatment, for instance to an archer in the Honfleur garrison who had managed to acquire a house which he let out for rent, but was admitted to have fulfilled his military duty properly. In one roll of 1434, nine archers were marked 'promoted foot lances' with a consequent increase in pay. The proportion was officially dropped at certain times to two archers for one lance, though the old balance of three to one or even four to one was constantly attempted.

There are precise notes of days served and not served, and of men punished, even executed for grave offences. Martin Lenfant, a Norman archer, lost his pay for missing a muster. An archer in the garrison at Conches was found to own land, by the king's gift, worth 10 livres a year, and his wages were accordingly reduced. Another bright rogue called Dancaster, a descendant perhaps of the captor of Guines, loaned a horse to a soldier for the muster day only, but when his accomplice Thomas Houvy's deceit was discovered his pay was not rebated because it was found that he had served properly and stood his watches. No doubt both he and Dancaster were dismissed with a caution.

Absences with good excuse were allowed, and escort duty was always recognised, though sometimes rebatement had to be fought. In March 1438 John Hastings went about his duties with two archers as an escort, who thereby missed two musters. It took him some time to coax their pay out of the Treasury. The next year when he went off with three archers he explained their absence from the February muster in advance to the commissioners, and there was no further difficulty. Men who went off on private adventures were not rebated as long as it was felt their adventures had been conducted against the enemy, even if to their own personal profit. The wounded were paid full wages too: '*Geffray Vrecknock* [of Brecknock] *archier fut malade*' for three weeks in March 1430 and was paid in full. Another archer was paid for '*le petit temps*' that he was a captive at Fresnay though prisoners were not always treated so well. One man failed to avoid a loss of pay for absence by pleading that he had 'gone to Le Mans to watch a fight between an Englishman and an Armagnac'. There were complicated and careful arrangements for the payment of archers transferred from one command to another, and the enormous detail of the army administration is shown by the fact that between 1429 and 1441 there are records of nearly 900 separate musters. It is all a far cry from the earlier days when, for example, in 1421 a muster roll for Falaise was submitted countersigned by two men-at-arms and two archers.

The military still protected its own against harsh civilian rules, when it was possible. Lord Scales' garrison was not properly mustered for the quarter September 29–December 28 1432, simply because the commissioners did not turn up till December 31. The Treasury refused payment, but Lord Scales went into action and obtained a warrant for payment because the system had been nonplussed by the officials themselves.

There are many details which help to build up the picture of an archer's life. By 1429 a small proportion of Frenchmen were allowed in companies both as men-at-arms and archers, so long as the latter were passable as '*archiers biens tirans et habillez souffisant*'. In the muster of the Pontoise garrison of June 1430 one third of the lances were French, and one eighth of the archers. Five Englishmen and three Frenchmen were married. This last piece of information preceded a general, extremely detailed, and secret muster of all the forces in Normandy the following year with a view to stopping soldiers carrying on trade in their garrison towns. Organisation was finally catching up with comfortably situated soldiers like the landlord in Honfleur.

The carefree days of pillage and living off the land were long past. Now musters began to be taken by the month, and payment in cash was monthly. The port authorities on both sides of the Channel were given powers to arrest deserters; those who changed companies without authority were liable to arrest as well, and the penalties in either case varied from simple return of the offenders to their units, to imprisonment and forfeiture of wages. When excitement and success and chances of booty were still in the air it was a different matter; now an archer who had received 'the king's wages' in whole or in part for his service, and who had 'departed and gone where he would' was to be apprehended and 'punished as a felon' and 'held for inquiry before the justices of the peace, except that he hath reasonable cause showed by his captain and by him'. What used to be a question either of private contract or compulsory levy, the latter being a form of recruitment that had not worked successfully for a long time, was now a public duty, the evasion of which constituted a crime against the king and the country. The organisation of war had moved into the modern context.

At the same time the novelty was gone. There grew up in France whole generations of captains, men-at-arms and archers, professional and mercenary soldiers, who seldom saw their homeland until they returned to it in defeat. They were paid well, and some of the magnates made great fortunes.

It is no great surprise that when they were finally chased home to an England where power was divided between kingmakers and factions the result was an appalling civil war. The nobles took sides and any of them with enough money could very easily raise large forces of trained soldiers who knew no other method of life. But, before we return to divided England, let us look at the last of those great archer armies of England in defeat across the Channel. Success in major engagements had come to them when they could stand on the defensive against massed attack, first by cavalry and then armoured men on foot in dense, ill-ordered battles. Archery was of course used in attack, but demanded the concerted organisation of foot and archers, or horse and archers. On the march, or while encamped, or if surprised by an imaginative and mobile enemy their strength was much less formidable.

On June 18 1429, an English army was in retreat. The pursuing French were unsure what to do. Jeanne d'Arc cried to their leaders, 'You have spurs! Use them!' They dashed after the English, under Talbot, Scales, Fastolfe and Sir Walter Hungerford (who had wished for 10,000 more archers at Agincourt), but did not make contact with them. The English were about 3,500 strong, and rather short of archers. The French came after them in three battles, led

The longbow in siege warfare (Bibliothèque Nationale, Paris).

by La Hire, Xantrailles, Alençon and Dunois, and Jeanne, who to her fury was kept in the rearguard. At two o'clock in the afternoon, French patrols had still failed to locate the English, when suddenly one of the French scouts put up a stag which ran off to the right and vanished among the copses on a ridge ahead. Within a moment of its disappearance there were loud 'holloas' from beyond the ridge. The English had seen the stag and now the French had heard the English. The French advance guard spurred to the ridge top, in the dip behind which they saw the small English force. They charged straight down on 400 archers who had no flank protection, and overwhelmed them. The archers, surrounded, shot as they could but they were totally crushed by the speed and extent of the attack. The rest of the French came on fiercely, and the whole English force was rolled up and rolled over within a few minutes. Talbot was captured. Fastolfe got away with some archers in a fighting retreat, to be unreasonably accused in Paris of having lost the war in a single afternoon. Why this defeat at Patay?

Henry V was dead. Jeanne d'Arc was alive, the inspiration of the French, and there were simply not enough archers in the English force, nor were they in a position from which they could shoot steadily into their attackers. They were unprotected on their flanks and they were in a dip. The waves of cavalry rolled down on them and drowned them.

In April 1450 as the French slowly forced the English back and back from their conquests, an army of 3,800 men, 1,500 of them archers, under Sir Thomas Kyriel formed up to face a slightly smaller French force, which had with it two guns.

Kyriel's formation was an imitation of Agincourt. A single line of men-at-arms with interspersed wedges of longbowmen, and wing archer formations. So short were the English of men-at-arms that their line had to be backed by billmen using a sort of lance-cum-pike-cum-battleaxe (one innovation the English had achieved) and some archers. This gave the bowmen the advantage of shooting the whole length of the line, but it depleted their main, enfilading bodies. There had been time for a careful preparation of the ground, which was backed by a ridge; stakes were driven in on the front and potholes had been dug with swords and daggers. Not only was Agincourt recalled, but Crécy; there was a windmill on another ridge to the south of the English position, where Kyriel had a small contingent.

At about three o'clock the French approached to within two bowshots, 600 yards or so, and halted. After a pause the French advanced straight ahead on foot, and suffered badly from the archery. Flank attacks of cavalry were repulsed as well, and Kyriel had occasion to think the victories of the past were to be repeated here, at Formigny.

Next, the French brought up their two guns, 'culverins', and started shooting at the English archers, who, in a fury, charged the guns and captured them, and began to drag them back to the English lines. The French broke off and withdrew, defeated. Had Kyriel attacked in force now he might have routed them completely, but he did not, and as he stood in his position, unmoving, a second French army under the Constable of France appeared in the distance, first sighted a mile and a half away by the guard at the windmill. It was not a large force, probably consisting of 1,200 lancers and 800 archers, whom we must take to be longbowmen, not 'arbaletriers'. They would have been supplied from the corps of archers now privately and officially raised, with royal encouragement, and known as the 'francs archiers', uniformed and well-trained men who claimed certain tax exemptions and other privileges. In every way, by 1450, the French had learned from the successes of their enemies in the past.

The Constable galloped off to find Clermont, the French commander in retreat, and persuaded him to rally his dispirited troops. Then he returned to his own force and began to advance against the English left flank and rear, extending his own left flank to join Clermont's re-formed right. Kyriel had to swing his left round far enough to allow his front, in the shape of a bow, to face the new combined attack, a manoeuvre which would hardly have been achieved without some confusion, even had it not been tried at the end of a battle. There are indications that Kyriel's was not the best-found or best-organised army. A force that marched without any digging tools may not have carried enough arrows for the archers to have many left to face this second fight. Whereas the French archers had full quivers.

The English, according to a French writer 'held themselves grandly', but they were pushed back and broken up piecemeal. Five hundred English archers, surrounded in a garden by a little river, fought to the death of the last man. Five-sixths of the English were killed. They lie buried in grave pits in a place called, until recently, Le Champ Anglais.

Three years later, the great veteran captain, John Talbot, Earl of Shrewsbury, well over 60 years of age, was sent out to Bordeaux with 3,000 men to the assistance of England's allies in Gascony. Normandy was already lost, and Guyenne was all but conquered by the French. But when 'Le Roi Talbot' arrived he was greeted with delight, and worked wonders with his small force. He marched rather against his better judgment to the relief of Castillon, beleaguered by the French, surprised a large force of French archers in their quarters at dawn, put many of them out of action and pursued the rest to the French camp three-quarters of a mile away. Talbot followed on towards this powerfully fortified French position, without his artillery, and found himself up against more than 150 guns, a large army, and a number of 'francs archiers'. Deceived by reports of the movement away from the camp of a body of horse (which in fact had been ridden off to make room in the constricted space for the archers who had been chased there), Talbot decided to attack. He directed the operation, unarmed, from the back of a little white cob. After release from his second capture three years earlier he had promised the French king never to wear armour against France again. So instead of a helmet he wore, as he rode out that day, a cap of purple velvet.

To the roaring of the guns, in the smoke-filled river valley Talbot and his men were slaughtered. His cob was brought down by a cannonball, a French battleaxe cleft his skull, and his body was found after the battle *'les cuisses et les jambes transpercées de flêches'.*

Castillon was really the final act of the Hundred Years' War, and of the long and happy alliance between Gascony and England.

In 1957, another John Talbot, Earl of Shrewsbury, visited Castillon with his wife. When they were at Villeneuve, before reaching Castillon, intrigued by some old buildings under a bridge they asked their history. The French replied that they were workshops and smithies 'built by the English, during the war'. 'The war?' *'Bien sure! La guerre de cent ans.'* When it was discovered at Castillon who the visitors were, the battle was explained in great detail, and they were shown exactly where the first Earl had died. The guides regarded his death and defeat as a tragedy. 'The war' to them ended in 1453, and disastrously. The Gascon memory of the old alliance with England is long, and grateful.

★ ★ ★

Except for their hold on Calais, the English were gone from France but, in spite of the horrors and the casualties and the burdensome taxes of the long war, England emerged from the years of strife a great nation of warriors, with a high, nationalist spirit that even the next 30 years of civil fighting did not eradicate.

Now in England the longbow was turned against the longbow, and the resulting slaughter of brothers and sons and fathers was appalling. The cannon and the handgun were still principally weapons of siege, and except in particular circumstances were no threat to the importance of the archer. The great battles of the Wars of the Roses, Towton and Barnet and Tewkesbury, were fought without the important use of artillery, much as Shrewsbury had been fought half a century earlier. At Northampton, though the Lancastrians, like the French at Castillon, stockaded themselves with a large number of guns, their method was useless because heavy rain soaked the gunpowder and hardly a piece was finally discharged.

Because the archers fought on both sides, and in a measure cancelled each other out, the longbow was not the deciding factor in these wars, but it was the main contributor to the masses of dead and wounded.

Sir John Smythe wrote more than 100 years after Bosworth. (We shall meet him again in the great controversy about the respective values of the bow and the gun, but two things he said are revealing as we look back over the wars in France and forward at the struggles of the wars of York and Lancaster.) He said:

'Many times French captains and gentlemen attribute all the former victories of the English against themselves ... more to the effect of our Archers, than to any extraordinary valliancy of our Nation ... They did think that the English Archers did use to poison their arrowheads; because that of great numbers of the French ... wounded or hurt with arrows, very few had escaped with their lives; by reason that their wounds did so imposthume, that they could not be cured. In which their concepts they did greatly err; because in truth those imposthumations proceeded of nothing else but the very rust of

Longbows, crossbows and hand guns together (Bibliothèque Nationale, Paris).

the arrowheads that remained rankling within their wounds ... Not only the great but also the small wounds of our arrows have been always found to be more dangerous and hard to be cured, than the fire of any shot.

'Besides all which it is to be noted, that horses in the field being wounded, or but lightly hurt with arrows, they through the great pain that upon every motion they do feel in their flesh, veins and sinews by the shaking of the arrows with their barbed heads hanging in them, do presently fall a-yerking, flinging and leaping as if they were mad, in such sort, as be it in squadron, or in troop, they do disorder one another, and never leave until they have thrown and cast their masters.'

So it was at Crécy, and at countless other battles, and so it would be through the civil war of the late 15th century, though often as in France the men-at-arms dismounted, left their horses in the rear and threw out their flanking archers as they advanced or stood to battle. In England it was far easier to amass great forces of archers than it had been in France. The towns and villages were full of them, returned from the war. The campaigns were short and swift, the distances of the march often small. Great sums of money did not have to be spent on maintaining transport, or on feeding men over long periods. When the battles were over, most of them returned home, until they were called out again. The nuclei of these archer armies were veterans kept in the pay of magnates on either side, whose numbers were swelled with others called out by the old system of commissions of array. Naturally many people evaded these levies, especially in the towns, where men called out by one side and the other failed to turn up and fight for either, waiting until the bloody issue was decided, before declaring their loyalties.

There were also many thousands raised by the system of 'livery and maintenance', whereby, as in the old way of feudal allegiance, captains and lesser magnates undertook to raise forces for those greater masters whose cause they followed, to wear their livery and fight under their banners in return for the great ones' protection and 'maintenance', or patronage. For instance in 1452 Walter Strickland, a Westmorland landowner, contracted in writing with the Earl of Salisbury, to follow the earl with his own men, who were listed as 69 'horsed and harnessed' bowmen, 74 billmen, equally provided with arms and mounts, 71 unmounted archers, and 76 unmounted billmen. The contract included a disclaimer of disloyalty to the sovereign against whose armies Strickland's force would serve. Such a method was in common use, based on the claim of those who fought against the king that they were in arms purely to deliver the king from those who misled him.

It was this form of recruitment that took over from the shire arrays, as the war continued.

At battle after battle the longbowmen volleyed their thousands of arrows into each others' lines and into the men-at-arms. When the barrage and the slaughter were intolerable both sides closed, to hack and thrust until the issue was decided.

On Palm Sunday, March 29 1461, something like 25,000 men were killed and wounded in a day-long battle fought between enormous forces in a snowstorm, at Towton Heath, two miles south of Tadcaster. The little river Cock ran red with blood for miles. The Lancastrian army chose the ground. Edward IV, aged 19, with his Yorkist army reached Saxton on the Saturday evening. They drew up for battle, and spent the cold snowy night where they were. As the Yorkists advanced to battle the next morning in a sudden blizzard of snow, the wind was behind them, blowing snow in Lancastrian faces. As soon as the Yorkist bowmen were within extreme range, the extra range that the wind would give them, they stood, in the earpinching storm, and loosed one high flighting volley at the enemy, then backed up hastily into the obscuring snow. The Lancastrian archers immediately replied, but they were blinded by the driving snow, the wind caught their arrows and steepened their fall, and the volleys were short.

I have no grounds for this next assumption, but I imagine that the Yorkist archers, shouting and crying out as if struck by the flights of arrows they watched falling to their front, deliberately encouraged more and more discharges. It is fact and not imagination that when the useless barrage petered out, forward came the Yorkist bowmen again, pulling Lancastrian arrows from the ground, advancing step by step, shooting down the snowflown wind the enemy's own arrows. This wild arrowstorm as thick as snow, in snow thicker than arrowstorm, went on until the archers ran out of arrows and the lines crashed together. For the rest of the day, the most frightful battle ever fought on English soil went on, with unparalleled violence and hatred, and unequalled slaughter.

At Edgecote Field near Banbury in July 1469, the Earl of Pembroke had more spears and billmen than archers; his adversary from the north, who had a large number of bowmen, shot Pembroke's force to pieces and rode down the remainder with cavalry.

In the heavy armour that had now developed, once a battle was joined the fully armed man had small chance of escape, so the losses among the nobility and the leaders of both sides in the civil war were enormous. Heavy, efficient and complicated as a man's full war harness was, with its steel boots, greaves to guard the legs below the knee, cuisses for the thighs, its breech of mail, the hanging tassets, breastplate, vambraces for the lower arm, rerebraces for the upper arm, pauldrons, gorget, helm and gauntlets, yet still the steel-headed arrows, driven true, could pierce not only mail but the plate armour itself. Some arrows were deflected, of course, but even then, in glancing off a curve or slope of steel, might find a chink in another man's armour or pierce his mail breech. Some parts of a good armour would withstand arrows, and undoubtedly the harness of this time afforded greater protection than ever before, but it was not arrowproof completely. The thicker the armour the slower its wearer. The thicker the arrowstorm driving at such slow-moving massed targets, the more vulnerable was each joint or overlap, by hanging plate or moving part.

In 1472 Edward IV had issued a statute from Westminster: 'Our sovereign lord the King ... hath perceived ... that great scarcity of bowstaves is now in this realm, and the bowstaves that be in the realm be sold at an excessive price, whereby the feat of archery is greatly discontinued and almost lost.' To cure this deficiency, every merchant bringing goods into England in any 'carrack, galley or ship of the city or country of Venice, or of any other city, town or country, from whence any such bowstaves have been before this time brought' must also bring 'for every ton weight' of merchandise 'four bowstaves, upon pain of forfeiture to the king for every default', 6s 8d. Sheriffs, mayors, bailiffs and port governors had to appoint examining experts to certify that proper amounts of staves came in, and to classify and mark them for quality 'after the manner as such staves in times past were wont to be marked'.

A hundred years later, a list, in the Hatfield Papers, examines the continental provenance of yew staves. First there was the Bishopric of Salzburg. The staves were sent down the Rhine and the Main to Dort, and then shipped to England. Such bows in 1574 fetched £15 or £16 a hundred. Secondly they came from Switzerland, above Basle. These were £3 or £4 cheaper than the German staves. Thirdly they came from 'Revel, Dansk, Polonia, and all countries east of the Sound', and sold for no more than £4 or £5 a hundred 'being of hollow wood and full of sap by reason of the coldness of those countries'. Fourthly they came from Italy, via Venice, and these were of the 'principal finest and steadfastest woods by reason of the heat of the sun, which drieth up the humidity and moisture of the sap'.

The stipulations for imported staves include orders that 'each bowstave ought to be three fingers thick, and squared, and seven foot long; to be well got up, polished and without knots'.

Edward IV tried again and again to maintain the quality and numbers of English archers. 'Every Englishman, or Irishman dwelling in England must have of his own a bow of his own height, made of yew, wych or hazel, ash, auburne [laburnum], or any other reasonable timber.' Butts had to be

From Joan of Arc to Roger Ascham

Longbowmen against longbowmen. Such confrontations lasted throughout the Wars of the Roses (Bibliothèque Nationale, Paris).

maintained and used on feast days in every township, and those who failed to shoot both up and down the ranges were to be fined ½d.

Unlawful games came in for attack again in 1470 'as dyce, coytes, tenys ... and many new ymagyned games' and 'every person strong and able of body should use his bow, because that the defence of this land was much by archers'.

That Edward's methods met with success can be seen in all we know of the Wars of the Roses. At Tewkesbury the king had the advantage in the number and weight of his guns, and the mass of his archers, who poured showers of arrows into the Lancastrian entrenchments.

On July 18 1473 the Commons granted the king £51,117 4s 7d in full payment of wages for 14,000 archers ... 'towards the payment whereof every county, city and town is severally taxed'.

On November 30 1474 the Commons granted to the king 14,000 archers to serve him at their cost for a year.

On June 20 1475 Edward IV landed in France with '1,500 persons in full armour, and each with several horsemen in their retinue, and 15,000 archers on horseback'. Louis of France wisely abstained from conflict and at Amiens bought off the invaders with money, promises, and wine and venison pies for the whole army. The yeoman, if he could not draw his bow and send his grey goose shaft hissing, at least dined well that day.

The orders and statutes poured out. The king's 'subjects in every part of the realm have virtuously occupied and used shooting with their bows' ... but now the king's subjects, because of the scandalous profiteering of the bowyers, 'be not of power to buy the bows'. Therefore, no bowyer or merchant could sell yew bows for more than 3s 4d. Penalty 20s.

Richard III in 1483 again had to control prices. This time, because of high charges by the 'seditious confederacy of Lombards', staves were costing £8 instead of £2 a hundred, and were not properly 'garbled', or sorted for quality. He also increased the import duties. With every cask of Malmsey or Tyre must come ten bowstaves, 'good and able stuff'. The penalty was 13s 4d for every cask of those wines without the proper number and quality of staves.

The last battle, and the death of the last Plantagenet king, at Bosworth, saw an interruption of the wholesale slaughter of Englishmen by Englishmen; a pause which lasted for some 150 years. On an August day in 1485 each army deployed archers to about a third of its total, in a fight lasting only an hour, and engaging only half the numbers of those present. Richard Plantagenet, betrayed by his allies, galloped with a bunch of friends to certain death, the last medieval warrior, the crown hacked from his helm. Among the victorious Henry Tudor's army was a corps of 2,000 mercenaries from France armed with handguns. It is not thought they were responsible for the victory. It is true that the muskets of Waterloo 330 years later were less efficient than the longbow in speed of shooting and in accuracy, but there were no longbows at Waterloo. From the last decade of the 15th century we watch the slow abandonment of the longbow in favour of the long pike and the musket. But for all that, Henry VII controlled the price of longbows once again, to 3s 4d each.

★ ★ ★

James I of Scotland, who had been a captive in England from 1406 until 1424, and then reigned until he was murdered in 1437, like the French at that same time tried to introduce into his country the use of the long war bow. He ordered 'that all men busk [practise] themselves to be archers from the age of twelve years, and that in each 10-pound-worth of land there be bow marks ... Yeomen of the realm between 60 and 16 years shall be sufficiently provided with bows and sheaves of arrows.' But in Scotland the orders were no more successful than they proved in France. There were longbowmen, but in the great days of English archery neither French nor Scots could match them, and then the gun displaced the archers as it did in England.

Henry VIII was on the throne in 1509, James IV was king of Scotland, and François I was king in France. Henry was 18, an athlete, like Henry V and, like him too, son of a usurper and anxious 'to busy giddy minds with foreign quarrels'. He was a longbow man, and with his great stature and strength could outshoot his own archers of the guard. 'No man drew the great English bow with more strength than the king, nor shot further, nor with truer aim.' At the enormous pageant in France, the Field of the Cloth of Gold, Henry shot alone before the great of the assembled armies and 'repeatedly shot into the centre of the white at twelve score yards'.

Pageant apart, Henry VIII invaded France three times. His first attempt to engage himself in Europe was limited to sending 1,500 archers to support his father-in-law, Ferdinand of Aragon, against the Moors. There was nothing new in this. Both Richard III and Henry VII had sent archers to Spain, and to Brittany. Some of those who went in the train of Earl Rivers, in the year of Bosworth, were described by a Spanish chronicler as 'loud and ungoverned in their drinking ... liable both to noisy revel and sudden quarrel'. They were otherwise proud, silent, contemptuous men, who had the highest opinion of themselves and their skills, 'marvellous good soldiers in the field, skilful archers' who disdained to rush forward in battle, or make dashing attacks like the Spanish or the Moors, but who fought with deliberation and obstinacy, and were slow to admit defeat. They would press into any opening made by the men-at-arms, shooting with great vigour and calmness, spreading death and destruction all around them. 'Withal they were greatly esteemed but little liked' by the Spanish, who found them staunch comrades in battle, but did not willingly seek their company in camp. 'They practised often with their great yew bows ... and shot with unerring aim at any mark set up for them.' So there was nothing new in the fact that Henry VIII's archers in Spain 'fell to drinking of hot wines, and were scarce masters of themselves'. A second force followed the first within a year, and followed their example too, drinking, eating garlic with their meat, and unaccustomed fruit 'which caused their blood so to boil in their bellies that there fell sick 3,000 of the flux, and thereof died 1,800 men'. Dysentery was not always a fever of the marshy lands of northern France.

Henry still eyed France, Scotland still eyed England, and France eyed both, one as enemy the other as ally. In 1513 Henry invaded France, and the same year and less than two months later Scotland invaded England. On September 9, Scots and English met at Flodden. Henry in France had captured two towns and won a minor battle, the Battle of the Spurs, so called because the French cavalry, routed on their maddened horses by the English archers, used their spurs to ride headlong away before they could be surrounded by the rest of the English force. They seemed to have remembered Jeanne d'Arc's cry at Patay, and adapted it to their own predicament.

Flodden was a major engagement, which started with a long exchange of artillery. Precautions to protect powder must have developed recently, for it rained throughout the battle, and the guns still fired with great effect. The result of the English gunfire was to bring a division of galled Scots charging down from their dominant position on a hill. The excitement and the apparent success of this attack brought King James' division down after them. The English cannon were accurate, and the archers from Cheshire and Lancashire (even now) huzzed and whirred their arrows as their forbears had done so often. The Scottish right, still on the hill top, suddenly found themselves attacked by a force who had climbed up to them through rain and mist, many of whom had taken off their shoes for better grip on the slippery hillside, and who let fly with arrows, driving the Scots back on their own reserves. The English archers followed up and shot at will into the confusion with devastating effect, until the bills and hand-arms went in to finish the rout. Back went the victorious English down the hill to where James IV fought among the best of his nation. He fell with an arrow in his head. Dusk and disaster to the Scots ended a battle which left 12,000 casualties among the defeated, 4,000 with the victors. It was the last battle at which it could be said that without the longbow the issue might have been different. The Scots had archers, Ettrick forest men, as at Falkirk, the 'Flowers of the Forest'. After Flodden they lay withered and gone. They had been outnumbered and outshot.

Why then was the longbow doomed? Because armour for vital parts was now being made thick enough to withstand the longbow arrows. Because it was harder to train archers than gunners, and harder to keep them trained to use bows of sufficient strength to be really effective. The gunners at Flodden certainly claimed the victory. Guns were newer, more dramatic, and fashion was the thing to follow. Guns for siegework were essential and, though they demanded a large ammunition train, they were becoming all the time more efficient, while it was a constant fight to raise archers in sufficient numbers, and supply them with enough weapons and ammunition.

But Henry VIII tried hard, 'especially,' wrote his early biographer, Lord Herbert of Cherbury, 'since the use of arms is changed, and for the bow (proper for men of our strength) the caliver begins to be generally received, which ... may be managed by the weaker sort'. Not only was the population falling, but the countryman was not so stout a yeoman as his ancestors. The tillers of the soil were increasingly driven from common lands which were being turned into sheep runs, and 'a few sheepmasters would serve for a whole shire ... and none left but a few shepherds, which were no number sufficient to serve the king at his need', and 'shepherds be but ill archers'. There was widespread bubonic plague in the 1540s and 1550s, and the even more lethal 'sweating sickness' seems to have killed about a third of the able-bodied population.

There was a lowering of living standards, a general loss of sheer physical strength among men 'pinched and weaned from meat', and physical strength, maintained by good diet and the robust work of the old

Above, left *Siege of Boulogne by Henry VIII, 1544. Longbowmen are there, though they are hard to find. Engraving by James Basire after S. H. Grimm from the original paintings at Cowdray which were destroyed by fire* (National Army Museum).

Above, right *The author with two* Mary Rose *bows in his right hand and a bough-wood longbow belonging to E. G. Heath in his left.*

agriculture, was a prerequisite of the longbowman. The wages of archers too were whittled away by inflation. '6d a day now will not go as far as 4d would before time, and therefore you have men so unwilling to serve,' wrote the author of *A Discourse of the Common Weal of this Realm of England* in 1549. Mercenaries were hired in increasing numbers to control the insurrections that resulted from the unquiet state of the country, and to which English soldiers were often sympathetic, and which they were reluctant to oppose. Further, the system of general recruitment that derived from the usages of the civil war, which themselves were a kind of revival of the old feudal system, as we have seen, were now more and more questioned or rejected by those who were ordered to arms. The social changes that were taking place, the inflation, the extravagance that was much complained of, the reduction in the size of households and the loss of belief in the ideas of duty and service, all worked together against the continuation of great numbers of strong and willing archers.

But Henry's efforts to maintain the longbow were fierce. In 1510, Piero Pesaro presented a letter from Henry VIII to the Signory in Venice asking for permission to buy 40,000 yew staves. This was against the laws of Venice, but the Doge granted the permission.

In 1528 a long proclamation was issued which reviewed the past glories of the bowmen of England, pointed out the abuses to which the laws and statutes to promote archery were now subject (this attack was no novelty), inveighed against the 'new fangleness and wanton pleasure that men now have in using of crossbows and handguns', against the idleness, the destruction of game, the usual (though now slightly altered) list of vain games, and so on. Then came the quick of the ulcer: 'no manner of person or persons must from henceforth ... have, or shoot in, or use any crossbow or handgun, nor keep any ... in their houses or any other places'. Persons were to seize any such weapons, or inform about their use. People who refused to give up such weapons for destruction did so on pain of death. Games players risked 'the king's indignation and imprisonment' if their 'tables, dice, cards, bowls ...' etc, etc, were not yielded up to be burned.

In 1529, Richard Rowley, a blacksmith, earned £6 13s 9d for delivering to Sir William Skevington 5,000 archers' stakes 'ready garnished with head, socket, ring and staple of iron, at 5 score to the hundred' [sic], and 2,500 sockets, rings and staples 'to garnish archers' stakes'. The Agincourt stake or a more sophisticated relative was still in use. So were 'leverye' arrows: Walter Henry, the king's fletcher, received from John Boyle, hardwareman of Sheffield, 192,000 at 8s a 1,000; and William Tempille, king's fletcher, had £23 5s 0d for 310 sheaves of new livery arrows at 18d a sheaf ... 'also for new nocking, new feathering, new heading, and new trimming of 500 sheaves of old arrows which came from the wars when the Duke of Suffolk was Captain General in France, at 9d a sheaf.

John Wilshire and Thomas Hixon, wardens of the Fletchers' Company of London, had £4 5s 8d for sorting out 'fectyff' or decayed, moth- or worm-eaten arrows in the Tower.

The Ordance Report for the Tower on September 21 1523, revealed that there were 'ready for use' 11,000 made bows, 6,000 staves, 16,000 sheaves of livery arrows (384,000), 4,000 sheaves of arrows with nine-inch fletchings (for close shooting?) and 600 gross of bowstrings. There were also 5,000 bits for carthorses, strakes and nails for 70 carts, 80,000 horseshoes and half a million horseshoe nails.

On January 17 1534 ambassador Chapuys wrote to his master, Charles V: 'two days ago the king ordered 30,000 bows to be made and stored in the Tower ...'.

An entry from the Privy Purse accounts for 1534 records that sums were 'paid to Scawesby for bowes, arrowys, shaftes, brode hedds, bracer and

shooting glove for my lady Anne [Boleyn]: 33s 4d', and 'the same daye paied to the king's bowyer for four bowes for my lady Anne at 4s 4d a piece: 17s 4d'. In 1536 the Queen was executed, and her body was put in an elm box made to store arrowsheaves.

In 1542 an Act established that no man who had reached the age of 24 years might shoot at any mark at less than 11 score (220 yards) distance.

But even the royal longbowman realised the value of the new weapons. Henry had taken heavy siege guns to France in 1523, among them his 'twelve apostles', which required 20 or more Flanders draught mares to pull each one. The three battles or 'wards' of his army had more than 60 guns each, varying in calibre from two to nine inches. There were also 'organs', multiple guns on a single cart mounting, which emulated the 'wholly together' shooting of arrows, and the king took increasing pains to encourage the development of the handgun. No doubt he believed the old and the new weapons should exist side by side, and that belief was really the basis of the argument that raged over the relative merits of bow and gun, from the troops that used them in war, to the carefully prepared and learned arguments of the experts on both sides.

Henry incorporated by charter the Guild or Fraternity of St George, for the better defence of the realm with the longbow, the crossbow and the handgun side by side, and in the charter the king revived the old rules of Henry I exempting from arrest or imprisonment any man who shot and killed or wounded a person running between the shooter and the mark. If a shooter saw an interloper he must cry 'Faste', which means to other shooters 'hold fast'. Hence one who shoots 'fast and loose', loosing his arrow in spite of the warning. 'Fast' is a shout that still echoes across the archery grounds of Great Britain.

By the end of the century the proportion of archers was very much reduced. Pikemen and 'harquebusiers' and light cavalry replaced them more and more. With the reduction in the use of the longbow, heavy armour would be reduced as well. First the leg armour would disappear, then the breastplate, and finally the helmet, but that time was not quite yet.

In 1547, at the beginning of the boy-king Edward VI's short reign, there was fought the battle of Pinkie, near Musselborough, with archers on the Scots and English sides, those in the English army contributing much to the victory.

In 1549 Bishop Latimer preached a sermon before the king, in which he said:

'The art of shooting hath been in times past much esteemed in this realm. It is a gift of God that he hath given us to excell all other nations withall, it hath been God's instrument whereby he hath given us many victories against our enemies. But now we have taken up whoring in towns, instead of shooting in the fields. I desire you my lords, even as ye love the honour and glory of God ... let there be sent forth some proclamation to the justices of the peace ... for they be negligent in executing these laws of shooting. In my time, my poor father was as diligent to teach me to shoot as to learn me any other thing, and so I think other men did their children. He taught me how to draw, how to lay my body in my bow, and not to draw with strength of arms as other nations do, but with strength of the body. I had my bows bought me according to my age and strength; as I increased in them, so my bows were made bigger and bigger, for men never shoot well, except they be brought up in it.'

Five years earlier, Roger Ascham, tutor to the young Edward, and also to Princess Elizabeth, had published *Toxophilus*, a treatise on archery, an exquisite examination of the history and practice of the longbow, which is the earliest surviving handbook for archers. Henry VIII gave him £10 a year for life, when the book was presented to him by the author.

CHAPTER 8

PLAYING BOWS AND ARROWS

Roger Ascham, in his treatise *Toxophilus*, covered the whole experience of longbow shooting, except for the making of tackle and bowyery, which he preferred to leave to experts. The treatise is written as a dialogue between Toxophilus and Philologus, archer and scholar, the latter very properly allowing the former the meat of the matter. Among the fascinations of the book is that Ascham was writing, admittedly during the military decline of archery, but 100 years before what I take to be the last recorded use of longbows in war. In quoting from *Toxophilus*, I have generally modernised the spelling, but retained the original in every other sense. Here is Ascham on choosing a good bow:

'If you come into a shop and find a bow that is small [slender], long, heavy and strong, lying straight, not winding, not marred with knot gall, wind shake, wem, fret or pinch, buy that bow of my warrant. The best colour of a bow that I find, is when the back and the belly in working ... do prove like virgin wax or gold, having a fine long grain, even from the one end of the bow to the other ... I would desire all bowyers to season their staves well, to work them and sink them well, to give them heats convenient and tillering plenty.'

He goes on: 'Every bow is made either of a bough, of a plant [sapling] or of the bole of the tree. The bough commonly is very knotty, and full of pins, weak, of small pith, and soon will follow the string ... for children and young beginners it may serve well enough. The plant proveth many times well, if it be of a good and clean growth, and for the pith of it, it is quick enough of cast, it will ply and bow far before it break, as all other young things do. The body of the tree is cleanest without knot or pin, having a fast and hard wood ... strong and mighty of cast ... if the staves be even cloven, and be afterward wrought, not overthwart the wood, but as the grain and straight growing of the wood leadeth a man. But ... you must trust an honest bowyer, to put a good bow in your hand ... and you must not stick for a groat or twelve pence more than another man would give, if it be a good bow.'

He follows with advice about what to do when you have your bow home: 'take your bow into the field, shoot in him, sink him with dead heavy shafts, look where he cometh most, provide for that place betimes, lest it pinch and so fret'. 'Cometh' means 'gives'; 'pinch' refers to the developing of tiny compression fractures, which can turn into 'frets' or 'chrysals' or collapsing of the fibres.

'When you have thus shot in him, and perceived good shooting wood in him, you must have him again to a good, cunning and trusty workman, which shall cut him shorter, and pike him and dress him fitter, and make him come round compass everywhere, and whipping at the ends, but with discretion, lest he whip in sunder ... he must also lay him straight if he be cast.' 'Pikeing' and 'whipping' refer to the cutting of the limb ends shorter, and making them slenderer, and so in their outward parts weaker, which, within reason, allows for a faster return of the limbs on release. A 'cast' bow is one that lies untrue, not straight, down its back. From the side a bow is often, of course, bowed; then it is said to 'follow the string'; too much follow indicates a poor weapon.

'Likewise as that colt, which, at the first taking up, needeth little breaking and handling, but is fit and gentle enough for the saddle, seldom or never proveth well: even so,' and here speaks the tutor suddenly: 'an easy and gentle bow when it is new, is not much unlike a soft spirited boy, when he is young. But yet as of an unruly boy with right handling proveth oftenest of all a well ordered man: so of an unfit and staffish bow, with good trimming, must needs follow always a steadfast shooting bow.'

Ascham goes on to the breaking of bows, listing many causes, the worst being when a bow is drawn too far, 'either when you take a longer shaft than your own, or else when you shift your hand too low or too high for shooting far. This way pulleth the back in sunder, and then the bow flieth in many pieces.' Many longbowmen have seen that happen, and heard the sharp crack across the shooting ground, or felt the sudden give in the hand, the whack on the head or in the groin.

There are 15 arrow-woods listed in *Toxophilus*, the most recognisable being brazil, birch, ash, oak, blackthorn, beech, elder and asp [or aspen], the most delightful being 'suger cheste', a wood not surprisingly used in the making of chests for the shipping of sugar.

'How big, how small, how heavy, how light, how long, how short a shaft should be particularly for every man ... cannot be told no more than your rhetoricians can appoint any one kind of words ... fit for every matter, but even as the man and the matter requireth ... Birch, hardbeame [hornbeam], some oak, and some ash being both strong enough to stand in a bow, and also light enough to fly far, are best for a mean [average].'

We have already seen that he preferred a good ash shaft for a war arrow, because it was fast flying and yet heavy enough to 'give a great stripe withal'. Of the shape of the shaft, he again says no rule can apply so fitly as that the shape should answer the purpose. Of the nock he says it must be narrow enough to hold the string, not to grip it too much, strong enough for the sudden blow of the string not to break the shaft, and smooth enough for it not to cut the string. 'The deep and long nock is good in war, for sure keeping in the string. The shallow and round nock is best for our purpose in pricking, for clean delivery of a shoot.' 'Pricking' is target shooting.

For the fletching of arrows, Ascham makes it plain that nothing but feathers will do, and that in his opinion the only feather is the goose. 'The goose is man's comfort in war and in peace, sleeping and waking ... how fit as her feathers be only for shooting, so be her quills fit only for writing ... the old goose feather is stiff and strong, good for a wind ... the young goose feather is weak and fine, best for a swift shaft ... the pinion feathers, as it hath the first place in the wing, so it hath the first place in good feathering ... you may know it ... by the stiffness and fineness which will carry a shaft better, faster and further, even as a fine sailcloth doth a ship ... The length and shortness of the feather serveth for divers shafts, as a long feather for a long, heavy or big shaft, the short feather for the contrary ... your feather must stand almost straight on, but yet after that sort, that it may turn round in flying.'

Of arrowheads he has this to say, among much else: 'Our English heads be better in war than either forked heads or broad arrowheads. For first, the end being lighter, they fly a great deal the faster, and, by the same reason, giveth a far sorer stripe. Yea, and I suppose if the same little barbs which they have were clean put away, they should fly far better. For this every man doth grant, that a shaft, as long as it flyeth, turns, and when it leaves turning, it leaveth going any further.'

I cannot end this glance at the first archer's manual without quoting Asham on common archers' faults, and on the requisites for good shooting.

'All the discomodities which ill custom hath grafted in archers, can neither be quickly pulled out, nor yet soon reckoned of me, there be so many. Some shooteth with his head forward, as though he would bite the mark; another stareth with his eyes, as though they should fly out; another winketh with one eye, and looketh with the other; some make a face with writhing their mouth and countenance so, as though they were doing you wot what; another blereth out his tongue; another biteth his lips; another holdeth his neck awry. In drawing, some ... heave their hand now up now down ... another waggeth the upper end of his bow one way, the nether end another way. Another maketh a wrenching with his back, as though a man pinched him behind ... some draw too far, some too short, some too slowly, some too quickly, some hold too long, some let go over soon ...

'Once I saw a man which used a bracer on his cheek, or else he had scratched all the skin off the one side of his face ... Some stamp forward, some leap backward ... of these faults I have very many myself, but I talk not of my shooting.' Those words of 1544 ring just as true now, and so do these, for every archer:

'Standing, nocking, drawing, holding, loosing, done as they should be done, make fair shooting.

'*The first point* is when a man should shoot, to take such footing and standing as shall be both comely to the eye, and profitable to his use ... A man must not go hastily to it, for that is rashness, nor yet make too much to do about it, for that is curiosity, the one foot must not stand too far from the other ... nor yet too near together ... for so shall a man neither use his strength well, nor yet stand steadfast.

'*To nock* well is the easiest point of all ... to set his shaft neither too high nor too low, but even straight overthwart his bow ... Nock the cockfeather upward always [the bow being across the archer's body]... and be sure that your string slip not out of the nock.

'*Drawing* well is the best part of shooting. Men in old time used ... to draw low at the breast, to the right pap ... nowadays we draw to the right ear ... Leo, the Emperor would have his soldiers draw quickly in war, for that maketh a shaft fly apace. In shooting at the pricks, hasty and quick drawing is neither sure, nor yet comely ... Therefore to draw easily and uniformly ... until you come to the rig, or shouldering of the head, is best both for profit and seemliness.

'*Holding* must not be long, for it both putteth a bow in jeopardy, and marreth a man's shoot; it must be so little, that it may be perceived better in a man's mind, when it is done, than seen in a man's eyes when it is in doing.

'*Loosing* [Ascham spells it lowsing] must be much like. So quick and hard that it be without all girds, so soft and gentle, that the shaft fly not as it were sent out of a bow case ... for clean loosing you must take heed of hitting anything about you. Leo the Emperor would have all archers in war to have their heads polled, and their beards shaven, lest the hair of their heads should stop the sight of the eye, the hair of their beards hinder the course of the string.'

Every archer ought to read Ascham, but there is no need to be an archer to enjoy him. His legacy is great. It includes the use of his surname for the narrow cupboards in which archers stand or hang their bows, and keep their tackle. They are called aschams and pronounced, as he would have, 'askhams'.

Ascham lamented the dearth of historical record. 'Men that used shooting most and knew it best were not learned. Men that were learned used little shooting and were ignorant of the nature of the thing, and so, few men have been, that hitherto were able to write upon it.' Since Ascham wrote that, innumerable men have written upon it, whether they were learned, or used shooting, or not. They have wanted to search beyond 'King Vortiger' (whom Sir Thomas Elyot had discussed with Ascham, reckoning that in his time the longbow was first introduced into England), and they have sought to perpetuate the use of the bow.

★ ★ ★

Giovanni Michele reported to the Pope, some years before the Spanish Armada, that the English 'draw the bow with such force and dexterity that some are said to pierce corselets and armour, and ... such is their opinion of archery and their esteem for it that they prefer it to all sorts of arms and harquebuses ... contrary however to the judgement of the captains and soldiers of other nations'.

The Elizabethan archer, contemporarily described, 'should have good bows, well nocked, well stringed, every string whipped in the nock, and in the midst rubbed with wax, bracer and shooting glove, some spare strings, trimmed as aforesaid; every man one sheaf of arrows, with a case of leather, defensible against the rain ... whereof eight [arrows] should be lighter ... to gall or astonish the enemy before they shall come within the danger of ... harquebuss shot. Let every man have a brigandine or a little coat of plate, a skull or hufkin, a maul of lead ... and a pike ... with a hook and dagger; being thus furnished teach them by muster to march, shoot and retire ... none other weapon may compare.' Others thought differently, and the battle of words about the bow as a war weapon was furious. After Ascham and Latimer, came William Harrison, in 1577:

'In times past the chief force of England consisted in their longbows. But now we have ... given over that kind of artillery. Certes, the Frenchmen and Rutters [Germans], deriding our new archery ... will not let, in open skirmish ... to turn up their tails and cry "Shoot, English!" and all because our strong shooting is decayed and laid in bed. But if some of our Englishmen now lived that served King Edward the Third the breech of such a varlet should have been nailed to his bum with one arrow, and another feathered in his bowels, before he should have turned about to see who shot the first.'

In 1575 Barnaby Rich published a *Pleasaunt Dialogue between Mercury and an English soldier*. The soldier was all for the longbow; he knew its past service and effect. The past is the past, replies Mercury, 'the order of the wars is altogether altered'. With light regard for what happened on so many campaigns in France and on the borders, he argues 'let a thousand archers continue in the field but the space of one week ... how many of those thousand men at the week's end were able to shoot above ten score [200 yards]? ... when every caliver ... will carry a shot eighteen score and twenty score, and every musket 24 and 30 score [600 yards]'. That poor argument is followed by a good one, that 'every bush, every hedge, every ditch, every tree and almost every molehill' is cover for a gun – the user can kneel, lie, do what he will, but the archer must stand up and be seen.

The soldier objects that if it is not a question of cover, but open fight, the speed of the archers' shooting, and the arrows 'so thick amongst them', will out-do guns. The answer is crushing: arrows are lobbed down and fall often between the enemy ranks; shot flies straight at them and must strike one rank or another. 'Now I perceive,' says the routed soldier, 'we may hang our bows upon the walls.' 'Nay, not so neither,' answers the kindly winner, 'no man doubteth that horsemen are serviceable for many causes ... so likewise archers may do very good service.' No doubt the crestfallen archer had his shoulder patted.

In 1588 among 6,000 trained men mustered against the threatened Spanish invasion there were no bowmen, and among 4,000 untrained, only 800 archers armed with what began to be disparagingly called 'the country weapon'. The year 1590 produced some hot verbal skirmishing. Robert Barret wrote in *Modern Wars* another altercation between a captain and a gentleman:

'*Gentleman:* Why do you not like of our old archery of England?

Captain: I do not altogether disallow them ...

Gentleman: Will not a thousand bows handled by good bowmen, do as good service, as a thousand hargubuze or muskets, especially among horsemen?

Captain: No, were there such bowmen as were in the old time, yet could there be no comparison.'

He argues superior range, which has to be admitted, superior penetration, which is arguable, and claims that a skilled musketeer can shoot faster than an archer by five to one. That would mean 50 shots in a minute, and is flagrant nonsense. But he bangs the gentleman down with 'Sir, then was then,

'*The Double-Armed Man*', *1625* (Radio Times Hulton Picture Library).

and now is now ... some have attempted stiffly to maintain the sufficiency of the bows, yet daily experience will show us the contrary.' The sop comes with 'well wishing in my heart (had it been God's good will) that this infernal fiery engine had never been found out ... you may note this by the way, that the fiery shot, either on horseback or foot, being not in the hands of the skilful, may do unto themselves more hurt than good'.

In the same year Sir John Smythe wrote a long and persuasive argument for the retention of longbows, charging Leicester and other advisors of the Queen with corruption and incompetence, and getting himself thereby into a good deal of hot water and into the Tower of London, for all that he was a first cousin to the late king. He and his allies claimed for the longbow superior speed of shooting; greater accuracy; the terrifying effect of great volleys of arrows; the dreadful results among cavalry; the fact that horses with gunshot wounds paid no heed once the first pain was over, and were soon trained not to fear the report of guns; the imperviousness of the weapon to weather; the ability of the archer to see his shot and so correct his aim which the gunner could not; the silence of archery for night attack; the effectiveness of arrows lobbed over high walls; the use of fire-arrows; the danger of handguns overheating when reloaded and shot as fast as possible, and so on. Replies volleyed back from Sir Roger Williams, Humphrey Barwick and others, claiming the greater impact of the bullet; greater accuracy; the decay of archery and the difficulty of maintaining a practical force of bowmen, laying stress on the amount a bowman must eat, or fail in his strength of shooting, claiming that bows *are* subject to bad weather, that any 500 musketeers are more use than 1,500 archers, and so on. Barwick issued a challenge, to arm himself with a pistol and let a band of archers shoot 10 arrows at him at 120 yards. Luckily for him, the challenge was not taken up.

Whatever the arguments, however Canute might brave the sea, the tide was set against the longbow. In 1595, on October 6, a request was made from the county of Hertfordshire for the replacement by muskets of the last '100 bows in the land, which for want of use are utterly unserviceable'. Three weeks later, on October 26, Queen Elizabeth wrote from the Privy Council to the authorities in Buckinghamshire:

'In regard of the purpose you have to convert the bows that are amongst the trained soldiers into calivers and muskets, as already you have begun, which your determination we do greatly allow, because they are of more use than the bows, and that bows in no other county be comprehended amongst the enrolled numbers ... therefore we pray you to convert all the bows in the trained bands unto muskets and calivers ... and to see that they may be trained and taught to use their pieces.'

This was the pupil of Ascham, who had learned to shoot in a longbow from him, and had herself re-issued the statutes of her father concerning war bows, and approved a proliferating legislation to promote archery practice, to curtail the encroachment upon and enclosing of archery grounds by building speculators, sandpit diggers and farmers, and to guarantee the maintaining of butts. But now she put her signature to an order which replaced the longbow in one of the last two counties that retained it, a position of honour or futility, depending on the opinion of each judge, shared with the county of Oxfordshire.

Even so in 1599 the return of stores in the Tower of London still included: 8,185 bows, 6,019 bowstaves, 196 gross of bowstrings and over 300,000 arrows.

The government's efforts to promote the bow now lost all semblance of being for the defence of the realm. As much as anything they were an attack on the mania for gambling that had gripped the nation, and on the use of guns for poaching, robbery and murder. Ascham had already known, as everyone else did, that legislation was hopeless. Men played with the laws of archery, he said, as those who put up a lantern outside their doors in response to civic orders, but never lit the candle within it.

The arguments and pleas continued for at least another 100 years, but the bow was dropped, and the main concentration of the military was on the perfecting of the musket and the training of soldiers to use it, which for all the jeering of the longbow partisans, required time and skilful instruction.

Some still swam against the tide. James Ferguson, bowyer to the Scots king, was sent into England to buy 10,000 bowstaves for military use in 1595. But by 1627 there were only four bowyers left in the City of London. Gervase Markham argued for the bow in 1634 much as others had done, adding, as a kindly thought for the manufacturers, if archery should be revived 'the now almost half-lost societies of Bowyers and Fletchers will get a little warmth'. Herbert of Cherbury declared 'bring an hundred archers against so many musketeers, I say if the archer comes within his distance, he will not only make two shots, but two hits for one'. In 1632 Charles I approved a system put forward by William Neade eight years earlier for a 'Double Armed Man' using both bow and pike, but nothing seems to have come of it. The useful range of a longbow, according to Neade, was from six score to eighteen and twenty score, or 400 yards, and he claimed six shots from a longbow for every one discharge of a musket. The king had already revived the archery statute of Henry VIII for the maintenance of the practice of archery, and encouraged the import of bowstaves. Also in 1632 the king ordered all impediments and enclosures of archery grounds to be removed, because 'the archers, using the commendable exercise and pastime, be very much hindered, letted and discouraged therein, and often times in great danger and peril for lack of convenient rooms and places to shoot in'.

At the beginning of the Civil War in 1643 *The Civic Mercury* reported of the king's troops at Oxford, 'they have set up a new magazine without Norgate, only for bows and arrows, which they intend to make use of against our horse, which they hear does much increase; and that all the bowyers, fletchers and arrowhead makers that they can possibly get they employ ... Also that the king hath two regiments of bows and arrows.' It goes on to say that therefore no arrowheads must reach the Cavaliers from London, and to advocate archery for Parliament men as well, and to warn that 'the flying of arrows are far more terrible to the horse than bullets, and do much more turmoil'.

In the issue book of the parliamentary Ordnance Department, there is an entry under April 26 1644:

'Delivered ... out of his Majesty's stores ... to Mr William Molins, comptroller of the Ordnance for the Militia of London ... to be employed in the service of the State, by warrant from the Lord General the Earl of Essex:

12,432 longbow arrows
   526 shooting gloves
   600 bracers
1,000 gross bowstrings
    64 quivers of leather
    28 bundles of bowcases.'

The bows had apparently already gone. It was a brave end. No one has recorded what damage was done by longbowmen in the Civil War, but it is an ironic thought that, if there had been as many longbowmen as there were

Gentlemen archers practise the gentle art of archery (Author's collection).

musketeers at the great cavalry battles of the Civil War, the outcome would almost inevitably have had to be decided by infantry, and both Prince Rupert's and the Cromwellian horse would have been in desperate straits.

In 1653 Thomas d'Urfey wrote:
'Let Princes therefore shoot for exercise.
Soldiers to enlarge their magnanimities,
Let Nobles shoot 'cause 'tis a pastime fit,
Let Scholars shoot to clarify their wit,
Let Citizens shoot to purge corrupted blood,
Let Yeomen shoot for th'king's and nation's good,
Let all the Nation archers prove, and then
We without lanthorns may find virtuous men.'

In 1670 Sir James Turner said with bitter regret, 'The bow is now in Europe useless.' From now on, the longbow was a weapon for sport, and much money was won and lost in gambling between longbowmen.

* * *

Longbow archery under the Protectorate was certainly allowed as a sport, though it had its awkward moments. The finest shooter in the north of England was John King of Hipperholme, Yorkshire. After a day of spectacular shooting on his part he was chaired by the crowd who had watched him, and carried about the town to cries of 'King ..., the King'. He was at once arrested by Cromwellian soldiers and tried in Manchester for high treason. But the Puritans, with a rare dash of humour, forgave him his name and he was acquitted.

When Charles II had been an exile in the Low Countries he became a member of the Guild of St Sebastian of Bruges. (St Sebastian, shot to death by arrows, is the patron saint of archers, much as if St Lawrence were the patron saint of steak grillers, or St Katherine of wheelwrights.) When Charles was king of England, one Edward Fawcett was employed as 'keeper of the longbows'. He had applied for a job in the household, having formerly taught the king to shoot, and provided four of his late father's bows, and all necessary tackle for him and his brothers.

'On March 1st, anno Domini 1661, four hundred archers with their bows and arrows, made a splendid and glorious show in Hyde Park, with flying colours and crossbows to guard them. Sir Gilbert Talbot was their colonel, Sir Edward Hungerford was their lieutenant colonel ... there were three showers of whistling arrows ... so great was the delight, and so pleasing the exercise that three regiments of foot laid down their arms to come and see it.'

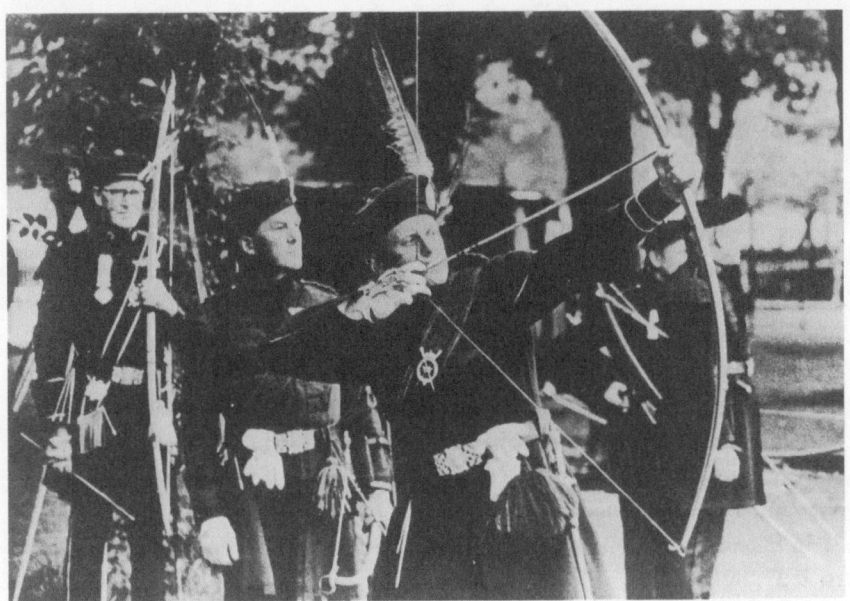

Royal Company of Archers (E. G. Heath).

The Royal Company of Archers was formed, or re-formed, from Robert Earl of Leicester's long disbanded Companie of Liege Bowmen of Queen Elizabeth. They were nothing to do with the Scottish Royal Company of Archers, the present Queen's Bodyguard of Scotland, which was a new foundation in 1676, and because of whom the English company changed its name to the Royal Artillery Company.

If Scottish archery appears to have been slighted up to this point, it is because, in spite of its known military effectiveness from time to time, it never achieved the national importance that it did in England, though as a war weapon it lasted longer. The Royal Company is the oldest surviving independently formed company of longbowmen in Britain. At its foundation, there was such a dearth of bowyers that an apprentice, Robert Munro, was sent to London to learn the craft. On his return he was made official bowyer to the Company, a post which has been constantly filled until the present day. William Law was bowmaker from 1900. He had learned his skills from Fergie, whose predecessor was Peter Muir, one of the finest of Victorian bowmakers. After Law came John Veitch, who was followed by Dowsen, who made the first longbow I ever used.

A royal charter was granted by Queen Anne in 1704 and, in return, the monarch was to receive a 'reddendo' of 'one pair of barbed arrows yearly'. In

1792, Robert Burns joined the Company and Sir Walter Scott followed him in 1821. In 1822 George IV visited Edinburgh and from that time the Company became the sovereign's bodyguard for Scotland. The royal tradition of shooting was revived when King George VI took lessons from the Royal Company, and when Queen Elizabeth II, as Princess, hit the gold with her first arrow.

Between 1703 and 1764 the Company shot every October for the Goose Prize. A goose was buried up to its neck and the bird's eye was the target. The first winner was Lord Tarbat who 'carried off the prize, the arrow entering the left eye ... so as she never moved after receiving the shot'. In 1764 mercy was extended to the goose, whose place was taken by a small glass ball. In 1791, two members of the Company fought a duel, loosing three arrows at each other at an unrecorded range. Neither is it recorded whether dodging was allowed, but it is known that no blood was spilt. The Royal Company now generally shoot at clout, either at 180 or 200 yards, and indoors during the winter at short ranges up to 100 feet.

In England the Honourable Artillery Company had been granted its patent in 1537, but it is possible a company of Finsbury Archers was formed even earlier, though their name does not exist in public record until the appearance in 1594 of a small book, called *Ayme for Finsbury Archers* which lists in detail all the marks that had been shot at by members of the fraternity in the past. The list includes one mark called 'Westminster Hall' to be shot from 'Lambeth' at 227 yards, another 'Whitehall' to be shot from 'Westminster Hall' at 222 yards, and 34 years later there is mention of a mark at nineteen score, or 380 yards. Few must have been the archers who could reach that, and few would they have been 250 years earlier.

Between 1684 and 1771 the boys at Harrow School shot for a Silver Arrow, and parents were to provide their young with bows and arrows. There are still two bows, 23 arrows, a wooden quiver, and a young archer's uniform at Harrow, among various other relics of the inclusion of the longbow in an academic curriculum.

By the 17th century, in general, shooting was done at 'pricks', or at 'clout' (the latter being 18-inch diameter white targets at 160 to 240 yards' distance); at 'butts', which were made of earth, and turfed and usually carried a target face of thick paper or canvas, sometimes marked with coloured, concentric circles, and set 100 to 140 yards apart, being shot at both ways, so that wind and light had to be gauged from either direction; at 'rovers', which meant at marks, chosen haphazard, and at any distances, except where there existed known marks as for the Finsbury archers, or at St George's Fields. Here in 1664 there were 108 marks existing and another 12 obsolete. There were Little John, Muzzled Bear, Cat and Fiddle, St George, Milkmaid, Grasshopper,

Brood Hen and Pigeon. In the middle of what is now Newington Butts Road, where undoubtedly there were butts, probably long after the shrinkage of the roving grounds, stood Dick Smith and the Fox's Stake. This kind of shooting is now called Field Archery.

The Finsbury Archers shot regularly three times each year, one of their meetings being the 'eleven score target', when competitors had to, as recorded in 1676, pay 20s either to Sir Edward Hungerford or to William Wood, who was an archer of great renown. One of his trophies was an enormous badge of 25 ounces of silver, the 'Braganza Shield', which is now in the possession of the Royal Toxophilite Society. His tombstone bears the epitaph 'In archery excelled by none', and on the day of his burial in 1681 three flights of whistling arrows were shot off over his grave. There is no record of where they landed.

Another fraternity was the Lancashire Bowmen, who held regular matches, and in May 1673 a Silver Arrow was first shot for at Scorton in Yorkshire, a prize which was later taken over by that county and which as the Scorton Arrow is still shot for every year, though most of the bows are not longbows. The distances for shooting at a four-ringed target varied, two dozen arrows being shot (the old sheaf), two at a time from each end, up and down the range.

Archers round London had a very hard time trying to keep hold of their ancient shooting grounds – their difficulties were constant and repetitive; farmers grew hedges, dug drains, removed their marks, and set their dogs on archers. In 1665 there was a petition from the 'Lords, Knights and gentlemen in and about London, who delighted in archery', for commissioners, 'such as had previously been appointed', to set up again the marks and stakes that had been pulled down by 'insolent farmers and field keepers'. There were petitions from the opposition as well. A brickmaker who dug in Finsbury, and had an 'estate of £1,200' there, complained that he had been kept from his property and trade for five years because of the archery laws, and as a result was in gaol for debt.

In the end the enclosers, the sand-diggers and the farmers with their mastiffs won, and the Finsbury Archers had to accept the hospitality of the Honourable Artillery Company to shoot on their grounds, which they gratefully did until 1677, when they were suddenly debarred from further shooting there. An appeal before the Lord Mayor was disallowed and that was the end of that. A certain Colonel Mew, of the Honourable Artillery Company, was apparently the cat among those last flights of Finsbury pigeons.

Difficulties notwithstanding, as Sir William Wood tells us in *The Bowman's Glory*, a thousand archers had marched in May 1676 from Bloomsbury, through Holborn, Chancery Lane and Whitehall, to Tuttle Fields, to shoot before the king and 20,000 spectators. Though they marched 'six abreast, yet when the van reached Whitehall, the rear had not passed through Temple Bar'. In 1681 archers marched from London to Hampton Court no less, where the king watched them 'near two hours', to their 'great joy, satisfaction and honour' and then 'permitted as many of them as pleased to kiss his hand, in token of his being well satisfied with that heroic exercise'.

During the next century the Finsbury Archers were again allowed the use of the Artillery Grounds, and they kept butts there until 1738. There were still targets to be shot at in Finsbury Fields during Easter and Whitsun holidays, until 1753, but within 20 years the enthusiasm of the Finsbury Archers seems to have burned out. Indeed the fires of archery in general burned very low. Archery as a sport seemed to be following the military bowmen into the shadows of the past. Bows were relegated to boxrooms, arrows to attics. Bowyers and fletchers and longbow-stringmakers were names only, perpetuated by the livery companies in the City of London.

It was a wind from constricted lungs that fanned the ashes into life again, and joined together a few survivors from the Finsbury fraternity to form the nucleus of a new society of archers. But it marked the beginning of a grand and lasting revival.

'About the year 1766,' wrote Thomas Roberts in his *English Bowman*, 35 years later, 'Mr Waring, who resided with Sir Ashton Lever, at Leicester House, and who may be justly styled the father of modern archery, having, by continued application to business, contracted an oppression upon his chest, arising principally from sitting too closely to his desk, and pressing his breast too much against it ... resolved to try the effect of the bow, in according himself relief. He accordingly made it a regular exercise, and in a short time derived great benefit ... and ascribed his cure, which was perfect, solely to the use of archery. Sir Ashton Lever, perceiving the good effects ... followed Mr Waring's example, and took up the bow; he was soon joined by several of his friends who ... joined themselves into a society, under the title of Toxophilites.' The Society has lasted until today, as the Royal Toxophilite Society, affectionately known as 'the Tox'.

Thomas Waring, who had learned bowyery from Kelsal of Manchester, one of a centuries-old family of bowmakers, became tenant of Leicester House, and there, until 1791, he opened his 'manufactury for implements' of archery, providing the members with bows and arrows to practise in the gardens, which extended from Leicester Square to Gerard Street. Their target meetings were held on bigger grounds, at Canonbury House, at Highbury, at the Honourable Artillery Company's grounds, at Blackheath, and at Mr Lord's cricket ground. By 1791 the Society had moved to a new headquarters near Bedford Square, which lasted them until 1805 when they lost the ground to builders and developers. From then until 1820, with a shrinking membership because of the wars, they shot at Highbury, moving the next year to Bayswater, where they stayed until they obtained six acres in Regent's Park, north-west of the lake.

There they had three pairs of butts on 'green, closely shaven turf ... except in the space between the targets, tastefully dotted with clumps of trees and flowering shrubs ... the whole reflects much credit on the taste of the Hon D. Finch, the Secretary, under whose direction it was planned, and whose judgement in ornamental gardening appears no way inferior to his skill as an archer'. There too they had their Archers Hall, 'an elegant building in the Swiss, or Rustic Gothic style' with Ladies' Rooms for the wives of members, who were allowed to shoot there occasionally, a range of 53 aschams in solid oak, 22 deer's heads and pairs of antlers, four large French windows and a bay window 'enriched with armorial bearings of patrons and various members'.

The elegance at Regent's Park lasted through public outcry at private occupation of part of a public park, and through the Great War, until the Society was finally given notice to quit in January 1922. For two years they wandered, but were able to settle in 1924 on ground in the old cemetery at Hyde Park Place, where the victims of the Tyburn gibbets had formerly been buried. Through the Second World War, a few members kept the Tox alive, filled in the bomb craters on the shooting ground, salvaged what they could from their damaged club house, and cleared the debris from the few rooms left safe enough to use. There, from the end of the war, they flourished until the builders once again chased such space-wasters away. Now, since 1967, they have their headquarters and their grounds at Burnham, in Buckinghamshire. There they hold more than 50 target meetings a year, including 13 longbow

Royal Tox Pavilion at Regent's Park with 'lady a rcheresses' (Author's collection).

George, Prince of Wales, in the uniform of the Royal Kentish Bowmen. This picture contains probably the best portrait of a longbow that exists (Gracious permission of HM the Queen).

rounds, and from there the Society and its officers wield great and beneficial influence in British and International archery, which reflects as much or more credit on the Society's Secretary, Lt Col Hugo Boehm, as ever was reflected on Mr Finch 175 years ago. His judgement in matters concerned with the organisation of archery appears in no way at all inferior to his skill both with the long, and as he would put it, the curly bow, the modern target weapon.

The Prince Regent had a considerable influence in the revival of the longbow. He shot with a longbow himself, was Patron of several societies, the Royal Tox among them, and the Royal Kentish Bowmen, for whom he designed a stunning uniform of green, buff and black. He reduced the varied and quirkish scoring systems of the past to a regular rule, and introduced the colours and values of the target rings, with one small exception, as they are today. 'The Prince's Reckoning' is still seen on standard target faces in Britain, America and other countries, the centre gold circle counting nine points; the next ring, the red, counting seven; the blue, which before 1844 was the inner white, five; the black, three; and the outer white ring, one. 'The Prince's Lengths' regularised the shooting distances, at 100, 80 and 60 yards, from which developed the York Round of 144 arrows, the major round shot by longbowmen, with six dozen arrows at 100, four dozen at 80, and two dozen at 60 yards.

1785 had seen the beginnings of another Company of Archers who still shoot the longbow today and, like the Royal Company, the longbow only: the Woodmen of Arden, whose lovely tree-lined shooting ground is at Meriden in the very centre of England. Members of the Royal Company and the Woodmen have the freedom of each other's companies, and from 1878 until today they have met alternately in Scotland and Meriden every three years, with gaps between 1912 and 1920, and 1938 and 1947 during the two World Wars. The matches are clout shoots, at good, old-fashioned longbow distances, nine or ten score yards. The Woodmen at their own meetings, called 'Wardmotes', shoot still up to twelve score on occasion.

The hankering for the military bow never quite left off. In 1798 Richard Mason, in the face of threatened invasion by Napoleon, published a book called *Pro Aris et Focis*, roughly speaking 'In defence of hearth and home', addressed to the British Public 'at the voice of the Country in danger' and 'on reflection of the causes which principally gave rise to the overwhelming power of France ... the number and excellence of her artillery'. He argued for the organisation of troops of triple-armed men this time, bearing longbows, pikes and swords, the pikes having two hinged legs which could be lowered, and screwed firm, turning the pike into a sort of 'palisado' or archer's stake, behind which the archer would stand and shoot. 'The model of the pike thus supported in defence, may be seen at Mr Thomas Waring's manufactory, Charlotte Street, Bloomsbury.'

He used all the well-tried arguments in favour of the bow, and against the musket, quoting a later authority on the latter point than had been available to earlier writers:

'Respecting the great Inefficiency of the firing of Musketry in modern War, and the consequent expence attending it, it may be judged of by what is stated by Marshal Count Saxe ... that on a computation of the balls used in a Day's Action, not one of upwards of eighty five took place [scored a hit].'

He went on to quote the battle before Tournay in May 1794, where, he says, 10,000 men were killed and wounded on the French side, by 40,000 infantry of the Allies, shooting at least 32 rounds per man, 'a total of 1,280,000 balls discharged to occasion the above loss of the Enemy, making 128 shots to the disabling one Object', even if none of the French loss was caused by 'the Bayonet, the Cavalry and Artillery', which he thinks would reasonably account for half the casualties, thus raising the expenditure of balls to 256 for the disabling of one man.

He claimed the effective range of musketry anyway was 'nine score yards' only, much below effective archery range, and that 'in the Discharge of a Body of well-trained Archers ... at least one Shaft in Ten would hit, so as to kill or wound ... Here then evidently appears an Advantage in Favor of the Bow, in point of Certainty of its Shot, of no less than upwards of Twenty to One!!! and as the Archer has the Power of discharging two Shots at least for one of his Adversary, the above Proportion is even doubled.' He concluded with detailed drill for the archerpikeman-swordsman but, alas for the longbowmen of England, his voice and his urgent capital letters were smothered in the roaring smoke and clattering cavalry of Waterloo.

In the fields of peace, the stately meetings of longbowmen proceeded; the banquets and balls were held for archers, and their ladies, among whom archeresses were increasing steadily in number. Princess Victoria became a Patroness of the Royal St Leonards Archers, and the Royal British Bowmen, shot in a longbow herself (many are the paintings and engravings to prove it) though as a child of 12 she found an archery meeting on the Isle of Wight 'pretty, but not half so grand and beautiful a sight as the fleet' in the Solent.

There was a figure in the shadows who would bring about remarkable change. As he comes into clearer view, we can salute the father of modern target archery, Horace Alford Ford.

Roberts could say, in the first years of the 19th century, that 'there are men now living who would not have disgraced the corps of Sherwood Rovers'. The Toxophilite Society 'can boast of one [Thomas Waring] who, shooting with two arrows at each end, put twenty successive arrows into a four foot target at one hundred yards; and twelve arrows into the compass of two feet at forty six yards, within the space of one minute'. Such old-world accuracy was to be scoffed at, such uncomely speed deprecatingly laid aside. Now all attention, all concentration, all skill and energy must be bent on calm, unwavering, unequivocal accuracy, at all costs.

Horace Ford was a natural player of serious games and instruments; bowls, cricket, billiards and chess, the violin and the piano. The son of a coal-pit owning solicitor in Glamorgan, he came to Brighton in 1845, where he started to shoot in a longbow, and was soon sufficiently enthusiastic to get up at four in the morning and shoot a round before breakfast, except on Sundays, when he had to content himself with gazing, bowless, but analytically, at the targets. Ford was tall and slim, not at all the broad-built archer who would spread his legs and open his bow with a wide muscular back.

In drawing the longbow, or any other non-mechanical bow, an archer uses well- controlled muscle power, the greater part of the work being done by the arm and shoulder muscles, but a firm and balanced stance is vital. The pull on the fingers of the drawing hand corresponds to the weight of the bow, the left hand holding the bow, and the left arm being fully extended, but not locked. The shoulder muscles and the muscles which join the shoulder blades to the trunk contract strongly, and the right elbow is fully flexed. The effect of the natural elasticity of the bow is to swing the bow arm across the chest so, to counteract this, the shoulder muscles have to develop a pull of about 300 lb (136 kg) force to draw a 60 lb (27.2 kg) bow, which is five times as great as the pull exerted on the arrow. The resulting force across each shoulder joint, for the draw of a 60 lb bow, is greater than three hundredweight (152.4 kg). The main advantage of modern bows is that their ratio between weight and efficiency allows for much lighter draw weights, to reach the same distances, than the relatively less efficient longbow. Ford's great contribution to modern longbow archery was to reduce the method of shooting to a clear rule and an economical style, which was in his case the result of thought, analysis and constant correction. The war archers of the past, at their best and for their need, must have applied themselves just as rigorously. Like Ford, their skills were the result of hard, painful and unremitting labour. By evolving a style and a technique to fit his physical shape, Ford was able to make scores that were thought beyond any possibility.

A friend said of him, 'blacks Mr Ford abhorred, and whites he seldom made.' In 1854 he was the first man ever to score more than 1,000 in a double York round. Out of 288 arrows shot, he had 234 hits and scored 1,074. In 1857 at Cheltenham, where by then he lived, he scored 1,251. Writing about

Viscount Sidney and Colonel Ackland from a superb painting by Reynolds in the possession of the Herbert family, showing a longbow and a composite bow (Permission of the Hon. Mrs M. Herbert.)

the achievement afterwards, he said: 'When it is considered that the second score, that of Mr Edwards was 786 only, and that the nearest approach to it ever made at the National by others than myself up to the present time, is but 902 ... I suppose it may be looked upon as very great shooting.' His record for a single York stood until 1943, but, given the facts that he shot both ways with a self yew longbow, which by its very nature lets down through the long

Above, left *A Finsbury archer, circa 1650. He has blunt target arrows in his sash, finger tabs tied at his wrist, a leather bracer and an abominable loose from 'ye pappe'. From Gervase Markham's* The Art of Archerie *(E. G. Heath).*

Above, right *Horace Ford* (E. G. Heath).

hours of shooting, and that a set of wooden arrows can never be as perfectly matched as the modern hollow metal shafts, it can really be said that he holds, probably for ever, the true longbow target record. Indeed in private practice he beat his own record.

'Under the risk of being considered egotistical ... I now give the following specimens of my private practice. I made with an Italian self yew bow of Mr Buchanan's and five shilling [⅞ oz or 24.81 gm] arrows of Mr Muir's ... 143 hits (out of 144), 765 score ... with a yew backed bow and the same arrows, 137 hits, 809 score.' On another occasion he wrote to a friend that he had made 141 hits, 799 score, and added, 'Mettez cela dans votre pipe and smokey le.'

'The archery of 1856 is not the archery of half a century ago,' he wrote in his *Archery: its theory and practice*. 'Scores that would then have been deemed impossible and visionary are now of every day occurrence ... First of all make up your mind to succeed ... and secondarily expect plenty of difficulties and discouragements ... it is not easy to become great in anything ... work hard and practise regularly ... study as well as practise ... do not fancy yourself a first rate shot when you are only a *muff*.'

'As it is certain that in no country has the practice of archery been carried to such a high degree of perfection as in our own, so it is equally undeniable that no bow of any other nation has surpassed, or indeed equalled the English longbow, in respect of strength, cast, or any other requirement of a perfect weapon.' When Ford wrote that, it was, in its way true. Now, of course, it is true no longer.

His advice about the details which make up the technique of good shooting are based firmly on the precepts of Ascham. 'An archer's general position ... must be possessed of three qualities – firmness ... to resist the force, pressure and recoil of the bow ... Elasticity, to give free play to the muscles and the needful command over them ... and Grace, which is almost the necessary consequent of the possession of the other two.

'The heels should be about six or eight inches apart ... it is neither necessary nor elegant for the shooter to straddle his legs abroad, and look as if he were preparing to withstand the blow of a battering ram ... The weight thrown equally upon both legs, the footing ... firm ... easy ... springy. The body upright but not stiff, the whole person well balanced ... the face turned round so as to be nearly fronting the target, with the expression calm, yet determined and confident', without 'frowning, winking, sticking out the tongue – the whole attitude suggestive of power ... command ... and will.'

During the action of drawing the bow, there must be no bending forward, he says, otherwise the shooter 'requires so much of his wits and muscles to keep himself from tumbling on his nose ... Not that he is to look as if he had a ramrod down his backbone', and if the head is too far forward, the archer will get 'such a merciless rap upon the nose as effectually to cure him'. He recommends that 'the pulling of the bow, and the extension of the left arm be a simultaneous movement ... that the aim be found by a direct movement onto it from the starting-place of the draw'.

Explaining that it is impossible at longer ranges to aim directly at the point to be hit, because of the curved trajectory of the arrow, he insists that aim should be concentrated on a point which will result in the centre of the target being struck, but which is not that centre itself, and that this 'point of aim' be directly looked at, not the mark to be hit. At the same time he ridicules the old

Grand Archery Meeting, Knavesmire, York, 1844 (E. G. Heath).

drawing of the arrow to the ear. 'Imagine a man being expected to hit accurately with a rifle, with a trigger at his ear, and his eye looking sideways at the barrel.' The whole length of the arrow must be directly beneath the aiming eye, which in 99 per cent of archers is the right eye.

As for loosing, he says Ascham describes that perfectly, adding that 'a slower flight and certainty' are far better than 'a rapid flight and uncertainty'. He wants 'great sharpness' of loose, but by that he does not mean a snatched string, or a rolling of the string, rather the allowing of an 'instantaneous freedom' of the string, without further pulling or jerking, which can only be achieved if the string quits all the fingers at the same moment. 'So slight is the muscular movement ... it is hardly perceptible ... yet so important is it, that the accurate flight of the arrow mainly depends upon it. The position of the left arm must be firm and unwavering, the attention never relaxed ... until the arrow has actually left the bow.'

Sadly, he speaks of clout shooting, the survival of the old archery, as 'a refuge for the destitute. In former times when the bow was *the* weapon of war, great force of shooting was the grand desideratum – precision being less required than penetrating power.' I bow my head like an indicted felon whenever I read this summing up. Distance shooting now, says the stern judge, 'as a test of skill is simply ridiculous ... the most excruciating muff ... may strut along the stage of his archery existence with a comfortable idea of his superiority over the poor, weak, benighted short range man – "he never shoots such paltry distances as sixty yards": he knows better – he would be found out if he did'. I comfort myself with a lingering thought that Ford might have been wiser than to scoff at what he did not do himself. 'Roving' he dismisses;

Lady archers at the Olympic games, 1908 (Radio Times Hulton picture Library).

'Flight' shooting for maximum range he describes briefly, and allows that there was a man at the turn of the century 'who shot repeatedly, up and down, in the presence of many spectators, a measured length of 340 yards, with a self yew bow of 63 lb'.

As for the past, 'it mattered little whether the archer hit the man he aimed at or the one next him, provided the arrow had sufficient power to penetrate the finely tempered armour then in use – shooting at a body of men, however few they might be, to miss all was an exceeding improbability … Each will form his opinion as to the skill attained in the first ages of the English longbow … according as his taste lies in the direction of the probable or the marvellous; and there the matter must be left.' Thus Horace Ford lays the old shooting to rest.

He was the target archer absolute. I have always thought it sad that the great ones of targetry will scarcely ever allow the skill of the hunter, whose instinctive shot, no less practised, subject to no lesser disciplines than his own, will strike a great stripe in the tiny lethal area that will bring down his quarry. The military archery of the past they bow to, perforce, but will whisper that it was easy, that the only necessities were strength and the discipline to obey those officers and sergeants of the army who directed the annihilating barrages in such mass that in the areas covered neither horse nor man could live.

We are near the end of the story of the longbow. Much of its later history concerns America, and it was in America that the longbow was first subjected to scientific scrutiny, resulting in an evolution which is examined in the next chapter. In the last chapter we will look at the longbow hunter, the bow itself and the wood of which, at the best, it is made. In the technical appendices the 'probable' and 'marvellous' are examined, to try and find where the truth lies.

The Royal Tox, the Woodmen of Arden, and the Royal Company exist still, and shoot both short and long, and there are companies and societies where addicts of the past can try to shoot as the old longbowmen shot, and find out more about their methods and techniques, sometimes even wearing appropriate costume. Also there are quite a few meetings where stubborn longbowmen may take their 'crooked sticks' and be sure of finding a cordial, if amused welcome and a target or two reserved for their rather less serious approach to the game. Ford would not have approved of their attitude, but he might have been glad there were some still shooting. Also, there exists one other society, devoted only to the longbow, which is open and welcoming to anyone who would try his hand with the wooden weapon: the British Long Bow Society, founded in 1951. Like the Royal Tox, it numbers among its members many women. The ladies of the longbow in England have shot

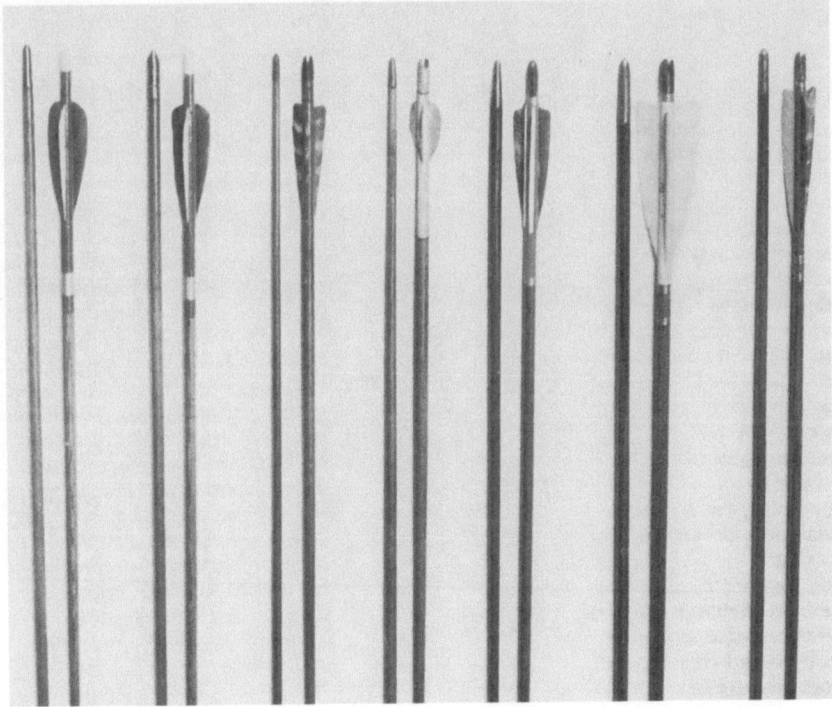

A selection of target, clout and flight arrows, 19th and 20th century, showing various pile, shaft and fletching shapes.

their lighter bows usually at shorter distances than the men for a century and a half, bringing to archery charm and delicacy, great determination and often great accuracy. Members meet several times a year to shoot York rounds and at clout, and among them I have heard much wisdom concerning the bow and arrow, and seen much skill both in the making and shooting of them. Two bows that are frequently shot with the Society I know to have been fashioned out of stakes of yew rescued from a humbler life, one in a rose pergola, the other from a fence. The yew is a wonderful wood, and the bowmaker a man of love and devotion.

CHAPTER 9

THE AMERICAN WAY

Horace Ford was a man I believe the Americans would have loved for his dogged and successful struggle with the ineluctable difficulties of the longbow, though they might have distrusted a certain narrowness in his attitude to archery beyond the limits of shooting at a target. It was time that America should begin to exert her influence on the bow and arrow, and, characteristically, the way she did so was to alter the whole concept of archery throughout the world. While it is true that there are at present as many bowmen in Great Britain as there were in the days of military archery, in the United States of America archers are counted in millions, a sizeable proportion of whom are 'bow hunters', and practically all of whom use the modern composite recurve bow, which is the weapon evolved by the Americans more than by any others. It could not really be said that the day of the longbow was quite over until the bowyers, archers and physicists in the United States put their heads together, determined to get more efficiency out of the old yew-stave. Their examination was the death-warrant of the yew bow in sport, as surely as gunpowder and the shooting barrel had caused the demise of the war bow.

Long before the European races began to spread across the American continent there were innumerable tribes of widely differing cultures across those vast expanses of land who used the bow and arrow to get much of their food, much of their clothing, and as an important part of their armament in war. From the astonishingly developed Incas, the Mayans and Aztecs to the most primitive nomad tribes, all, it seems, had reached the Americas in successive waves of migration from Asia, first across the narrow land that joined Asia to North America, and later across the sea gap that we call the Bering Straits. Most of those migrating peoples had brought with them some form of bow and, as they moved or settled, their weapons were slowly adapted to, and altered by the materials available, the needs of the tribe, the region they lived in and its wild life. Where fine timber grew there would be fine wooden bows; where there was no growing timber, there would emerge, or be retained from past experience in the lands of their origin, ingenious composite weapons with very little wood in their construction, sometimes none, but horn and sinew and rawhide instead.

As all hunters, the people of the Americas relied on the skills of stalking, tracking and trapping to get them close to their quarry, so the range and power of the weapon were not so important as the craft of hunting. Probably in some early stages of migration from Asia, before the sea burst through the isthmus, horses came with the moving tribes; yet the horse had died out in North and South America and remained extinct until the Spanish Conquistadores brought their ponies to the south. So for thousands of years the stalker on foot relied on his knowledge of animal and terrain to make his kill for meat, hide, sinew and horn, glue, grease, warming fur, decorative hair and feathers. When the horse from the south spread gradually to other parts of the continent the American Indian used its speed to close with the fleeing quarry. In the forest areas, where the horse was less useful, the old woodcraft remained and the kill was by stealth; sometimes poison was used, where, as in parts of South America, it could be found in root, plant and tree.

Some tribes or parts of tribes would become particular experts as bowyers or fletchers, and would travel and sell their wares or barter them for goods

American Indians hunting buffalo. Drawing by George Catlin, 1794–1872 (Radio Times Hulton Picture Library).

and beasts. Occasionally captured weapons and their makers would help to spread and intermix different methods, and improve the manufacture and use of bow and arrow. But wherever it was found, however well or ill-made, however developed or primitive a weapon, the bow was treated as an object of very great importance, even as a thing of magic, often as a symbol of power, or prowess, or manhood. By the same token, though there were tribes who allowed women to use the bow for hunting and fishing, it was more often taboo to women and children. Let no female nor unskilled hand touch what was so vital to the survival of the tribe! The cultures flowed and changed across the continent: the hunters of the open lands with stronger bows for greater range, and those of the forest areas with quick light bows for shooting among trees and undergrowth where arrows would be deflected and easily lost.

An uncertain, but beguiling flash of light is thrown into the shadows of medieval history. During the reign of Henry II in England (who had much trouble with the Welsh, and was often galled by the effectiveness of the longbow in Welsh hands), Prince Madoc, seeking respite from the incursive English, is said to have sailed to America with a band of followers and there mingled with the native Indians, leaving traces of physical appearance, culture and language that survived into the 19th century until smallpox wiped out the Mandan tribe from the banks of the Mississippi. Perhaps the Welsh took with them not only their fair skin and their language, but the skilled and aggressive use of the Welsh longbow that so impressed the English. If it were so, the imported bow from Britain first came into North America about 1170, and then not again for several hundred years.

When the Europeans began to pour into the Americas, they found bow-hunters and archer warriors everywhere. But in North America, as the European spread gradually westward from his first footholds on the Eastern seaboard, it seems that the memory of the western European longbow, not the adaptation of the native American bows, was the main influence that led to the growth of archery. This in turn led to the development of the modern composite bow, though that bow itself owes much to the ancient composites that came with some of those much earlier immigrants from Asia.

Modern American archery may be said to have become recognised and first organised by the formation of the United Bowmen of Philadelphia in 1828, after a discussion 'over a good bottle of wine' between Dr Robert E. Griffith and Mr Samuel P. Griffitts. They started to make and shoot, and break, bows of their own pattern until they decided that they would do better to order a longbow and some arrows in England. It turned out that the $90 they paid was well invested, and the longbow that crossed the Atlantic served as model for many other bows shot in Philadelphia by the 25 members of the club.

The rules of shooting largely followed the traditional English regulations, but the targets were smaller and the coloured rings in different order. Their bowstrings were of silk, the arrowshafts white holly fletched with feathers of the blue heron, the swan and the American eagle. Like so many archers through history, they were gradually forced from their shooting grounds by the expansion of their city, until the secretary's entry in the record book for 1859 exclaimed 'No grounds, no shooting!'

The Civil War put an end to elegant archery for many years, but in the process of dying the seeds of rebirth had been scattered. In 1844 James Maurice Thompson was born on his father's plantation in Cherokee Valley, Georgia, to be followed five years later by his brother William Henry. 1844 was, coincidentally, the year of the first Grand National Archery meeting in England, so it turned out to be an important year in the history of the longbow, and of its literature. The two boys 'played bows and arrows' as one hopes boys always will, but at that point Providence took a hand in the shape of one Thomas Williams. Maurice Thompson takes up the story: 'He was a sort of hermit whose cabin stood in the midst of a vast pine forest that bordered my father's plantation ... We found to our chagrin that ... all we had learned must go for naught, and an art must be mastered, the difficulties of which at first seemed insurmountable.'

The brothers enlisted in the Confederate Army. It is hard to see how Will got in; he can only have been 17 when the war ended in 1865; but somehow they were together when peace came and, as soon as Maurice recovered from a severe chest wound, the two of them set off home on foot. They found that during General Sherman's famous march through Georgia their house and plantation, like many others, had been destroyed. They had no money and they were allowed no firearms, being recent rebels against the victorious government. So they took to the woods with longbows and arrows, and lived like the bowmen of Sherwood. They made their weapons with knives, smoothed them with stones and polished them with sand. As time went by they were able to get better materials, finer tools, and their skills increased. Maurice's talent with words earned him money in 1877 and 1878, when *Harper's Magazine* published stories of the brothers' hunting lives. In 1878 these stories were collected together in a little book, *The Witchery of Archery*. It is unobtainable in England (there is no copy in the British Museum), and almost so in America. Here is Maurice Thompson again:

'So long as the new moon returns in heaven a bent, beautiful bow, so long will the fascination of archery keep hold of the hearts of men.

'Music and poetry sprang from our weapon. The bow is the old first lyre, the monochord, the initial rune of fine art ... The humanities grew out from

The American Way

Archery as a flower from a seed. No sooner did the soft sweet note of the bowstring charm the ear of genius than music was born, and from music came poetry and painting and sculpture.

'When a man shoots with a bow, it is his own vigour of body that drives the arrow, and his own mind controls the missile's flight ... Among wild peoples, a chieftain is invariably chosen on account of his ability to draw a mighty bow.'

He tells the story of a tramp who came on the brothers shooting one day, and stood watching. He said, 'On Brighton sands I have seen good shooting. I have shot there myself.' One of them handed him bow and arrows: 'For a moment he stood as if irresolute, then quickly fixing the arrow on the string, drew and let fly ... and he hit the gold in its very centre. Neither poverty, nor shame, nor hunger, nor dissipation, nor anything but death can ever quite destroy the merry, innocent, Arcadian, heathen part of our nature that takes to a bow and arrow as naturally as a butterfly takes to a flower.'

How pleased he would be to know so many hunters in his country use a bow and arrows:

'It is not sport to sling a handful, say from 300 to 700 pellets at a bird ... At 25 yards ... your gun will cover two feet square thickly with shot. See what a margin for successful inaccuracy ... The shot gun will exterminate game. It already has exterminated it in many large regions. The very sound of a gun is terrible to all wild things.'

He goes on to talk of hunting with the bow:

'First you must know how, under all circumstances and over all kinds of ground surface, to quickly and accurately measure distances with the eye; secondly you must be quick and noiseless as a cat in your movements; thirdly you must draw uniformly, that is put the same power in every shot, no matter how near or far ... draw to the head of your arrow every time you draw.

'Two years of sincere, systematic attention to the tried rules of archery will render you an expert, ready to knock down a flying grouse, a wood duck, and able to pierce a deer through the shoulders at a hundred yards. You will then be found in the jungles of Florida following the hounds after a deer, a bear or a panther, and handling a ninety pound bow and three ounce broadheaded hunting shafts with all the ease and power of a Tartar chieftain ... but first you must be content to chase the woodpecker ... see that one yonder on that slender stump ... only twenty five yards away. Try him with a light pewter-headed arrow. You pull very steadily and strong, loosing evenly and sharply. Away darts your shaft. Whack! What a blow on the stump exactly where the bird was! but too late to get him. He heard your bowstring, and quick as a flash he slid round behind the stump, and when the arrow struck he flew away.'

Not all birds are as sly as the woodpecker:

'What a glorious weapon the longbow is ... There was I within eighty yards of a great snowy heron, with two shots at it, and still it sat there! What if I had been armed with a rifle ... I shot twice, thrice, four, five times, the arrows whisking past ... my arm had got steady now, and I drew my sixth arrow ... a quick whisper ... a "chuck", once heard never to be forgotten. The feathers puffed out and sailed slowly away in a widening ring. The big wings opened wide and quivered a moment, then the grand old fellow toppled over and came straight down with a loud plash into the water.'

Two alligators made for the bird at the same moment as Thompson in his skiff:

'I gave one of them a bodkin point in the throat, causing him to turn somersaults, and beat the water into a stiff foam.'

Birds on the wing, quail, duck, grouse, all got with sheer skill, animals of all kinds, a 300 lb bear his largest prey. He ends his book with a curious and fascinating appendix on military drill with bows, and finishes:

'The marching commands and orders are those of the United States infantry. There can be no military display finer than that of a well-drilled, uniformed and equipped archery company. And at such short range as is needed in times of riots in our cities, no company would be more dreadfully effective.

American Indians hunting stags, disguised as stags (Mansell Collection).

A well-trained archer will discharge thirty arrows in a minute and every arrow is death.'

That astonishing claim is from a trained archer who knew and experienced the need for fast shooting, so we should believe that what he claims was true, at least for him. It is a claim worth putting side by side with the shooting speed of Howard Hill, perhaps the greatest of modern bow-hunters, and with some of the figures attributed to military archery earlier in this book.

The brothers went on to be honoured in every degree of archery, Maurice becoming President of the American National Archery Association when it was formed in 1879. Although he said, 'after all a bowman's skill is nothing if it is confined to a fixed range', he was a supporter of Horace Ford's shooting methods, and wrote once of United States archers:

'They have no idea of what aiming, as applied to archery means, in fact most of our archers do not aim at all, but use the hunting method of fixing the eye on the target's centre, and drawing to the ear, the arrow being directed by guesswork.'

Will Thompson's highest recorded scores for a double York round are one of 860 and a later one of 940, but remember he was a superb field archer and hunter as well, which Ford emphatically was not. Shooting at clout or rovers for Ford were mere excuses for shooting badly. Yet the Thompsons shot superbly both in the field and at the target, though they never ceased to warn the true hunting archers that the target was a danger to their very different, instinctive shooting.

After a promising resurgence, archery suffered a falling off during the last years of the 19th century. Equipment was difficult to come by and harder to make; the lawn tennis craze and the bicycle struck America together, and baseball, football, rowing, followed each other as alternative and often less exasperating outdoor exercises than the use of the bow and arrow. There continued to be a few devoted archers who kept the sport alive, but it was not until the turn of the century that the next tide began to make. At low water, the first movements of the flow are tiny, but they presage a flood.

In 1910 Harry Richardson won the York Rounds in Chicago, then the headquarters of archery, with 231 hits, 1,111 scored; that remained an American record until 1927, and Ford was the only Englishman to beat it. The same year, Dr Robert Elmer, a physician in general practice who decided upon archery as an exercise that would be less time consuming (he imagined) than golf, 'visited a sporting goods store, and with the usual luck of a novice, received from them a fair lemonwood bow and some miserable children's toys which they called arrows ... shot at a tree just eight yards away, and missed it by three feet'. By the following year he was shooting in the national tournament, with

Will Thompson (E. G. Heath).

better arrows that cost him $3 a dozen. Dr Elmer not only shot, he also applied a scientific mind to the questions of bowyery, and left a body of writing about archery that stands very high in the international literature of the sport.

In 1911 he met James Duff, who had come to the United States from Edinburgh where he had learned bowmaking from Fergie, the bowyer to the Royal Company, and later worked with Buchanan, the finest bowmaker in London. At about this time also, across the land in California, another doctor Saxton T. Pope had, thrown into his lap as it were, one of the most extraordinary chances to investigate the methods of the Indian bowmakers and bowhunters of the past, and to link the two archery cultures, the Indian and the European. In 1908, in the Mount Lassen district of California, there were still a few survivors of the white man's massacres of the Yahi or Yana Indians, known to the Californians as the Mill Creek Indians, perhaps a dozen of them out of more than 3,000. Saxton Pope wrote:

'One evening as a party of linemen stood on a log at the edge of a deep swift stream debating the best place to ford, a naked Indian rose up before them, giving a savage snarl and brandishing a spear. In an instant the party disbanded, fell from the log, and crossed the stream in record-breaking time ... this was the first appearance of Ishi, the Yana Indian.'

Ishi evaded the white man until 1911 when, emaciated and sick, he was captured, and eventually handed over to the Department of Anthropology at the University of California. He was clothed and fed, 'learned that the white man was good' and lived happily for five years, a stone-age man, fisherman and hunter, among the Californians of the 20th century.

Pope became an instructor in the University Medical School and so met Ishi, learned his language word by word, listened to his stories and became his pupil in bowmaking and hunting. 'He was a wonderful companion in the woods, and many days and nights we journeyed together.'

Sadly, Ishi died of tuberculosis in 1916.

'As an Indian should go, so we sent him on his long journey to the land of shadows. By his side we placed his fire sticks, ten pieces of dentalia or Indian money, a small bag of corn meal, a bit of dried venison, some tobacco, and his bows and arrows ... He looked upon us as sophisticated children – smart but not wise ... his soul was that of a child, his mind that of a philosopher. With him there was no word for good-bye. He said, "You stay. I go." He has gone and he hunts with his people. We stay, and he has left us the heritage of the bow.'

Saxton Pope, from his long association with Ishi, concluded that Ishi's bow and arrow making was the best by far of any examples of Indian bowmaking to be found in American museums. His bows were short and flat in section, cut from mountain juniper, as so many early bows must have been when the ice of the Glacial Age retreated and juniper began to grow early among the returning shrubs and trees. They were not longbows, and so do not properly fall within our scope, except that they were wooden bows, like the fore-runners of the developed longbow, and also shared a kind of ancestral kinship with the modern flat-section bow. For that reason, and because Pope's writings on archery are so excellent, and yet so hard to come by I make no apology for quoting freely from the 1925 edition of his *Hunting with the Bow and Arrow*.

Pope describes Ishi's bows as weighing about 45 lb (20.4 kg), being some 42 in long (106.7 cm), capable of shooting 200 yards (182.9 metres), made of juniper, backed with sinew, broadest at the centre of each limb, and of an elliptical cross-section. 'At the center of the bow the handgrip was about an inch and a quarter [3.2 cm] wide by three quarters of an inch [1.9 cm] thick, the cross-section being ovoid. At the tips it was curved gently backward and measured at the nocks three quarters of an inch [1.9 cm] by one half an inch [1.27 cm]. The nock itself was square shouldered and terminated in a pin half an inch [1.3 cm] in diameter and an inch [2.5 cm] long.'

Ishi would split a limb from the tree and use the outer part of the wood including the sapwood, as does the longbowmaker. He reduced it by scraping and rubbing on sandstone, and made the recurved tips 'by bending the wood backward over a heated stone'. It was then bound to a wooden 'former' and left to season 'in a dark, dry place'. After seasoning, he backed the bow with sinew from the leg tendons of deer, held on with a glue made from boiled salmon skins. The tendons were chewed, to separate the fibres and make them soft, and then glued to the roughened back of the bow. 'Carefully overlapping the ends of the numerous fibres he covered the entire back very thickly. At the nocks he surrounded the wood completely and added a circular binding about the bow. During the process of drying he bound the sinew tightly to the bow with long thin strips of willow bark. After several days he removed this bandage and smoothed off the edges of the dry sinew, sized the surface with more glue and rubbed everything smooth with sandstone. Then he bound the handgrip for a space of four inches [10.2 cm] with a narrow buckskin thong ... the bowstring he made from the finer tendons from the deer's shank', again chewed and twisted into a cord, with a loop at one end and a thong for tying at the other. 'Drawn to the full length of an arrow, which was about 26 in [66 cm], exclusive of the foreshaft, his bow bent in a perfect arc slightly flattened at the handle ... According to Ishi, a bow left strung or standing in an upright position gets tired and sweats. When not in use it should be left lying down; no one should step over it; no child should handle it, and no woman should touch it. This brings bad luck and makes it shoot crooked. To expunge such an influence it is necessary to wash the bow in sand and water.

'By placing one end of his bow at the corner of his open mouth, and tapping the string with an arrow, the Yana could make sweet music ... to this accompaniment Ishi sang a folk-song telling of a great warrior whose bow was so strong that, dipping his arrow first in fire, then in the ocean, he shot at the sun. As swift as the wind, his arrow flew straight in the round open door of the sun and put out its light. Darkness fell upon the earth and men shivered with cold. To prevent themselves from freezing they grew feathers, and thus our brothers the birds were born.'

Of all the arrow woods that Ishi used, he preferred wych hazel, cutting the stems with a peeled diameter of three-eighths of an inch (1 cm) at the base, binding them in bundles and leaving them to season. Then he picked out the best, and straightened them: 'this he accomplished by holding the concave surface near ... hot embers, and when warm he either pressed his great toe on the opposite side, or he bent the wood backward on the base of the thumb'. This thumb method is exactly the one we use to straighten longbow arrows today. 'The sticks thus straightened he ran back and forth between two grooved pieces of sandstone ... until they were smooth, and reduced to a diameter of about five-sixteenths of an inch [0.8 cm]. Next they were cut into lengths of approximately 26 in [66 cm]. The larger end was now bound with a buckskin thong and drilled out for the depth of an inch and a half [3.8 cm] to receive the end of the foreshaft.' This was of hardwood, tapered to the front and glued into the hole that had been drilled with a sharp bone. Measuring the joined shaft from his breastbone to the tip of his forefinger, 32 in (81.3 cm), he then cut the foreshaft. The rear end was notched for the string and the fore-end, or foot, to take the arrowhead. 'At this stage he painted his shafts. The pigments ... were red cinnabar, black pigment from the eye of the trout, a green dye from wild onions, and a blue ... from a root ... mixed with sap or resin ... applied with a little stick or hairs from a fox's tail drawn through a quill.' This aboriginal decoration and identification which Ishi painted sometimes in alternate rings, sometimes in dots and snaky lines, corresponds exactly with European cresting, put on for the very same purposes.

'In fletching arrows Ishi used eagle, buzzard, hawk or flicker feathers ... by preference he took them from the wings.' He split the quill and scraped the pith with obsidian (a kind of volcanic glass) until the rib was thin and flat, then he soaked the feathers in water and ruffled them against their lie, and with a thin piece of chewed sinew bound them, three to an arrow, as we have them, starting at the nock end with the first, as we should say the cock feather, at right angles to the line of the nock, and setting the others equidistantly in position to take the binding, as he slowly rotated the arrow shaft. He only occasionally bothered with glue. 'Two kinds of points were used on Ishi's arrows. One was the simple blunt end of the shaft bound with sinew used for killing small game and practice shots. The other was his hunting head, made of flint or obsidian. He began by taking one chunk of obsidian and throwing it against another.' Then he would choose a broken piece about three inches (7.6 cm) long, two inches (5.1 cm) wide and half an inch (1.3 cm) thick, and 'protecting the palm of his left hand by a piece of thick buckskin' he would use in his right hand a sort of horn-tipped chisel to press the obsidian and flake it, like a flint knapper, until he had a symmetrical head. 'Flint, plate glass, old bottle glass, onyx – all could be worked with equal facility ... These heads were set on the end of the shaft with heated resin and bound in place with sinew which encircled the end of the arrow and crossed diagonally through the barb notches ...', such heads had 'better cutting qualities in animal tissue than has steel'. Thus armed, and carrying up to 60 arrows in an otter-skin quiver over his left shoulder, Ishi would go hunting, just as Indians and primitive man had done for countless thousands of years.

He was no great shakes at the target. When Arthur Young, friend and fellow bowman with Saxton Pope, could shoot an American Round (30 arrows each at 60, 50 and 40 yards), and score 626 by the 'Prince's Reckoning', and Pope 538, Ishi's best round was only 223. But at tracking, calling game, waiting, and instinctive shooting Ishi was a master. 'Ishi could smell deer, cougar and foxes like an animal ... he could imitate the call of quail to such an extent that he spoke a half dozen sentences to them.' He would call deer by bleating like a fawn, in the season, or by bobbing up and down with a stuffed buck's head on his own to excite the curiosity of both buck and doe to come and investigate. 'First of all he studied the country for its formation of hills, ridges, valleys, canyons, brush and timber. He observed the direction of the prevailing winds, the position of the sun at daybreak and evening. He noted the water-holes, game trails, "buck look-outs", deer beds, the nature of the feeding grounds, the stage of the moon, the presence of salt licks ... the habits of game, and the presence or absence of predatory beasts ... He would eat no fish the day before the hunt, and smoke no tobacco, for these odours are detected a great way off. He rose early, bathed in the creek, rubbed himself with aromatic leaves ... washed out his mouth, drank water, but ate no food. Dressed in a loin cloth, but without shirt, leggings or moccasins, he set out ... As he walked, he placed every footfall with precise care; the most stealthy step I ever saw ... for every step he looked twice.' At 50 yards he could drive an arrow through the chest wall of a buck, half its length.

'In his youth Ishi killed a cinnamon bear single handed. Finding it asleep on a ledge of rock, he sneaked close to it and gave it a loud whistle. The bear rose up on its hind legs and Ishi shot him through the chest. With a roar the bear fell off the ledge and the Indian jumped after him. With a short-handled obsidian spear he thrust him through the heart ...

'In all things pertaining to the handicraft of archery and the technique of shooting, he was most exacting. Neatness about his tackle, care of his equipment, deliberation and form in his shooting were typical of him; in fact, he loved his bow as he did no other of his possessions. It was his constant companion in life and he took it with him on his last long journey.'

After Ishi died, Pope wrote, 'From shooting the bow Indian-fashion, I turned to the study of its history, and soon found that the English were its greatest masters. In them archery reached its high-tide; after them its glory passed.' It is certain that a new glory has come to archery both in the field and at the target, its high tide being in America. There bow-hunting is allowed, as it is not in England; there the millions practise at the target with the new and brilliant bows of today.

Pope went on from his time with Ishi, to make and hunt with yew longbows, to kill big game in his own continent and in Africa, and to produce a second volume of hunting reminiscences, *The Adenturous Bowmen*, and a remarkable study *Bows and Arrows*. In the latter, he examined a great number of bows and arrows from all over the world, and carried out tests shooting every bow that was in fit condition and using flight arrows of bamboo made by Ishi, which gave the bows every chance to show their maximum cast. Bows of the Mohave, Navaho, Yurok, Yaki, Yana, Blackfoot, Apache, Hupa and Cheyenne Indians were tested and showed that without exception they were inferior to any well-made longbow of the European type, though as far as hunting is concerned, and in some cases hunting from ponies, they must often have been more than adequate, indeed fitter for their purposes. As a war bow used on foot the longbow would have beaten them every time.

The Yaki bow, for instance, which was all but a five-foot (152 cm) weapon made of osage wood and with a draw weight of 70 lb (31.8 kg) shot 210 yards (192 metres), and the Yana bow, possibly one of Ishi's manufacture, of red yew with the sapwood removed and backed instead with rawhide, weighed 48 lb (21.8 kg) and shot 205 yards (187.5 metres). The Blackfoot and Cheyenne bows, both made of ash, one of 45 lb (20.4 kg) the other of 65 lb (29.5 kg), only managed 145 yards (132.6 metres) and 165 yards (150.9 metres), and none of the other Indian bows even reached those distances. An Alaskan, probably Eskimo bow, which Pope described as being 'by far the best made of any aboriginal weapon in the group', was of an elementary composite type, made from Douglas fir, backed with bone lashed in position with a network of twisted sinew. It weighed 80 lb (36.3 kg) and shot 180 yards (164.6 metres). There was also a bow from Paraguay, nearly six-foot long, made of ironwood, which was 'stubborn, strong, inflexible and quick casting', weighed 60 lb (27.2 kg) and shot 170 yards (155.5 metres). But this bow was reworked. It was first straightened by heating and then the limbs were planed so as to distribute

Ishi, the Yana Indian (E. G. Heath).

the bend more evenly when drawn; it was shortened to 67 in (170.2 cm) and was fitted with horn nocks and a linen string, when it weighed 85 lb (38.6 kg) and shot the flight arrow 265 yards (242.3 metres), providing an excellent example of the difference between poor native bowyery and the much more efficient European methods of bowmaking.

But extraordinary penetration has been claimed for some Indian weapons, and sworn to by eye witnesses. During their Florida campaigns the Spaniards again and again found their breastplates, which would stop musket balls, penetrated by arrows from the bows of Creek Indians, Choctaws and Chickasaws. An Indian captive, made to demonstrate their shooting methods, shot clean through a heavy coat of mail, the arrow dropping to the ground beyond the back of the armour. He also completely penetrated two such mail armours, one hung on top of the other.

There is also an account by a Jesuit priest from Cologne of Indian archery in the 18th century. He speaks of them as Apaches but, since Pope found the Apache bows to be weak, short-range weapons, and since another tribe, the Yaki, used heavy bows in the Sonora area of Mexico, it is likely that Father Pfefferhorn was referring to Yaki Indians when he described the native archers as incomparable shooters who seldom missed their mark, loosing their arrows with 'more power and effect than a bullet from the best musket'. A Mexican soldier, caught in an ambush, had his shield, his rolled-up cloak, his leg and the leather saddle pierced by an Indian arrow which penetrated deep into his horse as well. When Indian arrows failed to go through plate armour they were shot at the eyes and faces of their enemies.

While Saxton Pope was busy elucidating the mysteries of American bows and fostering the use of the longbow, the rotund medical archer further east was plotting the longbow's demise. Dr Elmer had it in for the for British, 'a race notoriously bound to tradition', and berated them for being thoroughly stick-in-the-mud about their archery weapon.

He was of course right to examine the old weapon for imperfections. It had lasted almost unchanged and unchallenged until 1928. He wrote, 'How microscopic by comparison appears the modicum of time since then, and yet how much has archery advanced within it'. Basing his thought on the belief that the best bow is one in which every part works and none is idle, he concluded that unequal stress and strain within the wood undermines the true efficiency of the traditional longbow. 'Wood is not a homogeneous material like metal', he explained. 'We are confronted with the problem of designing such cross-sections that neither back nor belly will give way from overstrain.' Samuel McMeen (a physicist of Columbus, Ohio, who used to watch the Thompson brothers as a boy, and who became an archer in 1916),

Dr Clarence N. Hickman, PhD and Dr Paul Klopsteg, PhD were with Elmer among those who were responsible for the improvement of the bow by their observation and experiments. Hickman discovered that in a yew longbow of traditional design the essential fault was 'the unequal distribution of stress and strain which works some parts too much and others too little'. From mid-limb outwards longbows tend to suffer the slight compression collapses which show as pale hairline cracks in the dark wood of the belly, which longbowmen for centuries have feared, and called frets or chrysals.

Elmer and Hickman were concerned with adjusting the section of a bow so that the neutral axis would be more balanced within the stave. The neutral axis is the dividing line between those parts of a bow that suffer tension, and those that suffer compression, the back and the belly respectively, and in the ordinary longbow they found an unsatisfactory balance. So they worked towards the idea of a rectangular section bow, rather then the 'D' or Roman arch pattern, in order that the neutral axis could be more easily kept constant and the inequality of the proportion reduced. They knew that the strength of a bow twice as wide is doubled, and that a bow twice as deep is eight times stronger, but they felt even the rectangular longbow was a *cul de sac* in the search for efficiency. A certain Captain Cassius Styles of Berkeley, California, who used to make traditional bows now followed the experimenting doctors and, reducing the convex contour of the belly, calculated how to tiller a longbow so as to distribute the stresses, and in doing so achieved up to 30 per cent more efficiency. In other words he was a very good bowyer, but his method also included flattening the bow. I possess a fine hunting bow made for me by Gilman Keasey, which is a longbow, although it is not so high-arched as the traditional bow. Earl Ullrich, whose rare skills as a forester and bowmaker are looked at in the chapter on bowyery, believed that the medieval bow was likely to have been flatter in section than the later target weapon. Certainly a slightly broader and flatter shape would tend to be a much better survivor of the knocks and bumps and strains that bows must constantly have met in conditions of war. Against this stands the fact that the surviving late medieval bows and bowstaves are high-arched. I think all one can believe is that some were flatter than others and that a compromise must have been reached or attempted between maximum punch in the weapon and maximum survival properties.

All re-bowyery of the old longbow failed to satisfy the physicists, though it is worth pointing out that Robert Elmer shot marvellously with whatever bow came into his hand. 'Obviously,' said he, 'if a substantial increase of cast be desired, it must be found by changing the design of the bow and not by simply making the same kind of bow heavier.' He had 'a feeling that ultimate beauty and perhaps utility of design cannot be achieved unless we do as the ancient Greeks did and rely on geometry instead of numbers'. The Americans were striving to get out of the prison of the straight-limbed bow, and in escaping they developed a bow of strikingly oriental appearance or, as Elmer put it, 'Classical proportions'.

Two further facts were leading bowyers away from the longbow. Doubling the thickness of a limb only doubles the speed at which the tip moves forward from the fully drawn position when the bowstring is released in shooting, but halving the length of limb increases that speed of spring four times; and, the flatter the section of the limb, the shorter the limb can safely be made. They were moving inevitably towards a flat bow that the bowmaker Russell Wilcox called a duoflex, that is a bow which bends towards the shooter from the handle and gradually bends forward again or retroverts towards the tips, made in such a way that the bow wood works equally at every point. Wilcox used a 'bendmeter' which measured in thousandths of an inch the curve at any point on his bow-limbs, knowing that without such indications it was virtually impossible to be certain of all tiny local variations in contour. Dr Hickman had produced figures showing the stress in yew wood at the limit of the wood's elasticity was 10,100 lb/sq in (710 kg/sq cm), and the stress at breaking point 16,800 lb/sq in (1,181 kg/sq cm). The bendmeter proved marvellously effective in regulating such fibre-stress, which could be measured accurately at any point on the limb, and the bowyer could use these measurements to distribute the work evenly and efficiently throughout the weapon, leaving a section of 12 in (30.5 cm) in the handle too stiff to work, and so making the weapon steady in the hand.

Between the simple wooden and the true composite bow there had always been many kinds of intermediate weapon, in which the wood element was the major source of power, the power of the composite bow being in the other materials than the wood component, which, if present at all, only acted as bone for the muscles of the other parts. Many such intermediate methods, some invented in the past, some new, were tried now. Hickory backing or sinew to resist tension, osage bellies to resist compression, laminates with greenheart or beech as the mid-layer, were all tried, and in 1928 the fine bowyer Rouncevelle introduced fibre backing. Wilcox began to abandon yew wood, and for his duoflex bows to use a single layer of summer-growth hickory sapwood as backing for an osage belly, though by now because of the flat section, 'belly' had begun to seem an inappropriate name. At the base of each identical limb the measurements were 1⅛ in (2.9 cm) by ½ in (1.3 cm) thick, all sections being rectangular, and tapering to half those dimensions at the limb-ends. The stressing throughout the limbs was uniform, about 18,000 lb/sq in (1,266 kg/sq cm), a figure at which yew would certainly collapse. The limbs were curved by hot water and a mould, other bowyers using hot oils, dry heat or steam for the purpose.

The steel bow was another departure from the old wooden bow, but really made with the longbow shape in mind, curiously ignoring the ancient and efficient designs of steel bows from Asia. In 1872, before the Thompson revival, a patent was granted to Ephraim Morton of Plymouth, Massachusetts, for a wood-handled bow with limbs of steel rod, having two spirals in each limb. By 1927 there had been various experiments which led to a sort of hollow golf-stick bow, and aluminium began to replace steel. There were metal bows with a T-section, one with a U-section called the True Temper bow, and flattened tube sections of which the best were the 'Seefab' bow from Sweden, and the Accles and Pollock bows from Britain. Everyone was trying them but, though some of them had lively casts, they tended to break the string, and they seemed to suffer from metal-fatigue fracture, which made people think them dangerous. There are some archers who remain devoted to the steel bow, who shot better with them than with any other bow and who maintain they were no more prone to breakage than wooden bows. But it is one thing to be hit about the head and legs with a breaking yew bow – some of the force disappears in the splintering – quite another to bear the thwack of a metal spring that has parted with one swift unpredictable snap.

The physics and mathematics of the longbow belong in the technical appendices at the end of this book, and are for those who enjoy such matters. For those who do not, it is enough to see that, just as the short bows and flat bows in Europe were superseded by the development of the longbow, so now

First Tournament of the National Archery Association, Chicago, 1879 (E. G. Heath).

the longbow, in a technically advancing society, was overtaken by the flood of such an advance. It was no great step from the moulded and laminated bows of Wilcox and others to the moulded composite modern bows that have taken advantage of every improvement in the manufacture of plastics and glass fibre, and whose design has put into the hands of archers the most superbly efficient and accurate weapons. In spite of traditionalism, the longbow was doomed in England as well as America, and in Europe too. In fact wherever it had flourished as a survivor from its continental war days into an Indian summer as a target weapon, it could now no longer hope to hold its own.

Elmer, Klopsteg and Wilcox were among the great innovators but they were not alone; they were in the van of change, but not its sole arbiters.

Among those who took advantage of the changes was a remarkable American called Howard Hill, who appreciated just where change was valuable and where traditional design still had a role to play. His career as an archer suggests the whole span of time from the medieval bowman to the modern. When he was young he made and used superb longbows, of enormous power, just the weapons of the yeomen of the French wars, 70 lb (31.8 kg), 90 lb (40.8 kg), 100 lb (45.4 kg) draw weight, and onwards. In 1928, using a 172 lb (78 kg) bow he shot an arrow 391 yards 23 inches (358.1 metres).

For some years, he made every effort to master a sensitive bow as a hunting weapon. He used different weights and lengths of both composite and self-wood bows but, as he himself admitted, he never managed to be really accurate with them. He made an 85 lb (38.6 kg) split-bamboo bow and practised four times a week for two months with it, only to find himself one day sitting at the bottom of a tree in which there were hen grouse, having shot off 38 arrows for only four birds hit. He went back to his camp and picked up a favourite old longbow that he called 'Granpa'. With that bow after 20 minutes he had shot 11 arrows and bagged nine birds. On that expedition, he killed more than 100 grouse, two deer, a moose, a great bald eagle, several ducks, rabbits and squirrels with his dependable old 'Granpa'. He concluded that recurved or other sensitive bows were not suitable for hunting. A hunting bow had to be accurate, hardy, steady and gentle to handle as well as smooth on the draw, but still have enough cast and what he called 'follow-through' to shoot a heavy hunting arrow. He found that the only bow which was a better hunting weapon than the conventional English longbow was the American semi-longbow. Hill's modern longbows were shorter than the old English weapon, slightly wider and a good deal flatter, though not as flat as the short bows of the American Indians, and he believed that a semi-longbow with recurved ends was too sensitive to be a good hunting weapon.

In choosing a hunting bow, he suggested that from a basic weight of 35 lb (15.9 kg) a pound extra was needed for every yard shot, up to 60 yards (54.9 metres). He seldom shot over 60 yards unless he was after predatory animals or birds. In general he would use the heaviest bows that he could draw without strain, finding that they made his release smoother, and threw the arrows in a flatter trajectory, which meant that he could shoot under low branches with more accuracy and less danger of deflection. He was often asked in later years why he did not use a recurved bow and his answer was always that, under hunting conditions, he could not shoot a short recurved bow accurately. If Hill could not, then nobody could.

Hill laid great stress on the choosing of arrows for each particular type of game, and on matching them to the bow, by spine and weight, and he believed firmly in quick, instinctive shooting from a high draw, not to the chin, but to the corner of the mouth. He used to find his anchor point with the middle finger on the back tooth of his right lower jaw.

He describes in his book *Hunting the Hard Way*, a classic in the annals of the longbow, a remarkable confrontation with a wild boar. Count Ahlefeldt, who has killed wild boar with the longbow, reckons that it is nearly impossible to pierce the boar's frontal bone head on with an arrow. But Howard Hill's narrative claims he did, cutting through the skull right in the centre as the animal charged him head on. He confessed that his knees were shaking and that he felt more frightened than he had ever been in his life as the creature toppled at his feet.

There are many books in which the techniques and construction of modern bows, both for target shooting and hunting, can be studied. We can only bow to the superiority of such bows, indicate how and why that superiority was achieved, and show to what an important extent America contributed to the breaking out of archery from the confines of the traditional. In the year of the bicentenary of American Independence, it is only proper to end this chapter with part of a letter, dated February 11 1776, to General Charles Lee, who was in command in New York, and was busy constructing defence-works. This plea was written years later than any from the traditionalists on the European side of the Atlantic:

'Dear Sir,

The bearer, M. Arundel, is directed by the Congress to repair to General Schuyler, in order to be employed by him in the artillery service ...

They still talk big in England and threaten hard; but their language is somewhat civiler, at least not quite so disrespectful to us. By degrees they come to their senses, but too late I fancy for their interest.

We have got a large quantity of saltpetre, 120 tons, and 30 more expected. Powder mills are now wanting. I believe we must set to work and make it by hand. But I still wish, with you, that pikes could be introduced, and I would add, bows and arrows. These were good weapons, not wisely laid aside.

1st Because a man may shoot as truly with a bow as with a common musket.

2dly He can discharge four arrows in the time of charging and discharging one bullet.

3dly His object is not taken from his view by the smoke of his own side.

4thly A flight of arrows, seen coming upon them, terrifies and disturbs the enemies' attention to their business.

5thly An arrow striking in any part of a man puts him hors du combat till it is extracted.

6thly Bows and arrows are more easily provided everywhere than muskets and ammunition.

Polydore Virgil, speaking of one of our battles against the French in Edward the Third's reign, mentions the great confusion the enemy was thrown into, *sagittarum nube* [by the cloud of arrows], from the English ... If so much execution was done by arrows when men wore some defensive armour, how much more might be done now that it is out of use.'

The writer was Benjamin Franklin.

CHAPTER 10

THE WEAPON AND THE HUNT, THE WOOD AND THE MAKING

In Denmark, on the island of Fyn, the tall and turreted, pale rose castle of Egeskov stands on its own dancing reflections in the water from which it rises. Thousands of people from all over the world who go to visit the gardens, the museums of veteran motor cars and aeroplanes, the collection of carriages and harnesses, to listen to summer concerts or watch ballet and national dancing, find themselves at the home of a remarkable man, with a remarkable and beautiful wife, who together organise all the beauties and pleasures of Egeskov for the delight of those who go there. He has cut his own yew trees, made his own longbows and his own arrows, fashioned his own arrowheads, and both in Europe and Africa used this true equipment of the Middle Ages both to hunt game, and to discover a great deal that is of vital interest to the historian of the longbow. He is Count Gregers and she is Countess Nonni Ahlefeldt-Laurvig-Bille. From 1945 to 1970 he was Game Warden and game advisor to the kingdom of Denmark, appointed Master of the Royal Hounds in 1930, an office which, he says, has little to do with hounds, and Chamberlain to HM King Frederic IX in 1959. He is also the author of several books and a dictionary on game and game preservation, on bows, bowhunting and bowmaking, as well as more conventional big game hunting, and ethnological investigation.

At the end of his professional career, when he was 65, he gave up shooting entirely, 'not for physical reasons, but just because my desire to hunt had petered out. I felt I had had my share. Or, to put it in other words: some people grow up slowly – others never do. I think I grew up rather late, but with no regret!

'Hunting traditions run in the family through generations,' he writes. 'Like most boys I played with so-called bows, generally made of a simple, untapered hazel stick, its ends connected by a bit of string or wire. Arrows were thin, fairly straight sticks of hazel, ash, bamboo or rush, without feathers to steer them. As arrowheads I was supposed to use bottle corks, for safety's sake, but in fact often fitted dangerous nails fixed with wire. Fortunately without serious accidents, I broke a number of window panes and missed quite a few sparrows. My first experience with a real bow was when my father gave me a rather poor laminated metal bow and badly fletched arrows, to practise in the sand dunes of western Jutland, which was fairly safe as long as you did not shoot too high over the dunes. I was 10 or 11 years old and well acquainted with stories of Red Indians and Robin Hood, William Tell and Einar Tambeskjelve.

'Then, in about 1928, I came across an article about Saxton Pope and Arthur Young in an American periodical. They killed moose and grizzly bear with longbows and sharp steel arrowheads. Frankly I rather doubted if it was true or not. But I had to have a go, to convince myself. I made a longbow, all by myself, out of a straight-grown ash stave – all wrong in choice of wood and in proportions – with a narrow rounded back, and belly far too high stacked, but beautifully tapered, and with roedeer horn nocks set on at the tips. I still have it. It worked, but it was stiff and unpleasant and kicked like a mule, and no animal came to grief.'

Count Ahlefeldt met the pioneer of modern Danish bow-hunting, Carl Dreyer, and through him was able to broaden his knowledge of the weapon they inherited from their Viking ancestors. Dreyer gave him a 30 lb (13.6 kg) lemonwood bow and a bunch of hunting arrows.

'Much later he gave me a 70 lb (31.8 kg) osage hunting bow made by Saxton Pope. It was baptised 'Stradivarius' and with it I killed my first and only red deer stag in 1935, but that, to quote Kipling, is another story.

'Dreyer used longbows of yew, lemonwood and osage, and arrows from such craftsmen as Ayers of London, and Duff and Saxton Pope of the USA. Tackle was expensive even in those days; a good hunting or a target bow was as costly as a fair shotgun, and fine hunting arrows cost far more than 12-bore cartridges, about 10 shillings apiece, and you broke or lost quantities, especially in hunting. I could not afford that, so the answer was simple: either give up, or learn to make your own equipment.

'I started searching for straight grown pine, such as dry old floorboards, for arrows, and – much more difficult – for straight elm and yew timber to season for bowmaking. Naturally the arrows were ready first, while the bow wood was seasoning, and I used them in the bows that I had.

'At first I split and trimmed every shaft myself, but in the end I was able to train a professional turner in the village to make me large bunches of excellent, uniform, dry shafts as good as any I ever met. Fletching by hand was no great problem, though it took much care and time. Turkey and goose feathers were plentiful, and for the rest I only needed good waterproof glue, large headed pins to hold the flights on while the glue dried, and the time in which to learn. I made a useful discovery when I found out that the narrow part of a good goose feather, with no trimming, is quite enough for a fairly light arrow, and does not crumple in the quiver or in rain, which even the best-trimmed feathers are apt to do.

'Target heads I made with a ring round the foot to prevent the shaft splitting, and then drilled a hole in the head and inserted a round-headed screw of the right size and weight to balance the arrow. For light hunting arrows I cut heads from a thin steel plate, with a broad head and a longish tang all in one. Then this was driven into the tubed and drilled arrowfoot, and finally formed and sharpened with a steel file or a grindstone, making a cheap and efficient arrow, suitable for birds or hares and rabbits, which cost about a shilling apiece in those days. Larger hunting heads, for bows of 50 lb (22.7 kg) weight or more, and for use against roedeer and bigger game, were made for me by a local smith from thicker and harder steel (shovel blades are excellent) with a ferrule soldered on to fit the shaft.

'Whatever you intend to hunt (except fish) never use barbed arrows. It is cruel. If an animal is wounded, the arrow must be able to fall out by itself. If the head is barbed it will remain in the wound even if the shaft is broken off in the animal's flight, and however shallow the penetration the wound will fester. But even a severe wound that is not fatal will quickly heal as long as the arrowhead comes away.

'For fish it is different. I used to stalk pike from a very light narrow kayak, drifting carefully along and standing up ready with bow in hand. My arrow was a very long one, such as natives use in tropical lakes and rivers, with a length of thin fishing line loosely attached, and one blade of the broad barbed head filed down in order not to hurt your bowhand in drawing. You aimed low, and if you made a hit, and not a plunge into the water yourself, which often happened, the fish would flash away, but soon show itself again on the surface not far off, with the feather end of your arrow as a floating mark.

'Bowstrings are easy to twist and wax and whip, using Irish flax or shoemakers' thread, or you can buy them, even nowadays, quite cheaply. Only never forget that the string for a hunting bow must be fresh, well waxed and contain enough strands to be strong enough for the bow. A broken string in action often means a broken bow. The finest wax I used to make by boiling

50 per cent beeswax with a wee bit less resin, and a small lump of dry tar. When it becomes a brown liquid mixture you pour it into cold water (it will not break up in the water) and when it is cool you can mould it into lumps of the size you want.

'Always carry wax with you when hunting, a knife and a steel file attached to your belt or quiver. It is cruel to use a blunt arrow, and heads should be sharpened even after a single shot into the ground. A head with a somewhat rough edge from the file is far better for its purpose than a razor edge.

'Once your logs or staves for the actual making of the bow are ready, it is simple and easy enough – almost as easy as making a violin, but remember how many bad violins there are in the world, and how few Stradivarii!

'My own best bow took me five years from the time I cut the log in 1932 out of a bunch of three heavy old yew trunks that had grown together in the shade by the side of the forest, for no one knows how long. I split the log carefully in two, immediately, keeping only the half that had grown on the inside of the clump, without twigs or pins or any overgrown knots in it. It spent some three months in clear running water, to drive out some of the sap. Then it went to a dark, damp old shed, where it was kept well above the wet ground for more than a year. After that it was moved step by step to progressively drier surroundings. The last year of its seasoning I remember it spent in the open air and the wind under the roof edge of an old thatched, half-timbered building.

'All the time, off and on, I had tapered the raw bow down bit by bit, and peeled off the drying bark. It was without a single flaw in its whole length and as straight as a ramrod. How long it finally took me in the workshop, with tools, broken glass for scraping, sandpaper, steel wool, wax and varnish, and testing in front of the long mirror, I have not the slightest idea, but it was finished some time in the spring of 1937, the year I took it to Kenya to hunt big game. It pulled 90 lb (40.8 kg), and was my strongest bow, a little too strong for accurate shooting. It is a long time since I had the strength to use it, but I still have it. I was 32 in 1937, I daresay not quite a weakling, and certainly in the best possible training, but my 90 lb yew bow was definitely at the limit of my capability for accurate and swift shooting. I felt that 80 lb was really my personal maximum. Hill, in America, and others may have been incredibly strong, but when one thinks of the average, rather small men of the Middle Ages I personally doubt if the ordinary military bowmen could handle more than 90–100 lb weapons.

'I preferred another single stave self-made yew bow that pulled 60 lb (27.2 kg), which I made from wood grown in the south of Norway, where yew is rare, but has the advantage of very fine grain because of its slow growth.

This bow was exceptionally fine in grain and lovely in its colours, with a yellow back and a dark salmon-red belly, and it had two almost symmetrical natural buckles, one halfway along each limb, which made it very pretty, quite unique, and probably added somewhat to the cast. It is broken now (due to a broken string) but I still keep it carefully glued together for happy memories.

'All the other bows I took to Kenya were self-made except for a Seefab steel bow which the Swedish factory gave me to test and which I detested (as I detest, and find hideous to look at, all modern artificial bows with sighting gadgets and what not) but I did kill a Thomson's gazelle clean with it. One of my favourites was a 50 lb (22.7 kg) two-piece yew bow, spliced in the handle, of a fine shape, fairly short, and backed with well-polished rawhide. Its shortness and the rawhide backing made it ideal for rough wear and tear in the bush, quick to handle and sweet to pull.

'Otherwise I had one spare longbow of medium strength made of elm with a thin backing of ash. It was not a bad bow at all, even compared with the yew bows. During the unpleasant German occupation of Denmark in the Second World War, when most firearms were confiscated, such bows became quite popular. I was often asked to make bows and finally in 1941 published a small book on bow- and arrow-making. At the same time – for lack of available yew staves – I trained a local carpenter to make ash-backed bows of elm wood. He is now a very old man, but he remembers making and selling in one year 111 finished bows, as well as a considerable number of staves.

'As to arrows for the bows I took to Kenya, that was a question of its own. Every archer knows the tricky business of the 'spine' of an arrowshaft. A heavy bow, say of 80 lb (36.3 kg), needs a stiffer arrow to 'round the bow' than a weaker bow of 50 lb (22.7 kg) does. The stiff arrow shot from the weak bow is apt to fly left of the mark, whereas the whippier arrow from the heavy bow will wobble in its flight. This applies even more to heavy-headed hunting arrows than to light-pointed target arrows.'

Count Ahlefeldt is referring to the 'Archers' Paradox', which describes the action of an arrow when the string is loosed. Because of the inertia of the head and the shaft, the nock of the arrow, under the impulse of the string, moves before the head, which makes the shaft buckle slightly to the right, or 'inside' the bow. It will then correct the first buckle with another in the opposite direction, and so on, until the oscillation dies out, and the arrow flies on straight and true. An arrow of correct spine bends round the bow and straightens out again almost at once. Too stiff an arrow does not buckle enough and flies off leftwards from the bowhandle. Too soft an arrow buckles too much, and the whole series of counter-bucklings are too big and last too long for true flight.

'To cope with the problem, I made during the better part of a year more than 1,000 arrows and tested each one on the target at least three times, training every day in rain or sunshine, but avoiding stormy or frosty days. Frost makes bows brittle and apt to break. All the arrows that were not obviously too soft by the feel of them were fletched and tipped with medium weight blunt heads. So as not to become used to a regular distance I set my little cardboard target on a hillside clear of stones, that had a big stretch of grass in front, and stuck the arrows (50 or 100 for a day's training) in the ground at various angles and distances, say 35 to 80 metres from the target. Then I picked up and shot the arrows, walking backwards and forwards in no fixed pattern, using a bow of about 50 lb draw weight. A poor 'slave' (often my wife) had to run in from a safe distance after every shot and mark each arrow with a pencil according to my shouted instructions. I could always feel if the draw and the loose were good and steady, and so knew when a good arrow would have flown true. Bad shots got no mark; wobblers were out; nice shots had a cross marked on them; arrows that flew left earned a question mark.

'Only arrows that after repeated shooting achieved three crosses were selected for medium weight heads. Then all the question marks were tested again with heavier broadheads and shot from the 90 lb (40.8 kg) bow. Hard and strenuous work it was and I could not generally trust myself to shoot more than a dozen or two arrows steadily enough for testing with that big bow.

'Six boxes of 100 arrows each were finally packed and sent off to Nairobi. I could have done with a third of that number, but my expectations were high.

'When the "three Danish pioneers" – Dreyer, a mutual friend, Mr Pelsoe and I – got going in the 'thirties, there was no objection to hunting in any Danish game law, as there is now, but soon we were to be strongly criticised in

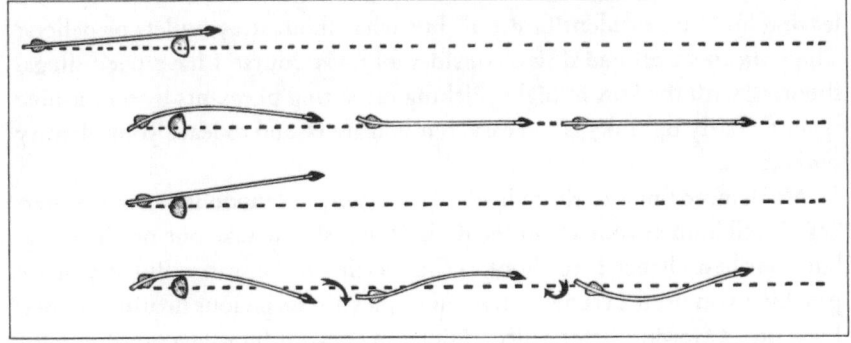

The Archer's Paradox: Stages of the arrow's movement past the bow handle (Drawing by Clifford Anscombe).

newspapers and by well-meaning animal protectionists who knew no better. Was it cruel, as they claimed? All hunting and killing of any animal may be considered cruel by some people, but I never met one of them who was a vegetarian for that reason.

'From a long experience of using both bow and rifle, and from seeing the reactions of animals hit by arrows, by shot and by rifle bullets, I know that killing an animal or even wounding it with a broadheaded arrow is less cruel than using a shotgun or a rifle, with pellets or a bullet.

'An animal hit by the blunt missile from a firearm will always show strong reaction and fright. That rarely happens with an arrow wound, as long as the head is not barbed. Hit in a vital spot, even shot clean through the body with an arrow, the animal will move off with no panic, with hardly any shock reaction, or none at all, but very soon it will stagger, fall and die very quickly from internal bleeding. In the case of a flesh wound the arrow will soon fall out (provided always that it is not barbed) and the animal will be none the worse for it. I have seen a red-deer stag, shot clean through the body so that the arrow buried itself in the ground beyond, go less than 50 yards (45 metres), without panic, before it fell dead. I saw a roebuck shot clean through the upper fleshy part of a foreleg with an arrow. He moved off quietly without fear and without limping. To see what happened I shot him with a rifle a week or so later. There was no sign of a limp when I saw him, and when I examined the wound, it was perfectly clean, dry and healed on both sides.

'We were also blamed for teaching poachers to use a new (!), silent and secret weapon. I was once interviewed by a policeman who had found some of my arrows sticking in the ground near a wood. I was able to draw the policeman's attention to the ease with which he had been able to track down the culprit.

'I suggested that any bow hunter who is good enough to poach for profit will also be well enough known in the neighbourhood, and he cannot help leaving his "card of identification", but what about stray bullets or pellets? The policeman calmed down considerably. Of course I have tried illegal shooting with the bow at night. Picking off sitting pheasants by moonshine against a fairly light sky is too easy, but you are bound to leave your identity cards about.

'Much more fun is to shoot mallard on a calm and moonlit patch of water. Stand well hidden yourself in the dark. You just may take one on the wing, but your best chance is to shoot as they settle on the water. But if you are poaching you need a retriever trained to pick up suspicious floating arrows. I had one, a Gordon setter called "Flint", my dearest hunting companion for more than 12 years. When I took up bow hunting he was quite unhappy at first and apt to go home by himself.

'I had him search and point, when we walked together, and then when the birds flew up and I loosed an arrow – well, it was mostly just a clean miss, and no bang at all – he would look at me sadly and thoughtfully; that is putting it mildly. When I bagged a bird on the wing he would proudly and happily retrieve it. Then one day, of his own accord, he started retrieving arrows, in the fields by day or from the pond at night; but I never managed to teach him not to chew them.

'Shooting at a bird in a tree or a sitting hare was not beneath the dignity of the "pioneers", especially after a good and fair stalk. The stalk was always the main thing, not the killing. I often caught, say, a roebuck napping at such close range that I knew I had him but, if I didn't need to kill him, I would just back out, perhaps taking a few photographs before I left. One thing is certain: bowhunting will never threaten the game population. We preferred competing with the natural alertness and cautious instincts of our game at close range, rather than using long-range firearms.

'Clay pigeon shooting was good sport, and we were often challenged by mildly mocking gun users. I would sometimes bet on hitting three or four pigeons out of a dozen and I often won. I would be left to choose my own stand, and I would go a fair distance in front of the trap, so that the target was thrown over my head. With luck and judgment I would be just where the saucer lost its speed and started falling, nearly above me and at short range. But be careful of your eyes, and your bow. I once had a good yew bow badly damaged by the sharp splinters of a clay pigeon showering down on it.

'But is shooting with the bow and arrow dangerous? Yes it is, when in the hands of the inexperienced, or of children. It is easy to make a poor bow and bad arrows, but it is terribly dangerous to children's eyes when they start shooting sparrows or playing William Tell. While I was chief of the Game Department I suggested a new measure which did put the brakes on. Target shooting on ranges was allowed, but a potential bowhunter had to carry a gun licence, and also a special permit from the Game Department, to be allowed to hunt on his own ground or in a defined area. To get that permit he had to be over 16, to have a testimonial from a well-known bowhunter that he had experience of target shooting, and that he would use bows of over 45 lb (20 kg) draw weight and unbarbed, sharp broadhead arrows only.

'In 1937 Denmark's oldest publishing house offered to pay my fare to Africa if I would go with bow and arrow in my armoury and write about my experiences afterwards. They did not have to make the offer twice. Actually the book was far more of a success than my rather restricted efforts with the bow. When I arrived in Nairobi I had to see Game Warden F. H. Clarke, to get the special permit for shooting game with a bow. His first question was

"Do you think it is cruel, young man?" My slightly angry reply was: "No, and if I did think so I would not use the bow at all", and I gave him all my reasons. Clarke laughed and pulled one trouser-leg well above his knee. He showed me a number of well-healed scars, saying: "These are from bullets, and they hurt like hell. Those others are from arrows, unpoisoned, thank God!, barbed Pigmy arrows. They did not hurt particularly, provided they went right through or could be pushed through the right way. If they struck bone they had to be pulled out backwards, and the only way to do that was for my native servant to twirl the arrowshaft between his palms, to free the barbs – that *did* hurt!"

'I got my permit all right, but not for the Big Five, elephant, rhino, buffalo, lion and leopard. Clarke explained: "It is not for your own sake; you probably know what you are doing. But if you were to wound an animal, it could turn nasty afterwards, and if somebody else – an innocent native for example – were to get mauled, not necessarily by your victim, but in an area where you had been shooting, I would be the person to carry the blame ... *so please* ..."

'A very wise and very British way of putting it. Had he threatened me with fines or expulsion I might well have ventured to break his ruling. Now it was impossible.'

The Kenya Game Department annual report for 1937 records: 'The visitor, who is a sportsman in the highest sense of the word, was good enough to remark afterwards that he is quite convinced that my ruling was a correct one and that it would be wrong to allow the use of the bow and arrow on rhino, elephant, buffalo or lion.'

'I know I could have killed at least one lion with an arrow. I met him without his ever knowing, broadside on at less than 25 yards (22.9 metres). He

An Impala shot in Kenya in 1937. Shot and photographed by Count Ahlefeldt-Laurvig-Bille.

scratched himself pleasantly and then sat down, while I drew the arrow but never let fly.

'My bow hunting experiences in Kenya were not so very different from those at home in Denmark. On one of our early days out I killed my first Thomson's gazelle with an arrow clean through the neck, so that it died on the spot. I was as pleased as a dog with two tails, until back in camp, having a drink over our camp fire, I noticed some rather peculiar behaviour among the native "boys". Our two gunbearers were obviously mimicking an archer, and the archer was me. They were all roaring with laughter. Not knowing enough Swahili yet, I asked Syd Downey, the famous white hunter and our guide, to translate for me. He was reluctant at first, but I persuaded him. "Well Count," he said, "if you insist. The boys consider you the biggest bloody idiot they have ever come across!" I said, "Thank you. Why?" and Syd replied, "Well, they're simple boys from the bush, but most of them have been on safari before. They know you have good rifles and they've seen you shoot well with them, so they wonder why the hell you bother to crawl about on your tummy, to get near enough to a gazelle to shoot it with a bow and arrow."

'One fine evening Syd and I were walking back to camp side by side on a narrow track. I had a light bow in my hand, braced, and with an arrow on the string. I suddenly noticed a faint wriggling on the ground where Syd's next footstep would fall. We were both wearing canvas shoes on bare feet. With one action I pushed Syd sideways into a thornbush and loosed an arrow at whatever it was. Syd made an unpleasant remark about thornbushes and at the same time asked if I was trying to shoot off his big toe. But he was happier when we discovered that I had neatly clipped the head off a deadly poisonous night adder.

'Because of my fair ability with the longbow I earned a nickname among the natives, which I found very flattering: "Bwana Mechali" – "Mr Arrow". I was greeted by that name, often by natives quite unknown to me, when I returned to Africa several times after the war. I was flattered because the phrase carries two meanings: one, that the man can shoot with an arrow, and two, that his way of thought follows the flight of an arrow.

'Perhaps the finest arrow shot of my life was at a Grant's gazelle. I could get no nearer than 70 yards (64 metres) and I shot him dead in his tracks with a broadhead arrow high in the shoulder, from my favourite 60 lb (27.2 kg) Norwegian yew bow. I was less fortunate with an impala. I saw him at a distance, nearly hidden in the grass and made a mental note how far to crawl. When I got up to shoot I realised to my horror that I had badly overestimated the distance. I could have touched the animal with one end of the bow. I aimed too quickly and shot badly and hit him far back in the kidney area. He staggered a short distance, and I immediately asked Syd to put an end to the ugly and pitiful sight, from where he had watched the stalk, with his rifle.

'When hunting anywhere with the bow, if there was not a rifle marksman with me, I always carried an accurate, long-barrelled .22 pistol in a holster, and I never entered a kill in my game diary that was not achieved with an arrow, before lead came into the question. Fortunately the occasions when I had recourse to bullets were rare.

'After 1937 I deliberately put the bow aside. I lacked time to practise with it. It was not until 35 years later, when Robert Hardy, author of this book and of the BBC film *The Longbow*, turned up at Egeskov to make part of the film here that I drew the bow again.

'To reconstruct the story of Einar Tambeskjelve, we sacrificed a yew bow. We put it in a carpenter's bench vise and drew it on a winch. Nothing below the handle was seen by the camera, only the upper limb against the sky. I shot from beside the camera, out of sight. My first arrow flew a little left, but the shaft and feathers slapped "Einar's bow". Either I was too near (15 paces or less) and the Archers' Paradox was operating, or the arrow was the wrong spine for the 50 lb bow. I corrected to the right and my second arrow grazed the target on its right side. My third broadhead struck just in the middle of the bow limb, which was tightened to 'full draw' on the winch. It broke "with a horrible noise" just as Einar's had in the old story. For some reason unknown to me that shot was not used in the completed film. Apart from that I admit I was quite pleased. I had barely drawn a single arrow for 35 years.

'I remember calling attention on that occasion to the fact that, though the longbow meant so much in British medieval history, it could not by any right be called a British invention. Caesar tells us that when he crossed the Channel in 55 BC he found the Britons armed with "short and crude bows". Neither do the early Vikings in the Icelandic Sagas tell us that the enemy they plundered on the British coasts used the longbow. They were beaten over and over again, indeed largely conquered by the longbow in the hands of the Vikings, from all over Scandinavia. From certain Vikings (referred to as Normans) the British were quick to learn the use of the longbow – the hard way!'

The Count regards his bowhunting adventures in Africa as 'humble experiences compared with those of other people who killed every variety of game there, such as Saxton Pope and Arthur Young at the beginning of our century, and much more recently the astonishing results of Bob Swinehart, and others. They must have used stronger bows, sharper and better arrows and have been both braver and stronger than I am,' he said. But it was he who put into my hands the details of Swinehart's safaris.

Bob Swinehart was a friend and pupil of Howard Hill. He is a man of great strength and skill with the bow, who from 1964 to 1966 four times took longbows on safari in Angola and killed all the Big Five. I do not know his complete armoury, but he had three longbows at least, from 75 lb to 100 lb (34.0–45.4 kg) and, though not the yew weapons of the traditional shape that would meet the conditions of the British Long Bow Society, they were certainly longbows (one made of laminated bamboo), and shaped like the now-traditional American longbow, with a flattish section of limb dipping from a deep handgrip.

Like Count Ahlefeldt and most other bowmen who hunt game, he was accompanied by a rifleman, and his own remarkable courage sometimes put his second in a tricky position, too close to dangerous game for the gunman to have time to shoot more than one bullet as a stopper, if the need arose.

In North America, Swinehart had already shot deer of all sorts, moose, bear, and sharks believe it or not, with the bow and arrow. Now he was up against the last remaining challenges, the last if you leave out the Yeti and the Loch Ness Monster. When, in man's insatiable curiosity, one of those is finally slain in the cause of science I had far rather see it done with bow and arrow than with a gun, or whatever else. It is so easy to kill or maim with a gun. The bow demands all that is hard-come-by: courage, patience, skill, strength, nerve, accuracy and wisdom. How many hunters

 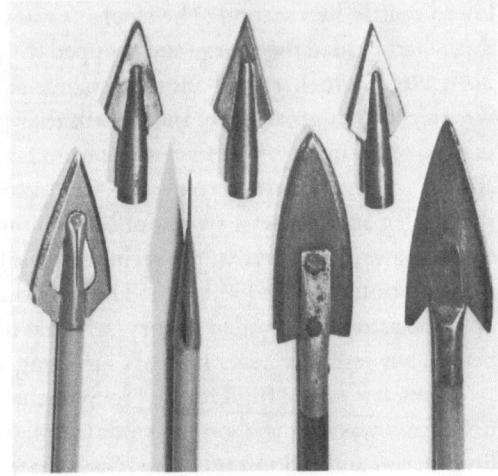

Above, left *Bob Swinehart draws his American longbow at an African buffalo* (From *In Africa with Bob Swinehart*).

Above, right *Various hunting broadheads* (Photograph by Graham Payne).

with rifle or pistol can make the weapon they use with such dash, or understand the chemical components that propel for them such lethal missiles? Most bowmen who hunt, not only can, but prefer to make their own tackle and when they have done that they have only started. There is much to be achieved before they can draw the arrow that will kill them merely a rabbit for their supper.

Swinehart shot his first buffalo with a 90 lb bow, running to within 35 yards (32 metres) of a stampeding herd. The broadhead arrow pierced the jugular vein of the one beast he singled out, and it fell dead 83 yards (75.9 metres) beyond the arrow strike. It is worth remembering that the African black buffalo is a much more dangerous animal than the buffalo or bison of North America.

His first attempt at a black rhinoceros was nearly a disaster. The arrow hit, penetrated deeply and would have proved fatal, but not before the animal charged. 'The back-up man had to pour a lot of lead to stop it.'

Leopard, Swinehart achieved with a 75 lb bow, not a longbow on this occasion, but a Ben Pearson glass fibre recurve 'Bushmaster', sending an arrow clean through its chest. The creature went 70 yards (64 metres) before it was dead.

A five-ton elephant bull was the next of the Five to fall. Swinehart stalked, to within 15 yards (13.7 metres) and then shot a heavy 40 in (101.6 cm) arrow from his 100 lb bow, which penetrated deeply. The first arrow was followed swiftly by a second. The elephant was tottering but suddenly charged. Swinehart evaded the charge and dropped the animal with three more arrows.

In 1966 Swinehart had another chance at a rhinoceros, after a long and weary chase in great heat, and a stalk that put him within 20 yards (18.3 metres) of his quarry. The first arrow from his 100 lb longbow glanced off the animal's chin as it turned to charge. The second buried itself deeply behind the foreleg and the beast swung off. The hunters found the dead rhino some distance away. One arrow, the second, had killed it, piercing the lungs, with a penetration of 20 in (50.8 cm). That was the first rhinoceros ever to fall to bow and arrow in recorded history, which is not to say that it never happened before, but certainly never in man's surviving knowledge.

A lion, last of the Big Five, fell to Swinehart the same year, on July 4. The first arrow was shot at about 35 yards (32 metres) and sank into his chest. The lion turned and hid in high grass. Swinehart closed to 25 yards, found him in an open patch of grass, dying but ready still to leap. He drove another arrow into the lion's chest. The animal vanished into the grass again, biting the arrowshaft. After a few more blind shots into the grass and a very careful circling stalk, they found the lion dead.

I do not expect these hunting details to suggest to very many that they should go out and tackle wild animals, bow in hand, even if they could get permission. There are few people brave enough to do so. Nor do I think it desirable that anybody who has the urge should be allowed to rattle about with arrows and bows among the wild life in Great Britain. But while it is a comparatively easy matter to obtain a licence to shoot at game with guns, while, as well as the careful experts, any kind of poor shot can blast way with little chance of a clean kill, and while for instance the deer population in the British Isles is greater than it has ever been, it seems to me sad that no bowman, no matter what his skill with his weapon or as a stalker, is ever allowed to draw an arrow in his bow at anything except vermin, unless he goes to Scotland or Northern Ireland where the prohibition does not apply.

Legend has it that when the Bill came before the House of Commons in 1963, the member who was to represent the interests of archers was in the lavatory at the moment the appropriate clause came up for debate, which was thus passed without argument.

★ ★ ★

Among his collection of bows and arrows, European, Asian and African, Count Ahlefeldt has some fine Japanese weapons and missiles. Any study of the longbow ignores the long Japanese bow at its peril, but a cursory look at Japanese archery is even more dangerous. It demands long examination which cannot come within the scope of this study. It might also be argued that the very construction of the Japanese bow puts it beyond our brief. It is a bow that has developed from the two traditions of composite Chinese weapons of the mainland, and the simple wooden bows used by the predecessors of the present Japanese and Ainu people. With the development of the weapon since the early centuries AD has grown a philosophy and a mystique attached to its use and the training for it which makes western archery over the same centuries seem a rough and ready sort of exercise. Anyone interested in archery should seek out some of the keys to the Japanese mystery.

Here, it must be enough to glance at the main types of bow; the *maru-ki*, a plain self bow of round section, usually over 7 ft (213.4 cm) and up to 9 ft (274.3 cm) long; and the *shige-tô-yumi*, a bow of the same sort of length, and of a mixed wooden construction, having either a central core of mulberry wood, cherry, or sumac, boxed in with strips of bamboo, and bound at intervals with rattan, or entirely made of bamboo laminates, sometimes set edge to edge, sometimes laid together as plates, strengthened at the sides with long strips of bamboo and, finally, most exquisitely lacquered. The average length

Above, left *Modern Japanese longbowman* (E. G. Heath).

Above, right *Japanese warrior of the 18th century, with longbow* (Victoria & Albert Museum).

of these bows is now about 7 ft 3 in to 7 ft 6 in (220.0 cm to 228.6 cm) but some of their surviving predecessors, which are of great age and often of great strength, are nearly 9 ft long (274 cm), or as short as any European longbow. One survivor, of 7 ft 7 in (231.1 cm), dates from AD 1363, and shows how a society that regards its weapons as almost sacred provides much better evidence for the historian and archaeologist than another, such as the British, whose attitude to weapons and armour has only recently been anything more than totally pragmatic.

The Weapon and the Hunt, the Wood and the Making

The shape of these Japanese bows incorporates a slight recurve forwards, more pronounced when unstrung than when braced, and the handle is well below the central point, about two thirds of the total bow length from the tip of the upper limb, partly so that the small Japanese can shoot kneeling and still keep the weapon's foot clear of the ground, and partly so that the bow can be shot from horseback. Naturally, this inequality in the limbs makes for very complicated and delicate application of balances and stresses by the bowmakers.

There are two shorter but similar types of bow, *the Bankiū* and the *Hankiū*, which means 'half-sized bow', and one quite different composite recurve bow of the Tartar or Mongolian type, the *Hoko-yumi*, with stiffened limb-ends and some metal in its construction.

The longbows are drawn well behind the ear, using a different hold on the string from the Western, and a different release, with a shooting-glove. So the delicate, beautiful arrows are correspondingly longer than ours in the West, more varied in design and more frightful in destructive potential. *Karimata*, 'the flying goose', is a forked head with a width across the points sometimes of 6 in (15.2 cm). *Watakusi*, 'the flesh tearer', is of various efficient and unpleasant shapes often viciously barbed, for making good the claim of its name. *Togari-ya* or 'pointed' heads vary from lance shapes to bodkins and to heart shapes, often fretted and decorated;

Yanagi-ha, 'willow leaf' heads, are roughly of the shape of the leaf and like other Japanese heads vary greatly in weight and size. There are also blunt heads for birds, just as there were in medieval Europe, and globular whistling heads of wood or horn for signalling and for ceremonial and religious use.

As the war use of bows faded so their ceremonial importance increased, and the decoration of arrows became more fantastic, but so much remains to be seen of the old weapons, so infinitely more than we have left in the Western world.

★ ★ ★

There is a member of the British Long Bow Society who has made a bow which corresponds to all the demands of a traditional English longbow as defined in the rules except that it is constructed, Japanese fashion, of bamboo strips glued together. He shoots with it, and calls it a reliable and effective weapon, but he is reluctant to pass on the secrets of its manufacture. Since the foundation of the Worshipful Company of Bowyers, and long before that, bowmakers have had their secrets, and suffered their secret troubles, which I believe stem from their work with the dark, sinister, ancient, sacred, poisonous yew tree. That is the wood for longbows; no other is as good. Osage comes near it, but is as hard to get. Dagame from the West Indies and South America, often called lemonwood, is usually obtainable from archery stores and good timber merchants, and is the best wood for a novice bowmaker to use. There are yellow lance-wood which is scarce, brittle for a workaday bow but which can give, at its best, a marvellous cast for long shooting; hickorys of various types, often used for backing strips; and red cedar which makes a good bow for hot or cold climates because it tolerates temperature extremes. Other woods include black walnut, red mulberry, common elm, wych elm, brazil, chittam or smoke tree, fustic, snake-wood, ruby, locust, sassafras, stopper-wood, beef-wood, greenheart, ironwood, amaranth, rosewood, cornelwood, crabapple, laburnum or *bois d'arc* which by its very sobriquet should rate high in the list, and so on; very many woods can be used to make a bow, but above them all, *le vrai bois d'arc* is the yew, *taxus baccata, canadensis,* or *brevifolia,* male, female or hermaphrodite.

That great old American forester Earl L. Ullrich of Roseburg, Oregon, who has been cutting bow timber and making bows since the first quarter of this century, wrote: 'The Almighty meant yew wood to be for our use, and we should not deprive the generations to come of such a prize ... We cannot replace the minerals and the oil taken from the earth, but we can grow trees if we have a mind to. This is our heritage; we must pass it along.' Most of the best yew longbows are made now of wood imported from Oregon (Ullrich has supplied a lot; some of it very fine). Before that, it was imported from Spain, as in the Middle Ages, but Spanish yew is very hard to come by now. When it can be found it is a lovely brown, close grained timber, ideal for bowyery. English yew can be good, the further north the better; Scottish therefore is apt to be better than English; Scandinavian yew as good or better than Scottish; Welsh slightly better than English when it grows on high ground; but the yew of all temperate climates tends to be open in the grain, and hence softer in action than its brothers of the hotter southern, or colder northern climates. In the days when longbows were made in huge quantity, the young yew trees were tended, generation by generation; suckers, pins and twigs were removed and everything possible done to encourage straight growth, to discourage the writhing and wind-twisting, the gnarling and fasciation to which the yew habitually resorts when left to its own devices.

Anyone connected with archery is asked about churchyard yews. One of my daughters was told at school the other day that yews were grown in churchyards so that folk taking sanctuary could nip out and make a quick bow or two for their defence. My own belief is that on the whole yews stand about churches because often churches were built on what was sacred ground to the 'old religion', and to the old religion the yew was a sacred tree, a symbol of

The Weapon and the Hunt, the Wood and the Making

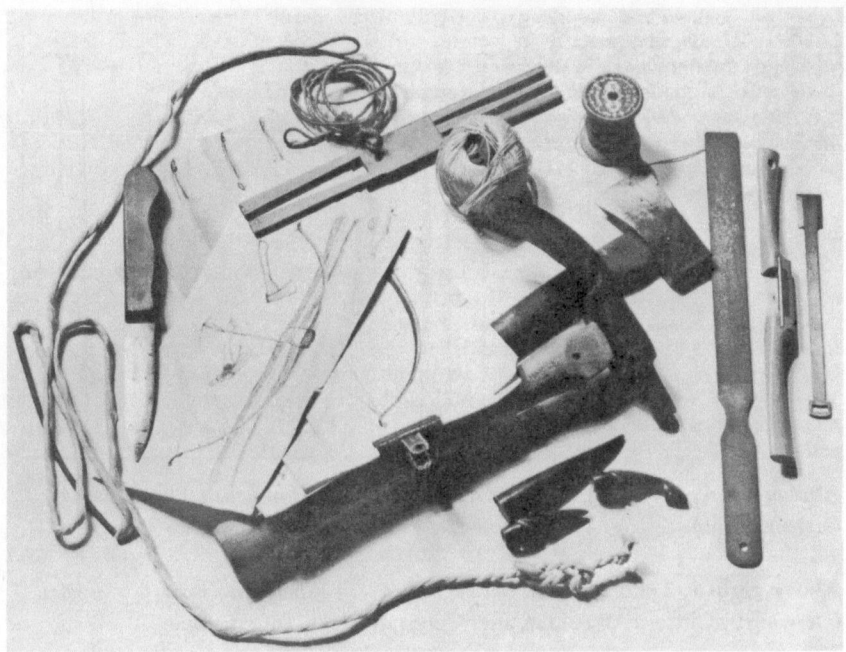

Some of the longbow maker's tools (Photograph by Graham Payne).

eternity. The Christian religion took over the idea, as it did so many others, and the dark, melancholy trees were accepted as suitable around the new places of worship, symbols still of death and eternity.

Were bows made from them? From time to time, no doubt; but it is worth noting that when Henry V, before the campaign of 1415, sent Nicholas Frost, his principal bowyer, round England to gather yew wood for bows, he expressly forbade him to take timber from any ecclesiastical land. On the other hand, medieval butts were often close by the church, and many churches bear on their lower stone courses grooves where arrowheads were sharpened and cleaned during practice.

If you want to make a longbow, first try with dagame, making sure that your stave is flat grained and lying straight; but once you feel you have mastered the principles of bowmaking, then go out and beg, borrow, I will not say steal some yew. Best of all from the point of view of sheer satisfaction, select a tree which offers promise of straight timber, cut it, season it, work it and tiller it, and shoot in your own, your truly own longbow.

It is easier to find a log of yew that will afford two sister-split billets, than one good stave, and most longbows now are made from such paired billets, jointed in the handle. Of seasoning I would only say that the weight of

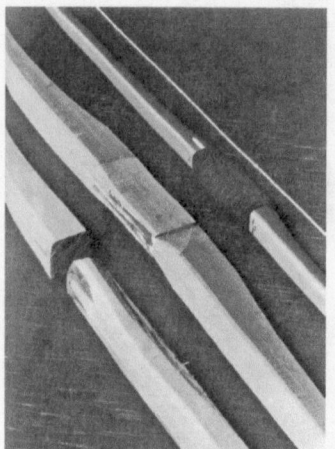 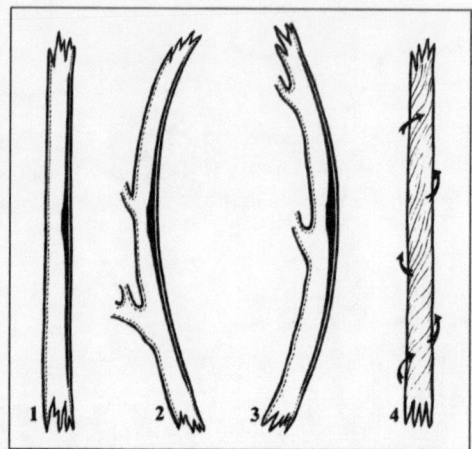

Above, left *A pair of billets, the jointed stave and the finished bow handle* (Photograph by Graham Payne).

Above, right 1, 2 and 3 *How the bow lies in the tree. 4 Wind twisted wood, to be avoided* (Drawing by Clifford Anscombe, after Count Ahlefeldt-Laurvig-Bille).

bowyers' opinion goes against 'water drying' and suggests that careful air-drying is the best, though slowest way of going about it. Through the three or more years of seasoning, work your timber down, as Count Ahlefeldt did, until you have a straight clean stave, or twin billets measuring together something over 6 ft (182.9 cm), some 2 in (5 cm) square, and showing, with any luck, an even layer of sapwood on the outside. When the billets are dry enough to work, butt them end to end so that one is upended on the other and the peculiarities of one limb will tend to be repeated in the other, with the pale sapwood on the back, or forward-facing part of the bow stave.

Square the ends to be joined and mark off 4 in (10.2 cm) on each butt end. Within that 4 in mark draw out the 'W' shape of a double fishtail joint, or a single 'V' joint, which with modern glues is quite sufficient. Saw out and clean the joints until without glue they fit perfectly together – a plane blade tapped by mallet is the only effective method of reaching the apex of the inside of the joint; no chisel is slim enough. Then glue and clamp the billets together and make sure that they lie straight, as a stave. You must adjust the joint before gluing until they do lie straight. Then mark out the centre line on what will be the back of the bow, using a string tacked into the limb ends, rubbing the string with a coloured chalk and snapping it on to the stave so that it leaves a line from end to end. Mark that line firmly with pencil or thin felt-tipped pen.

Apart from marking, you must leave the sapwood completely alone. Sometimes it is necessary to remove a little, but it is a ticklish job and requires that you do not transgress the grain. Almost always I have regretted reducing sapwood, because to avoid breaking the grain is almost impossible, and where the grain of sap is broken you are very likely to get lifting splinters, sooner or later.

On the whole it is best to set the two limbs together so that the back is as flat as possible, but, depending on the lie of the grain and the contour of the billets, it may be possible to set them so that they incline slightly forwards from the handle. This is called 'setting back in the handle', and it means that the eventual string-follow of the limbs is counteracted before the limbs are shaped. It can also contribute to good cast, but you must take care that this forward set of the limbs is not severe – eye and experience alone can judge it to perfection – or the finished bow will break, because you are asking for tension and compression beyond the limits of the timber.

Next, set the handle in a wooden vice, padded with felt or some material that will through the working processes allow a firm grip but will not bruise

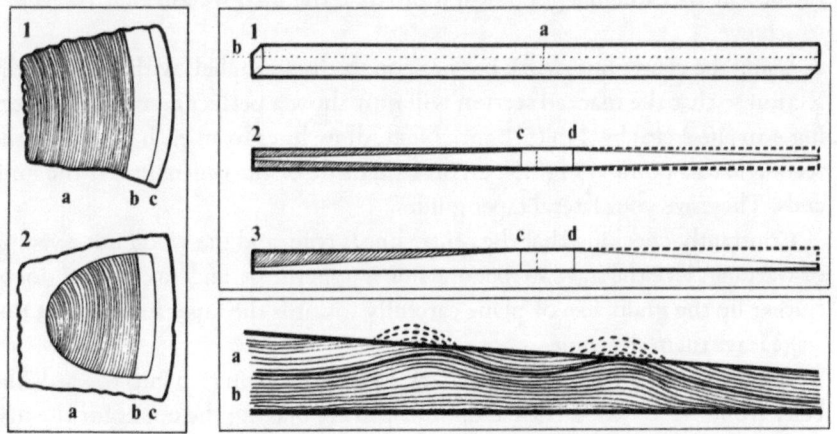

Above, left 1 *Section of a billet of yew wood showing* **a** *heart wood,* **b** *sapwood,* **c** *bark.*
2 *Section of finished bowstave lying within the billet* (Drawing by Clifford Anscombe).

Above, top right *Marking out the bowstave.* **1 a** *Midline.* **b** *Centre line.* **2** *and* **3** *show handle section* **c** *to* **d** *and the wood to be pared away in shaping the bow.* **2** *is a view from above;* **3** *from side* (Drawing by Clifford Anscombe).

Above, bottom right *Side view of exaggerated grain wave in a yew stave,* **a** *being sapwood,* **b** *heartwood. The dotted lines show where wood must be left proud* (Drawing by Clifford Anscombe).

or squash the wood, and in the 4 in handle section mark a line right round the stave 1¼ in (3.2 cm) from the top, and 2¾ in (7 cm) from the bottom of the section. This is the centre of the bow; but to allow the grip of the bow hand and the arrow-pass a proper compromise, the handle is more below than above the centre line of the bow.

You will have to decide before this which is going to be the upper and which the lower limb. When deciding bear in mind that the lower limb is a little shorter below the handle than the upper and must consequently be a little stronger.

You must now make the limbs of equal length, measured from the mid line. Decide on the overall length of bow you want. If you stand 6 ft (182.9 cm), then, as a rough rule, give yourself a 6 ft to 6 ft 4 in (193 cm) bow. If you are 5 ft (152.4 cm) give yourself a 5 ft 3 in (160 cm) to 5 ft 6 in (167.6 cm) bow. But length depends also upon the weight you are aiming at. If you want to make a 100 lb (45.4 kg) bow, which you almost certainly will not, then allow plenty of length. You can always pike a bow, cut it shorter, but you cannot make it grow. When deciding on the weight, remember that when you first string the bow it will weigh a good deal more than the final weight that you mean to achieve.

Using the centre line, mark out next on the back parallels within the handle section so that the marked section will now show a perfect rectangle measuring 4 in (10.2 cm) by 2 in (5.1 cm). Next, draw lines from each corner of the rectangle to a point ¼ in (0.6 cm) on either side of the centre line at the limb ends. These are your lateral taper guides.

Constantly checking that the centre line is true, and the wood not twisting or warping, vice the stave so that one side is uppermost, and, taking care not to feather up the grain, axe or plane carefully towards the taper lines, but at this stage leave them showing.

On the sides of the limb ends mark a generous ½ in (1.3 cm) towards the belly from the sapwood back and, until you are making the nocks for the first stringing, never cut the wood within some 3 in or 7.6 cm or these marks. Then, having decided on the depth of the handle, and marked it, draw a taper line on each side of the bow from that mark on the belly to the ½ in marks on the sides of the limb-ends. That straight line will never be the real line of taper, because of grain cadence, so, once you have drawn it lightly, re-draw, free-hand, a proper taper line that takes account of the grain, and do not cut beyond it.

At this stage I generally mark rings at 3 in (7.6 cm) intervals along the limbs, and put a depth and thickness figure against them as a rough guide. They keep disappearing as you work, but I keep replacing the rings at least, and constantly checking them against a measurement list on the wall above the workbench. Measurement can only be rough, as far as yew is concerned, because the shaping of the limbs is dictated by grain and contour, by knot, pin and wander.

Now you are into the long process of 'feeling' the stave down towards a workable bow. In every tree there stand so many potential staves or billets. In every stave or joined stave there lies its optimum bow. Your job is to release that bow, as Prospero released the imprisoned Ariel, pinned within the trunk of a tree.

You must use drawknife and spokeshave, scraper, glass, knife, sandpaper and wire wool to slim and taper each limb from the handle to the tips, dipping where the grain dips, letting it rise where the grain rises, leaving proud wood around knots or pins. If these are not in the centre of the limb width, leave them proud and forget about them, but if they are central, or penetrate quite through the limb, drill them out carefully and tamp in a plug of yew wood, which bowyers call a 'dutchman' or a 'dutch thumb', mindful of the story of the boy who plugged the dyke.

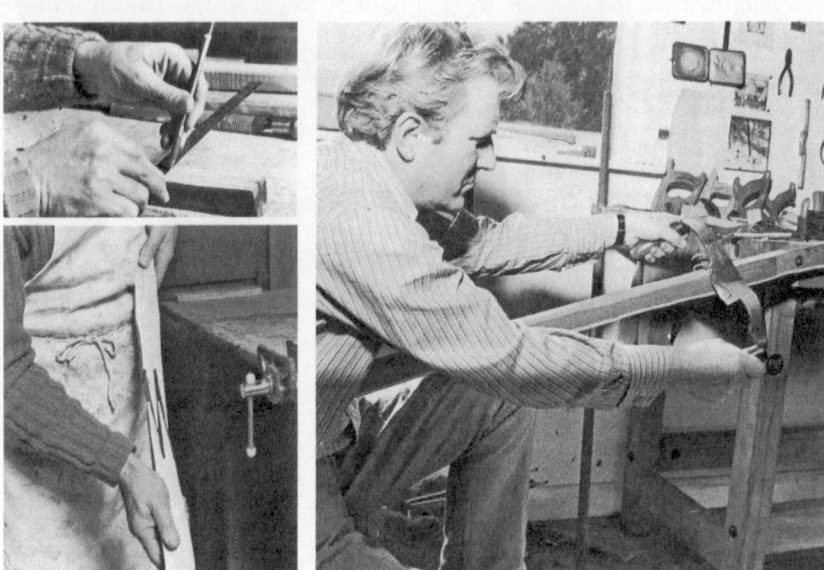

Above, top left *Marking out the double fishtail joint* (Photograph by Graham Payne).

Above, bottom left *The billets are joined* (Photograph by Graham Payne).

Above, right *The author at work on a stave with a drawknife* (Oxford Mail & Times Ltd, photograph by Bill Radford).

Remember, as you reduce the limbs, that you are aiming for a 'D' rather than a rectangular section. You must decide how high or low stacked the bow is to be, and constantly, at each stage of reduction, round the belly to the shape you want. Remember also that the grain further down the belly should, ideally, show a regular pattern, always dropping a layer, and only being allowed to regain layers when you are intentionally leaving proud wood for safety.

When you think the bow is capable of its first bending, cut nocking slots at about 45° to the long axis of the bow at either tip, with a rat-tail file, and make all smooth. Fit a very tough thick bowstring (strong enough for at least 100 lb [45.4 kg] pull if you are making a 50 lb [22.7 kg] bow), put the bow in a tiller, your heart in your mouth, and ease it back an inch or two.

What is a tiller? It is any device, plank or stave, to hold the bow firm by the handle while the string is drawn to a series of measured notches or, in a more complicated pattern, wound on a ratchet, so that you can retreat from the weapon and gauge its bend. You can do without a tiller, if you have a long looking glass in which you can watch the reflection of the bow as you draw it progressively from its earliest inches to the full draw of your own particular arrow length, 26 in (66 cm), 28 in (71 cm), or 30 in (76 cm), whatever it may

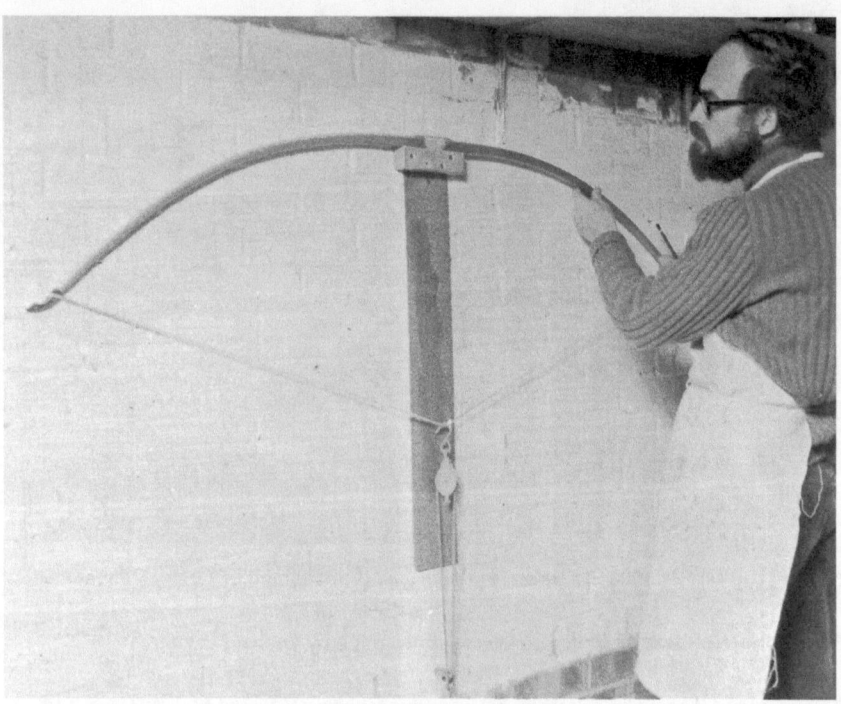

Stuart Homer with a bow on the tiller (Photograph by Graham Payne).

be. You can 'see where it cometh most, and provide ... betimes'.

You want your bow to describe an arc of a circle, or 'come round compass', but with a flattish middle to its arc. To make the handle steady and allow the wood to work gradually from a few inches outside the handle is a difficult trick to learn. The old bowyers did it by simple tapering from a thick handle. Later bowyers, Buchanan particularly, allowed the limbs to work closer to the handle by dipping the contour of the belly steeply from the handle. 'Dips' must be carefully controlled and smoothly shaped or they will force the wood to a sudden work-bend which will eventually endanger it. When the curve of your full-drawn bow has pleased you finally, then polish it with the finest steel wool until it gleams, and is satin to the touch. File the tips to a cone and fit on horn nocks, which you can buy if you persevere, or make yourself from the tips of cowhorn, buffalo, stag, or kudu, whatever you fancy or can obtain, and that may require perseverance too. Not long ago a visit to an *abattoir* would net you enough horn for many bows. Now we have bred out horn from our cattle, or we cut it away at birth, so your nocks may require a pilgrimage.

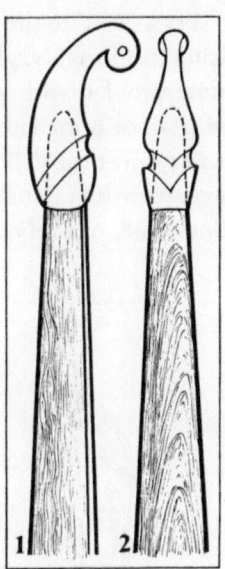

The application of the upper nock on the bow tip. **1** *From side.* **2** *From belly* (Drawing by Clifford Anscombe, after Count Ahlefeldt-Laurvig-Bille).

Far left *Work starts on a horn nock and* **left** *the nock nearly finished* (Photographs by Graham Payne).

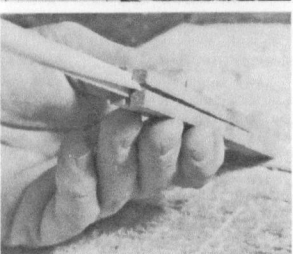

Far left *The arrowshaft is quarter-cut and fitted to the hardwood footing and* **left** *a four-point footing completed* (Photographs by Graham Payne).

Fix a 'riser' to the flat back of the handle so that the grip agrees with your hand comfortably, glue it on, whip the whole handle with stout thread and waterproof it with glue or varnish. Then let in an 'arrow-plate' of mother of pearl or horn and work its surface till it is sheer to the side of the bow. It will protect the delicate yew wood from the constant abrasion of the passing arrowshaft. Finally protect the bow with layers of wax polish, or boiled linseed oil, or apply carefully dried and smoothed layers of varnish or french

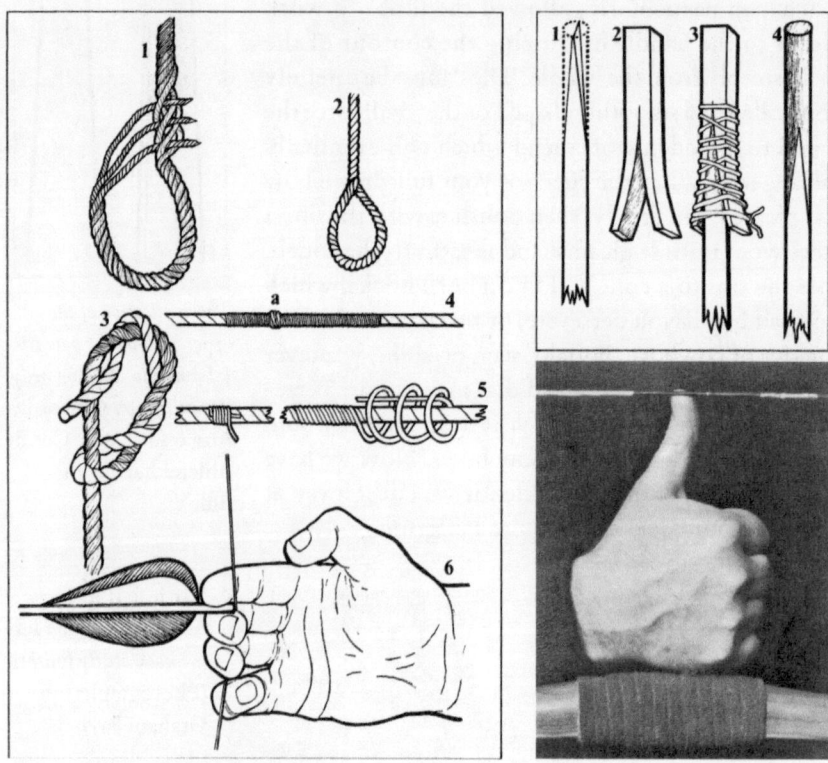

Above, left *The bowstring.* **1 and 2** *The lower loop.* **3** *The timber hitch or bowyer's knot for the lower loop.* **4 and 5** *Whipping at the centre of the string where the arrow nock will fit* **6. a** *in the centre of the whipping shows the slight over-whip below the position of the arrow nock* (Drawing by Clifford Anscombe, after Count Ahlefeldt-Laurvig - Bille).

Above, top right *The various stages in two-point footing of an arrowshaft* (Drawing by Clifford Anscombe, after Count Ahlefeldt-Laurvig-Bille).

Above, bottom right *The bow is strung, the handle bound and measured by the fistmele* (Photograph by Graham Payne).

polish. I unashamedly use a modern varnish, always to the scorn and horror of old Ned Thompson who until he died was bowyer to the Woodmen of Arden. He thought that no manner of good; 'a bow must breathe,' he would say. I would hesitate to disagree with anything the old master said, but I cannot rid myself of an instinct that suggests the more complete the protection the longer will that last small percentage of moisture which a bow needs endure within it.

Now all you need is a string for the bow, a tab for your fingers, a bracer for your bow arm, and arrows. If you cannot buy arrows to suit you and your bow, then take advice. There are many woods to make them, many ways of shaping them, many different ways and types of fletching, many different feathers to choose. Weight and 'spine' you must investigate, and 'footing' and 'cresting' but, complicated as it may sound, the problems are not great, and half the joy is in the learning and the solving. Otherwise there are many archery stores in the world and they will help you to proper matched sets of longbow arrows for target or field shooting or for hunting.

You have your longbow in your hand, arrows at your side or on your back. If you follow the advice of the *Roi Modus* in a book printed in 1486 in France, your bow will not be less than 5 ft 6 in long (167 cm) and it will be strung with silk because silk is stronger and more durable than any other string material. Your bow will be braced, the distance of the string from the belly, or the 'fist-mele', being the width of a palm and two fingers, and in your quiver will be 30 in (76 cm) arrows with light heads and low-cut feathers, and perhaps a few with broadheads and taller feathers. You will set your first arrow true in the bow with the cock feather outwards, your finger will test the sharpness of the arrow blade, your well-trained dog will wait at your side, where your file will hang, and in your pouch you will have a spare string. So, in clothes coloured to match the greenwood you will wait for your quarry, and when it comes you will let it pass, because the arrow will do more damage diagonally than shot straight through.

Of course if your quarry is a target, the 'prince's colours', a white paper rabbit, or cardboard elephant, you will do better if you shoot straight on. There is more chance of a hit. Whatever you shoot at, however well or badly you shoot, the satisfaction you will have from shooting your own arrows in a bow that you have made yourself will be very great indeed, and you will feel like the yeoman whom Chaucer described 600 years ago and more, when men really knew longbow shooting, when

'... he was clad in cote and hood of grene;
A sheef of pecok-arwes brighte and kene
Under his belt he bar ful thriftily:

'(Wel coude he dresse his takel yemanly:
His arwes drouped noght with fetheres lowe),
And in his hand he bar a mighty bowe.
A not-heed hadde he, with a broun visage.
Of wode-craft wel coude he al the usage.
Upon his arm he bar a gay bracer,
And by his syde a swerd and a bokeler,
And on that other syde a gay daggere,
Harneised wel, and sharp as point of spere;
A Cristofre on his brest of silver shene.
An horn he bar, the bawdrik was of grene;
A forster was he, soothly, as I gesse.'

CHAPTER 11

FORWARD TO THE PAST

And all would have remained as it was; our knowledge was very unlikely to have been extended or deepened; the old arguments would have rocked to and fro; much that is claimed in this book would continue to be hotly contested; the old sources would have been re-examined, old deductions and guesses repeated, new guesses and deductions put forward; the various lobbies of historians, of archers, of enthusiastic amateurs would have held ground, or slightly lost it; all would have remained the same – and then suddenly something occurred, so unexpected and so revolutionary in its possibilities that although we have become used to it all, it is still difficult to appreciate its full significance.

The wreck-site in the Solent of Henry VIII's great ship the *Mary Rose*, sunk during battle with the invading French fleet in 1545, had been re-discovered, and for some years the patient and skilful work of investigating the ship had been going on, under the inspired command of Dr Margaret Rule. The site had been protected from amateur and professional robbers, and from time to time the Press offered stories of artefacts being brought to the surface – guns among them, just as there had been when the Deane brothers first investigated the wreck in the 1830s. The Deanes had also brought up longbows. The Anthony Anthony Roll of the King's Ships, at Magdalene College, Cambridge lists 250 longbows of yew, and arrows and bowstrings in the *Mary Rose* inventory. Margaret Rule was hoping to find some example of everything mentioned in that Roll, on board the *Mary Rose*.

One day in 1979, a diver surfaced alongside *Sleipnir*, the control vessel anchored over the wreck site, with a long slightly bent pole tapering at each end, waterlogged, black and covered with small marine accretions. Was it a longbow? If it was not a bow it was a very odd pikestaff indeed.

At the Mary Rose Trust Headquarters Margaret Rule's desk was piled with books on every subject with which she was likely to have to make herself familiar, during the long process of exploring the *Mary Rose*. By chance among them was the first edition of *Longbow*, with its references to those first bows from the *Mary Rose* now kept by the Royal Armouries at HM Tower of London. Margaret got in touch with me, told me of the bow, of arrows that had been found, and of her hopes that she might find more longbows, and more arrows. With my friend and colleague, Professor Peter Pratt, I went down to Portsmouth. There was not the slightest doubt: before us lay a longbow of 1545, all the more authentic to my unaccustomed eyes because it immediately called to mind an early photograph of the Hathersage weapon 'Little John's Bow'. It was knobbly, where the bowyer for safety had allowed extra timber about the knots and pins; even in its uncleaned state there was faint differentiation between the sapwood of the back and the heartwood of the belly; and it was massive; it looked as if it must in its prime have had a draw weight well over 100 lb (45.35 kg).

There was at the back of my mind a faint disappointment that it appeared a rather primitive weapon, though that was outweighed by the fact of its existence, and its completeness, and the promise it gave of yielding secrets. We could not know that of all the 138 bows that later came to light this first one would prove one of the very few that had a rough-hewn appearance.

Much, much more was to emerge, of much greater quality and in much better state of preservation, but the lifting of that first bow, through the surface of the Solent, was to me as magical as the appearance of Excalibur held above the mere.

As more bows and arrows were discovered in the murk of the silt that covered the *Mary Rose*, and were brought to the Trust workrooms, Peter Pratt and I were joined in our examinations by John Levy, Professor of Wood Science

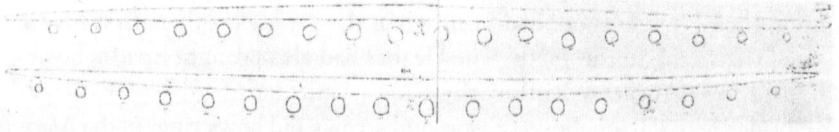

Profile drawing of a Mary Rose bow at the Royal Armouries in HM Tower of London. This bow, drawn by Debbie Fulford of the Mary Rose Trust, is one of those brought up by the Deane Brothers in the 19th century (Imperial College of Science and Technology, London).

at the Imperial College of Science and Technology. He was already advising the Trust on the timbers of the ship's hull, and brought his vast knowledge and experience of timber to our team, which was now officially appointed as the consultant body on longbows and arrows to the Mary Rose Trust. Our job was to conserve, examine, research and make detailed measurements of all the bows and, when thought advisable, to make experiments on them to determine their strengths now, their degrees of degradation, and so their likely strengths when new. From the beginning, through Professor Pratt we were able to count on the assistance of Dr B.W. Kooi of Gröningen University, author of a highly technical work *The Mechanics of the Bow and Arrow* who has made himself responsible for what we called the computer profiling of the bows. We were also able to recruit the services of Roy King as bowyer to our enterprise and John Waller as a practical archer of thorough experience in the making, the shooting, and the history of longbows. Roy has made very many fine bows, and has as much knowledge of yew wood, its strengths and weaknesses as anyone in this country. He has combed large parts of the country for yew timber suitable for bowmaking, and from a piece which shows the faintest promise he will elicit the best weapon that could possibly emerge.

By now we knew that every bow so far brought up was made of fine-grained yew, and when the first of two complete bow boxes from the orlop deck was discovered, and then its companion was brought to the surface, it became obvious that we were dealing not only with a reasonable proportion of the 250 bows enumerated in the Anthony Anthony Roll, but with bows of the finest quality imaginable.

When in 1981, the last bows were up, John Levy, in his first report to the Trust, wrote: 'The wood from which the bows were made has been identified as yew (*taxus* sp.) The staves from which they were formed were cut or cleft radially from a log across the sapwood/heartwood boundary, so that, when finished, each bow had sapwood on the back (the convex side), whilst the belly consisted of heartwood.'

When the bows from the chests were still wet, they had almost the appearance of new wood, though it was only after drying that the pale sapwood took on the true colour that one sees today in a yew bow made fifty or a hundred years ago. The deep reddish brown of the belly lightened as the timber dried through the months, and it was only later gentle oiling and waxing that restored to them something of their youthful look.

John Levy went on to explain that at the time he wrote the report it had only been possible to examine the anatomical characteristics of broken fragments of bows found on the decks. Later, after breakages during experimentation (of which more anon) the interior wood structure of less degraded

Longbow men in Henry VIII's army at Southsea in 1545. From the Cowdray engraving which shows the sinking of the Mary Rose (Mary Rose Trust).

bows became apparent, but this did not materially alter his conclusions. He wrote, 'to all external appearance these fragments looked very similar to the wood of the complete bows found in the bow chests. Cross sections showed a clear distinction between sapwood and heartwood. In most instances there is evidence of some degradation of the cell structure in the sapwood, whereas the heartwood is substantially without any sign of degradation, apart from a very thin surface layer, rarely more than 1mm deep'. Whether there was vital degradation present, and its degree, could only be revealed by the processes of testing, and that was to come. John Levy continued: 'The cross sections show the growth rings and from the narrow radial increment between each, it is clear that the trees from which the staves were cut grew slowly. This is likely to have been the deliberate choice of the bowmakers, since in general slow grown softwoods have greater strength properties than fast grown.' John Levy concluded that from the average curvature of the growth rings of the samples examined the diameters of the logs from which the staves were cut or split would have been some eight inches or more.

The first thing was to clean the bows of salt, so they were immersed in cascades of fresh running water until they appeared free of chlorides. Then they were washed, one by one, with the greatest care in a mild soap solution, and again rinsed in fresh water. It early became clear the amount of work that would have to be done on the bows necessitated their release from the Trust's store into the care of the consultants. Initially, all examination and conservation was carried out at the Trust's Old Bond Store in Portsmouth, but because of the difficulties inherent in travel, lack of storage space, and the steady increase in artefacts arriving from the sea bed, a large number of the bows was released to the consultants in January 1982. A few were retained at Portsmouth for exhibition in Southsea Castle, from which in 1545 Henry VIII watched his great ship founder. Twenty-two bows remained in the Old Bond Store and were not released to the consultants until September 1982. These were the first 22 bows recovered, all during 1979 and 1980. During their life at the Old Bond Store some of these bows were treated with polyethylene-glycol (PEG) as a conservation agent, and many of them were meticulously measured by Debbie Fulford using a method which the consultants were later to develop to a greater degree of accuracy.

The consultants' team at this time was fortunate in two respects. They had access to a secure chamber where the temperature and humidity could be controlled, and where special racks were at once constructed to allow all the bows to be free and well aired. The other piece of good luck was that Mrs W. M. Garcin, at that time my secretary, who had typed the original edition of *Longbow* and done endless work towards its completion, was prepared to undertake the daily weighing and cataloguing that now became essential as the bows began to dry out. Through that months-long process she never lost patience with the task, and by the end of it all she had produced a very great number of extremely valuable figures, charts and graphs.

By the time the bows had reached stability, that is to say when their weights decreased and increased only in response to changes in humidity, they had lost almost exactly half their waterlogged weights, the weights first recorded when they were removed from their final baths of clear water.

During the drying out period, each bow was frequently wiped with a soft damp cloth, because as they dried they exuded the last vestiges of sea salt, which would cloud their surface with a sticky dust, easily removed, and which gradually lessened, and finally ceased altogether.

Also during those long months each bow was examined, and preliminary measurements were taken. A complicated system of description and identification was begun which would soon record the idiosyncrasies of each weapon: its vital statistics obviously, its appearance, its 'feel', and its type. It began to be clear that there were certainly two very distinct types, possibly but less clearly, three types. In many instances, among the best preserved, the extraordinary skill and confidence of the bowyers were apparent. Where the surfaces seemed almost like new wood, as they now did in many cases, the marks of the bowyers' draw knives, or 'floats', were plain to see, even to feel, like delicate fluting on a glass stem. So sure were they of their skills and of their timber that they clearly felt no need to work out those last straight marks of manufacture, as we do now. Now we sand and polish till the surface is like plain glass; the Tudor bowmakers knew when the necessary work was completed, and no doubt a protecting wax polish was applied soon after the last curls of shaving had fallen from the finished staves.

When all the bows were stable and clean, it was decided that some of them, the greater part, should be lightly oiled with a non-penetrating protective vegetable oil, and that when the oil coating was completely dry a beeswax polish should be applied. This was particularly done to 12 bows which were to tour the USA in the Trust's fund-raising exhibition. Some other bows were for the moment to be left untouched so as to reveal how their unprotected surfaces reacted. Now, some years later, it has been decided to oil and wax every complete bow, as a safe and handsome way of preserving them.

One of the most exciting and mysterious things to emerge from early examination was the existence, on many of the weapons, of 'bow-marks'. We shall perhaps never be quite sure what they are, but we shall concentrate on these marks in more detail later in the chapter. Before that it is necessary to return to some general background information.

The bows were found variously about the ship, on the weather deck, in which number must be included those that fell into the ship from the bow and stern castles, either on impact with the sea bottom or during later disintegration of the castles; in the gun deck, and in the orlop. In one chest were 48 bows, in the other 36, almost all in miraculously fine condition after 437 years' immersion. The rapid inflow of silt accounts for the good state of preservation, which is probably far better than would have been the case if the bows had been preserved in air, since much of the natural make-up of the timber was sealed in anaerobic conditions.

The particular components of Solent silt and its sulphur-reducing bacteria, while being extremely destructive to metal and certain other materials, favoured particularly the preservation of timber. Yew timber itself has an astonishing ability to withstand inundation. I have seen yew wood that has been immersed in bog water for a thousand years which, when dried, is sound within a narrow margin of its outer surface, almost as sound as new wood. Perhaps oddly one of the components of the silt which is most likely to have ensured the survival of timber is human effluent, flowing regularly from west to east through the Solent from the many towns and villages of the south coast of England.

Drawing of a sheaf of arrows in a leather spacer (from the wreck), and actual spacers and arrows found in the Mary Rose (Mary Rose Trust).

The first 22 bows, from the upper parts of the ship, the first to be dried, remain pale and dusty looking, showing little difference between heart and sapwood. On some there are the marks of oyster-spat, on some encrustations from long immersion contact with metal objects, and some are slightly or badly marked and eaten away by micro-and macro-biological agents, mainly gribble and toredo worm.

Various other bows and part-bows show a broad variety of the effects of degradation and juxtaposition, and of contemporary breakage. Since the same bow can show good preservation at one end, and bad deterioration at the other, it is likely the bad ends were not covered with silt until a later stage of the immersion.

Now we approach a very important matter: it was noticed at once that every bow, no matter what condition it was in otherwise, as long as one or both tips remained, showed at the tips a plain differentiation of colour for some 5 cm. This was clear evidence that the tips had originally been covered by an applied nock of some kind. Since horn is and was the most usual material for such applied nocks, and since horn has been proved to perish fairly rapidly in the conditions of the *Mary Rose* silt, it can safely be assumed that the nocks were of horn. This is borne out by the fact that of the thousands of arrows recovered, having a slot at the nock end which runs down towards the fletching and which would originally have taken a horn sliver for the purpose of strengthening the force-absorbing end of the arrow, only one or two still have the horn in place. Those few only remain as a result of being protected, for instance by a coil of tarred rope, from the effects of micro-biological and seawater decay. It is also notable that among all the *Mary Rose* finds, no horn buttons have survived, no horn panels for lanthorns, no horn handles, where it is obvious that there had been many such articles in 1545. Horn is certain, which is protein, which micro-organisms of the sea find delicious; and, as has been said, if not eaten it dissolves eventually.

Horn or bone nocks ('horns') are definitely and contemporaneously illustrated on longbows as early as the late 14th century. In the case of the *Mary Rose* bows, on the 'underhorn' timber, tillering nocks – slots used to hold the string loop or the knot during the manufacture of the bow – are to be found, one on the left-hand side of the upper tip, the other on the right-hand side of the lower tip. In many cases the tapering or coning of the tips for insertion into a horn nock has resulted in the reduction of the depth (and so the usability) of such slots; in some cases the slots have disappeared altogether. Such evidence is exactly in accord with normal experience in the making of horn-nocked wooden bows today, and such evidence applies to the whole range of the *Mary Rose* bows, including the two examples recovered by

the Deane brothers in the 1830s remaining in the Royal Armouries of the Tower of London. A number of the bows show a 'trimming' or whittling for a few centimetres down the limbs from the inwardmost marks where the horn nocks would have been, leaving the tip cones slightly proud. This suggests that individual archers worked at their bows after issue, 'whip-ending' them to increase cast.

It will be seen that in the opinion of the consultants these *Mary Rose* longbows, whether found at action stations or in boxes in the orlop, were finished weapons ready for use. It might seem unnecessary to make such an obvious claim, but it is necessary because within the archery world a good deal has been said and written expressing the view that the *Mary Rose* longbows are not bows, but bowstaves, unfinished and not ready for use. Apart from the very oddity of the idea that a ship of war, in time of war and actually in action against the enemy, should put to sea for action with no longbows but a large number of unfinished staves, there is the massive evidence of the bows themselves. What started the hare of this nonsense was our first published suggestion of the draw weights of these bows. It was hard to believe them ourselves, but as will be seen in Professor Pratt's contribution to this chapter, we have to believe them. Until it was incontrovertibly proved that our estimates were correct, few believed us when we came up with those first massive weights, arguing that the *Mary Rose* bows ran from about 100 lb (45.35kg) draw weight at 30 in to 180 lb (81.64 kg). Many said these weights were impossible, and that therefore the bows must be unfinished, carrying more timber and hence more weight than they would when completed. That they are completed is now self-evident, and I no longer meet people, at least not face to face, who are foolish enough to maintain the contrary.

As Professor Levy indicated, it was to be seen at once that all the bows were made from yew timber, each from a single, unjointed, unpieced stave. The quality of the timber, its density, the extreme fineness of the grain in most cases, suggested that we were dealing with imported staves of a straighter and finer quality than can readily be found in the soft climates of the British Isles. That most of it was imported from the Continent there is small doubt, and several documents from Henry VIII's reign record such imports either through Venice by the Doge's special permission or from elsewhere by special mandate of the Emperor Maximilian. Such timber would be gathered from those parts of Europe – Italy, Austria, Poland – where the yew grew high and fine-grained, and where for centuries timber had been felled and split into staves to supply our military needs. Henry VIII, as is known was a great encourager of the military use of the longbow, at the same time as he keenly pursued the development of gunpowder artillery. He sent his agents into Europe to choose the finest yew timber, selecting at a time thousands of the best staves which were then stamped with the Rose and Crown for export to England. The orders were almost always large; one part-order was for 40,000 staves to be sent to England through Venice, and the names exist of five bowyers who made up 600 of this particular batch of staves, into finished bows, for which they were paid altogether £200 13s 4d at a time when a master carpenter was paid fourpence a day and beef was about twopence a pound.

Yew wood will vary in colour from a dark, almost red heartwood to a lighter, orangey hue, and the sapwood from a pale cream to a yellowish buff.

Though darkened with age (a process of change which begins in any case very soon after seasoning) the *Mary Rose* bows exhibit just such a range of colours. The apparent remarkable preservation of the better protected of the bows is in sharp contrast to the poor and debilitated condition of the thousands of arrowshafts found in the ship. Most of the woods used in arrow making are plainly not good survivors of immersion or silt burial. Professor Levy writes: 'The arrows, many of which are shown to be of poplar or alder are, in general, in a very poor state of preservation. Both these woods are hardwoods and belong to the group described by Butcher & Nilsson (1982) as being susceptible to decay by soft rot fungi, caused by a group of micro-organisms that are still active even when the conditions for decay are marginal. It is therefore not surprising that these non durable species should be heavily decayed whilst the very durable heartwood of yew remains sound.'

The broad spectrum of measurements and hence of draw weights suggests variety of purpose in the use of the bows. In section they also vary, from the nearly circular to the almost rectangular or trapezoid. In between, the majority have a 'D' or an oval section with a rather rounded back and a deeply rounded belly. Longitudinally all the bows are of the expected longbow pattern, tapering from the centre to the limb tips, a shape inevitable for the even bending of a beam of wood. But unlike Victorian or later longbows made for sport there is no rise or stiffening at the centre to form a handle. This means that when drawn the bows really do come 'round compass' as Ascham described, and it means that I have had to rethink my claim that a bow which bends in the handle is awkward to shoot. Plainly, when bows were made in massive numbers by experts, for experts to use in war, in life and death situations, they were made to move in the handle and form a near semicircle at full draw, so that every part of the timber worked.

The bows are superbly even daringly crafted; knots are in cases excised with a brave disregard of weakness dangers, or rather perhaps with absolute knowledge and experience of dangers and safety margins; in other cases wood is left proud over pins, knots and swirls in the grain, exactly as a more cautious modern boywer would be likely to allow. Everything suggests sureness in the makers' craft; the finish, as we have seen, is not too perfect but has a fine dash to it. Instead of an immaculate smoothness there is that clear fluting left by the long, swift strokes of the float.

After long and frequent examination we came to the conclusion that the bows showed exactly what today's longbows show in the way of age and use. Those in regular use exhibit a slight or a marked 'string-follow', that is they remain curved towards the belly, or 'de-flexed'; one of the deck bows was certainly in use when the ship foundered and the string somehow survived long enough to

The one remaining shaft and part arrowhead from the ship which suggests the same type of head (type 16) as that illustrated in the colour section (Prof. P. L. Pratt, Imperial College of Science and Technology, London).

set the bow in the braced position for good. It now has a true fistmele bend. Others lie almost straight; but a majority of the boxed bows show a 'reflex', a bend towards the back, away from the natural bow shape. The only convincing reason for this is that the timber was selected in that form to achieve optimum straightness when the bow was well used, a straightness which means a longer and faster return of the limbs from full-draw to the braced position at which the arrow quits the string; the faster that return the greater the cast of the bow. It is the general Scandinavian experience with closely comparable yew longbows from an earlier period, which have suffered longer lake or marsh immersion than the *Mary Rose* bows in seawater and silt, and they show a general reflex towards the sapwood back which has led commentators in the past to suggest, quite wrongly, that the Scandinavians, such leaders of the development of the longbow in Europe, made their weapons back to front. The phenomenon is not explained by swifter shrinkage of sapwood, pulling the body of the heartwood towards it; in fact it is now known that cannot occur. Nor can it be ascribed to the swelling of the larger section of heartwood from the absorption of salts, since all bow pieces anatomically examined show bright natural colour and no sign of penetration into the centre of the wood. But it is true that yew wood when seasoned in staves can tend to reflex itself a little. It might be the result of pressure during immersion, though this can be discounted by the regularity of the reflexes, or it could possibly have been caused by heat treatment during manufacture, which is unlikely as it would tend to weaken both heart and sap; the most likely explanation is undoubtedly the choice of timber showing a

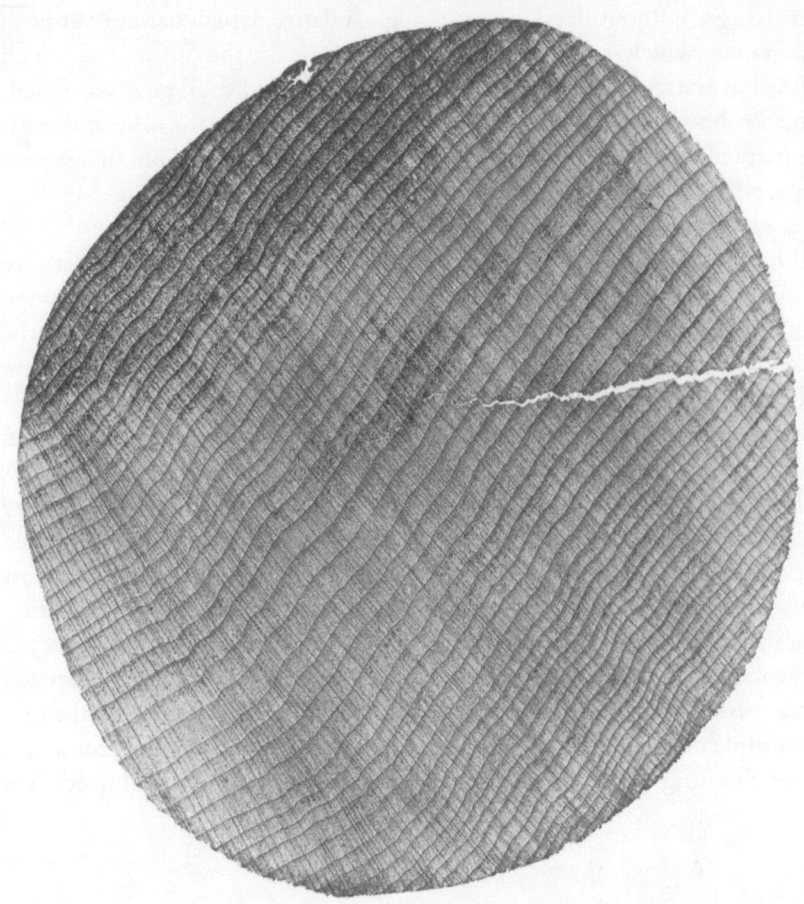

Sectional photograph of a broken piece of bowstave from the ship, showing growth rings and considerable degradation of the sapwood (the darker top section) in particular. This was photographed by Professor John Levy, one time Professor of Wood Science and Technology at the Imperial College of Science and Technology, and consultant to the Mary Rose Trust.

natural reflex, which would ensure comparative straightness in a used bow and which means a greater maintenance of strength and cast than in a deflexed bow. Nowadays we often set the two billets that are jointed in the handle, to make a bowstave, at a slight angle away from the belly to achieve the same effect. The Tudor bowyers and presumably their predecessors had such choice of fine timber they could select the naturally reflexed single staves. There is a significant number of *Mary Rose* bows that are almost straight, or only slightly deflexed, to support this theory, and only very few showing a marked deflex.

As to age, without dendro-chronological dating which has not yet been undertaken, which is in the working programme for the future, and which will tell us the age of a particular piece of timber, and the date it was felled, some few bows do give an impression of greater age than the others; there is one particular beauty made from not very fine-grained wood but of a superb shape, which somehow suggests that it was an old hand already in 1545. Roy King gave it the name, and we all call it Agincourt.

It has already been said that the *Mary Rose* bows are handle-less. There are no indications of any binding being put on them, and it must be assumed they had none. The approximate position of the 'arrow-pass' is just above the handle position (for even without a marked handle section, there is of course still a handle position), and it is in very many cases indicated on the *Mary Rose* bows by incised, pricked, and in some cases stamped marks. These marks can be roughly grouped, though some are unique examples.

They cover a small area on the left-hand side of the upper limb rarely more than 15 x 15 mm, though there are some examples of double, even triple marks. We have called them 'Bowyers marks', in the belief that their most probable explanation is that they were applied after manufacture and that their message, thus incised on a 'livere', or 'livery', or 'issue' weapon, conveyed to the user: 'This is the upper, this the lower limb, and here the recommended arrow-pass'. We may be wrong but this is what we have deduced. In the case of multiple marks it is possible that the archer added personal identification, or the choosing of an alternative position for the arrow pass, at need. It is

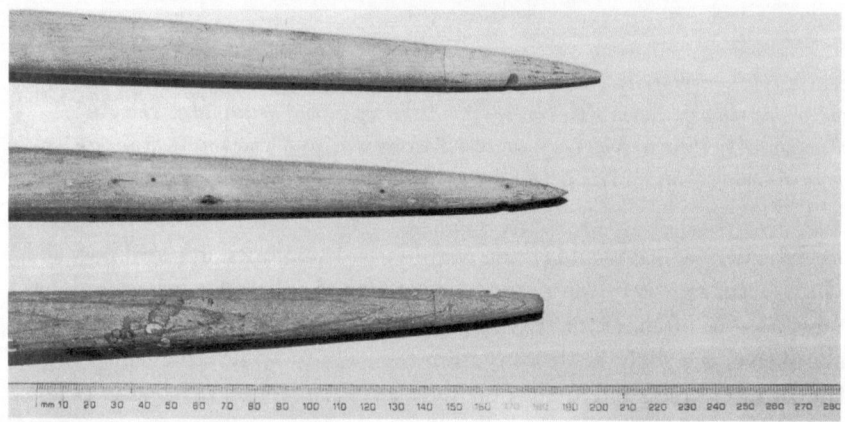

The tips of three longbows from the ship, showing the paler wood once covered by horn nocks, and underhorn tillering nocks (Imperial College of Science and Technology, London).

perfectly possible for instance to drop the position of the bow hand considerably without danger to the bow, and thus increase the draw-weight somewhat.

The marks consist mainly of groups of incised dots, as if made with a chisel corner (perhaps a float blade corner) arranged in pairs, threes, crosses, or little tree-like groups. There is a variety of circular marks: plain circles, circles with a cross, segmented circles, some apparently made with dividers, one or two possibly with a tubular stamp. There are variations on the cross: plain, and with dots in various arrangements. There are other linear marks, often in association with dots, sometimes whole clusters of pinpricks up to 30 or more in number. In general there seems a difference between the heavy marks which suggest a maker's advice or identification, and rather more random markings which could be personal additions. But those bows which are not marked at all, some fifth of the total, do not seem to be generally inferior, or different from the marked bows.

Analysis of marks

There are
- 6 crosses
- 5 circles in a variety of styles
- 4 dotted circles
- 1 circle containing a cross

About
- 40 marks either of a single point, or double, triple or quadruple points, probably made with the tips of 'float' blades
- 25 marks of five or six points in triangular or angular arrangements
- 5 with double types or tripled types of mark
- 10 six or seven point marks rather in the shape of fir trees
- 11 point marks between, or associated with lines incised on the bow side

There are also some 'sports':

- 1 thirteen point mark in a definite arrowhead shape
- 1 neatly incised 'H' mark
- 6 various complications of point marks around crosses or within circles
- 1 example shows a double pyramid with a five point angle mark with another double pyramid above that, making a total of 47 point marks on that bow

It is hard to find a standard in these marks. But then there was hardly a standard bow; yew does not yield to a standard; there are not two dozen bows at exactly 100 lb and two dozen at exactly 110 lb and so forth. There are more standard arrows at predictable lengths; the bow by its nature is personal and unique, first in the timber, then in the bowyer's hands, then in the hands of the archer.

★ ★ ★

Testing the Bows

By Prof P.L. Pratt, PhD, FInstP, CEng, FIM, FACerS, Professor of Crystal Physics, Imperial College of Science and Technology

Introduction

While the bows were being recovered underwater from the hull of the *Mary Rose*, a lot of thought was being given to the best way to find out how strong they were, how far they could shoot an arrow and how much damage could be done to their target. After conservation, the external appearance of many of the bows from the boxes in the hold was excellent and so an experimental programme was planned to measure the strengths directly and to compare these with theoretical estimates based on computer modelling by Dr Kooi. In the end the bows proved to be more decayed than they appeared and so new bows were made by Roy King as approximations to the design of the *Mary Rose* bows. These were tested successfully and used to verify the theory, thus lending credibility to the figures predicted for the *Mary Rose* bows.

These predictions were always higher than those we were able to measure on the medieval bows themselves.

What do we need to know to predict the strength?

To make use of Dr Kooi's elegant modelling technique we need to know the properties of the wood used for each bow and the details of their design. These include:

1 The weight and density of each bow.
2 The elastic stiffness of the wood.
3 The length of the working limbs and the shape of the unbraced bow.
4 The shape and the size of the cross-section along the length of the bow.
5 The bracing height.
6 The length, weight and elastic stiffness of the arrows.

Forward to the Past

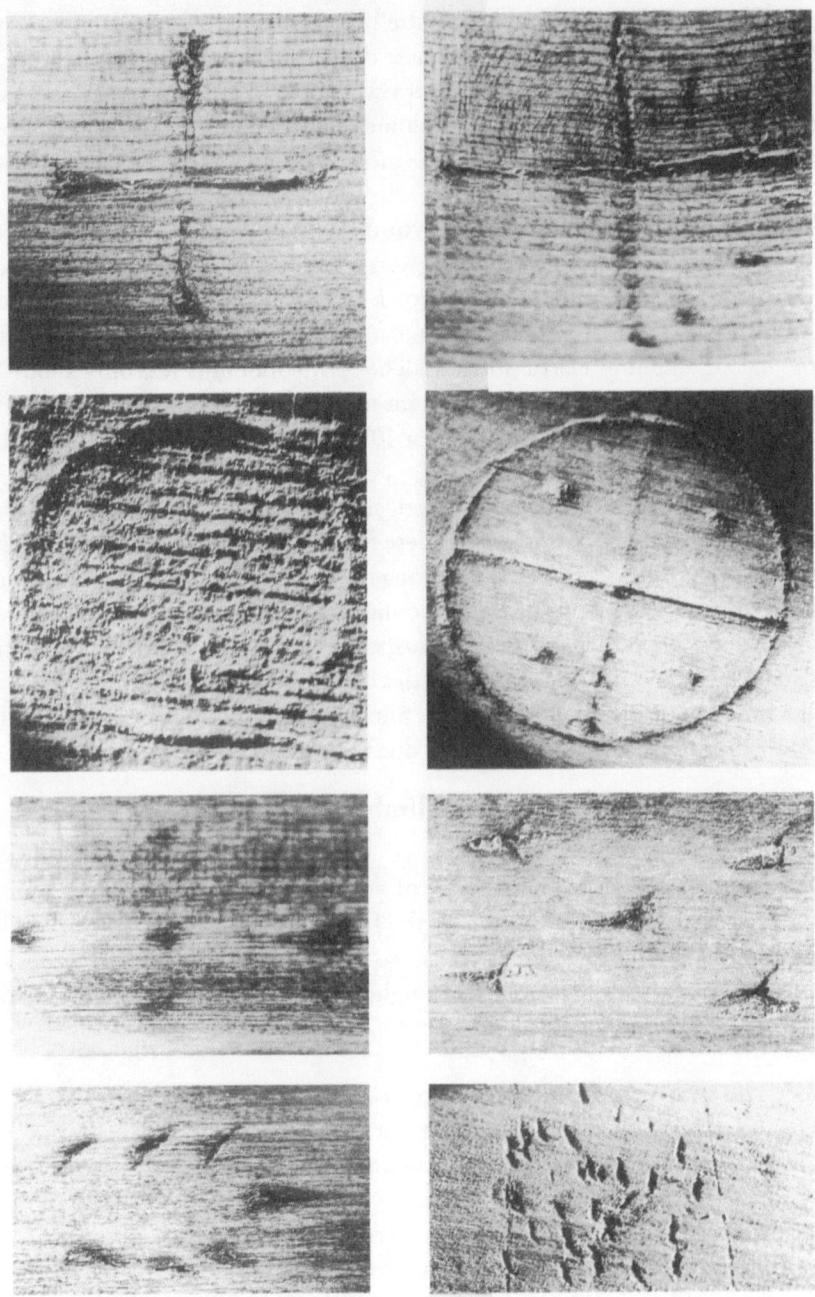

Eight different bowmarks from the many found on the Mary Rose *longbows* (Imperial College of Science and Technology).

1 The weight and density of the bows

Each bow was weighed repeatedly as it dried out so that its stable weight at a relative humidity of 60–65 per cent was known. The density was obtained from the weight by measuring the volume of water displaced by the bow from a length of plastic downpipe, after the method of Archimedes.

2 The elastic stiffness of the wood

The first fragment of bow we were given was about 25 cm long and appeared very degraded, especially in the sapwood. Internally, however, the heartwood was in a good state of preservation and a beam 1 × 1 × 10 cm was cut out for the measurement of elastic stiffness. The elastic modulus was only about 15 per cent less than would be expected for modern yew of the same density and a second fragment had a modulus near 10 GN/m^2, a figure typical of a similar beam of modern yew.

While it is easy to measure the elastic modulus of a square beam, it is more difficult to measure that of a complete bow. Two stages are involved. The distribution of the section stiffness along the length of the bow must be calculated from measurements of the shape and size of the cross section as outlined in 4 below. Then the bow must be suspended from the middle with one tip fixed while the other tip is loaded with a spring balance to determine the load as a function of displacement. The results for whole bows are expected to be lower than those for small specimens due to knots and other imperfections.

3 The length of the working limbs and the shape of the unbraced bow

For these bows made from one piece of yew with no thickening at the handle the length of the working limbs is the full length from nock to nock. The position of the vanished horn nocks was estimated from the tillering nocks. The shape of the unbraced bow, straight or curved forwards or backwards, was recorded photographically.

4 The shape and size of the cross-section

For the early bows, the shape and size of the cross-section of the limbs were obtained from the archaeological drawings prepared by the Mary Rose Trust. An example is shown in 3974 which has an almost rectangular or trapezoidal cross-section with flat sides. The increased width on the belly side of the neutral axis is exactly what is needed to prevent buckling in compression. For the later bows, pin gauges were used and the openings recorded photographically at 10 cm stations along the bow. Photographs of A3965 show the more common D-shaped cross-section with a narrow well-radiused back and the nearly circular cross-section towards the tips. For all the bows to which we have had access, these cross-sections have been analysed on a quantitative image processing system to generate the area, the position of the neutral axis and the second moment of area, or section stiffness, at each station. Apart from obvious knots and knobbles the cross-section was shaped by the bowyer to produce a linear increase of section stiffness from bow tips to a point near the middle of the bow. The position of the maximum section stiffness was always offset from the middle of the bow by a few inches, enabling the upper and lower limbs to be distinguished. These measurements confirmed the position of the bowyer's mark and gave an indication of the strength of the bow from the value of the maximum stiffness. The importance of the linear distribution of section stiffness is that all parts of the bow limb are equally stressed, so that the drawn bow 'comes round compass'; the excitement of it is that the medieval bowyer was able to produce a constant bending moment long before Newtonian mechanics were able to analyse it and to declare it desirable.

5 The bracing height

This determines the initial curvature of the bow before it is drawn as well as the moment when the arrow leaves the string during the release. This distance, from the string to the belly, the fistmele, is explained in Chapter 10.

6 The length, weight and stiffness of the arrows

The length of the arrow determines the length of draw of the bow. The distribution of the length of the arrows measured by the Mary Rose Trust shows two peaks, one at 29½ in corresponding to a 28 in draw length, and the other at 31½ in corresponding to a 30 in draw. The ratio of the longer to the shorter arrows is 18 to 6, averaged over some 300 arrows. The longer length was used for modelling.

The weight and stiffness are more difficult to measure because the non-durable hardwoods used for the arrows are badly degraded and the steel arrowheads largely rusted away. A major sample of the arrows was measured by the Mary Rose Trust, and the wood was finally identified in every case. For birch, hornbeam, oak or ash, the woods recommended by Ascham as 'heavy to give it a great stripe', the average weights of the longer shafts would have been about 60 g and the shorter about 35 g. The small English warhead, Type 16, to be found still in many museums, weighs 7 g and with the shorter shaft the total weight of 42 g is remarkably close to that of the Type 16 arrow found in Westminster Abbey in the Chantry above the tomb of Henry V. For the longer shaft a Type 7 armour-piercing long bodkin seems more appropriate. With an average weight of 13 g for Type 7 heads, the total weight of 73 g for the arrow is very close to that of a replica of a heavy bodkin made by Saxton Pope from drawings of a medieval arrow published by Hastings in 1831. For poplar or alder, woods identified amongst the *Mary Rose* arrows but disliked by Ascham for their weakness and lightness, these weights would have been less, 33 g and 58 g respectively.

Estimates of the strength of bows

Most of the factors mentioned in the previous section can be measured quite reliably on the *Mary Rose* bows, with the exception of the most important of all, the elastic stiffness. This is the most important because it is the scaling factor by which all other calculations must be multiplied. If we assume the value of 10 GN/m^2 given in the Technical Appendix 1 and in standard reference books for small specimens of yew, what value should we use for a real bow? We know that the larger the specimen the more knots and other defects it is likely to contain and therefore the lower will be the effective value of the stiffness. Based on the experimental value of the stiffness of one of our modern approximations to a *Mary Rose* bow, MRA1, we have chosen to use a conservative value of 7.6 GN/m^2 for our estimations of the strength.

This leads to the following table of draw weights, or strengths, for a number of bows of interest:

	lb	kg	density
X1–2 Tower of London	98	44.4	
X1–1 *Mary Rose* bows	101	45.8	
A812	110	49.9	
A3952	115	52.1	0.55
A1654	124	56.2	
A1648	136	61.7	0.63
A3975	137	62.1	0.53
A1607	185	83.9	0.62

Table 1 Calculated strengths of *Mary Rose* bows at 30 in draw.

This covers the range from the smallest and lightest of the *Mary Rose* bows, at about 100 lb, to the largest and heaviest at 185 lb. A1607 is so thick that it seems likely that it would break at a 30 in draw length; reducing the draw to a safer 28 in would reduce the strength to 172 lb. Even this range of draw weights from 100 lb to 172 lb would require a very different archer from the majority of those living today. Nevertheless it is very close to the range of 80 to 160 lb predicted on page 54 of *Longbow* in 1975 long before any bows had been recovered since the Tower bows in 1841. The figures could be corrected for differences in elastic modulus by making use of the well-established relationships between modulus and density, but we have chosen not to do this yet. The measured densities range from 0.53 to 0.76 g/cc implying a range of moduli from perhaps 8 GN/m^2 to 12 GN/m^2 for small specimens or 6 GN/m^2 to 9 GN/m^2 for whole bows. However, we do not yet know the extent of degradation of all the bows and this is almost certain to reduce their density.

An independent prediction of the draw weight of A812 at 76½ lb, based on simple measurements of breadth and thickness of the bow limbs, was published by Patterson in 1981. It was realised later that this computation was for only one limb of the bow and doubling the figure to 153 lb gave better agreement with Kooi's estimate of 144 lb. Both of these, however, used a high value for the elastic stiffness compared with our value of 7.6 GN/m^2 which gave 110 lb in Table 1.

Are these draw weights credible? They are large by present-day standards and they must be compared with experiments on the bows themselves and on modern approximations.

Experiments on *Mary Rose* bows

The last five bows in Table 1 were fitted with horn nocks and strings and gently exercised with frequent rests until they could be braced. Their force-draw curves were measured in a testing-machine with the bows drawn manually, and thus quickly, by a system of rope and pulleys. The target draw length was 30 in.

A1654
The first bow released for testing reached a draw length of 29½ in and a tip deflection of 14 in at a load of 79 lb. A faint click was heard at this draw and a transverse crack was found penetrating the sapwood in the upper limb.

A1607
This very large bow broke at 8 in tip deflection at a low load. There was local degradation inside the bow at the region of fracture. This bow came from the box that had broken open at one end, allowing attack to occur.

A3975
This bow exercised well, was braced to 7 in and drawn to 22 in at a load of 42 lb before the sapwood cracked.

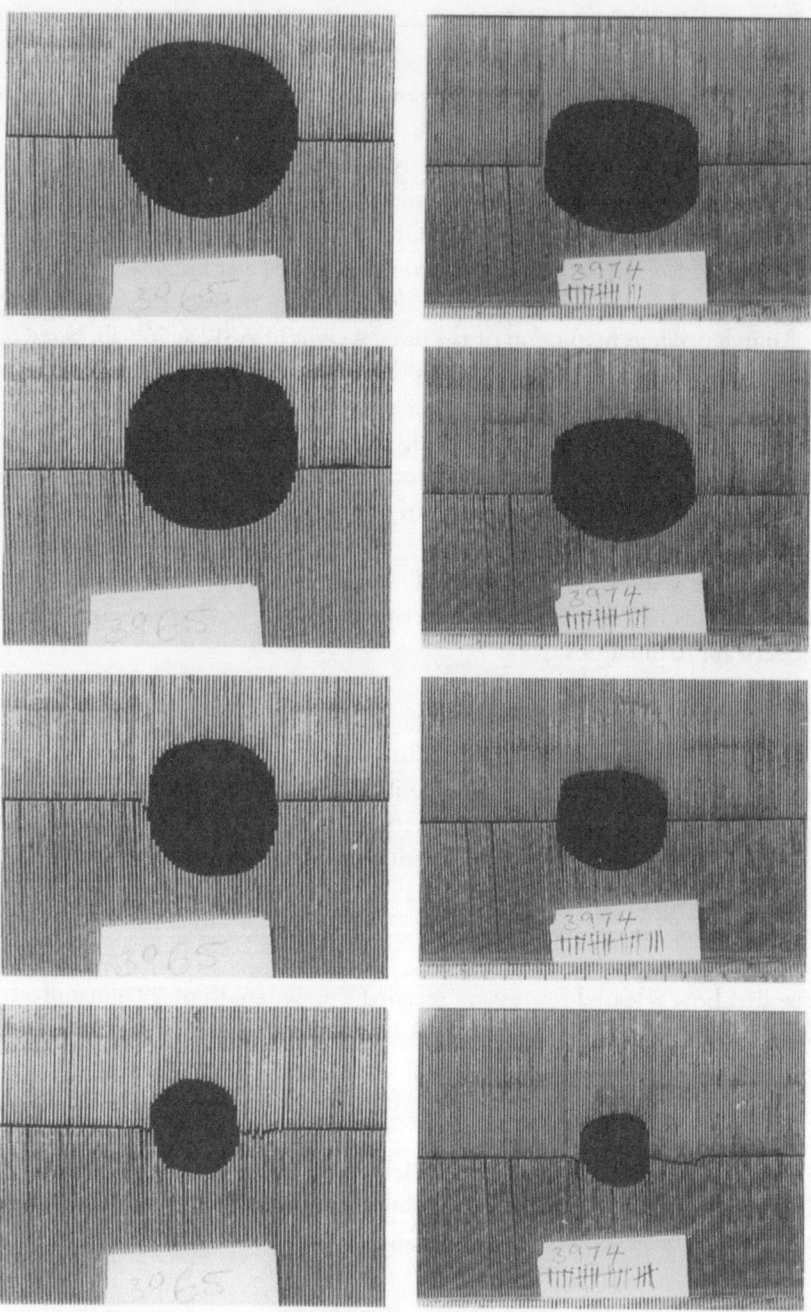

Cross section of two Mary Rose longbows, numbers 3974 and 3965, using pin gauges. The two profiles show two different kinds of longbow shape (Imperial College of Science and Technology).

A3952
This bow was found to have longitudinal cracks in the sapwood and, following the advice of Roy King, was not tested.

A1648
This bow exercised well, was braced to 7 in and drawn to 30 in with no sign of damage. The maximum load was 55 lb and this increased to 60 lb after further exercising.

Despite their outward appearance these bows obviously had become degraded especially in the sapwood. At this stage the elastic modulus of three of the complete bows was measured and found to be some 50 per cent lower than expected. In view of this confirmation of the degradation the decision was made to stop testing *Mary Rose* bows and to concentrate on the modern approximations.

Experiments on modern approximations

The earliest record of a modern bow made to the *Mary Rose* design was given by Saxton Pope in 1930. He made an approximation of a Tower bow, based on the very brief description in the Badminton Library volume on Archery. The draw weight of this bow was a disappointing 65 lb at 28 in draw and 72 lb at 36 in and only shot a light arrow 225 yards (206 metres). More recently Richard Galloway made an approximation of a Tower bow which weighed over 100 lb at 28 in draw; this is close to the predicted figure in Table 1. Either Saxton Pope's stave was not as stiff as Galloway's or more likely, since it did not break at 36 in draw, his bow was thinner. The thickness of the bow is very important because the strength varies with the cube of the thickness. A small increase in thickness gives a large increase in strength.

In our experiments we have used MRA 1, 2 and 3, bows made by Roy King to the general design of the *Mary Rose* bows. He is making the strongest bows he can from the best staves of Oregon yew we could obtain. His difficulty in making the bows is outlined in the notes he made at the time. Measurements of the distribution of section stiffness revealed that too much stiffness had been built into the centre section, where modern longbows have a thicker handle, and also into the ends of the bow, a valuable safety feature in the opinion of the bowyer. Dr Kooi's computed draw weight for MRA 1 was 128 lb, assuming a draw of 30 in and an elastic modulus of 9.5 GN/m^2. Subsequent measurement of the elastic modulus of the bow gave an actual value of 7.6 GN/m^2. On this basis the computed strength is reduced to 102.4 lb and this is in remarkably good agreement with the experimental strength of 102.8 lb, measured at the same draw length, in the testing machine at Instron's Laboratories in High Wycombe. It was this agreement which gave us justifiable confidence in the computer modelling and in its ability to estimate the strength of the *Mary Rose* bows.

Comparison of theory with experiment

Results for all the bows tested are summarised in Table 2.

	Theory, lb at 30 in draw	Experiment, lb
A1654	124	Cracked at 79, 29½ in draw
A1607	185, or 172 at 28 in draw	Broke at low load; degraded
A3975	137	Cracked at 42, 22 in draw
A3952	115	Cracks in sapwood; not tested
A1648	136	60, 30 in draw
Saxton Pope, Tower bow	98–101	65, 28 in draw
Galloway, Tower bow	98–101	100, 28 in draw
MRA 1	102.4	102.8, 30 in draw

Table 2 Strength of *Mary Rose* bows and modern approximations

Two conclusions can be drawn from the results of the tests. The computer modelling is capable of predicting draw weights in good agreement with experimental measurements so that the range of 100 to 172 lb for the *Mary Rose* bows is likely to be correct so long as the elastic modulus of MRA1 applies to medieval yew. There is one obvious difference between modern Oregon yew and that used for the *Mary Rose* bows, namely the growth ring spacing. On a microscopic scale the density of wood varies within each growth ring, being lowest for the early spring growth and highest for the late or summer growth. The more spring growth there is with large cells and thin cell walls, the wider is the growth ring and the lower is the overall density. The *Mary Rose* bows have a very fine spacing of the growth rings, on occasion too fine to count easily without a magnifying glass. One bow had 152 growth rings/inch compared with 25–50 for English yew. On this basis their elastic moduli could well have been higher than 7.6 GN/m^2 and the draw weights correspondingly higher. This difference probably is due to the high quality of slowly grown yew produced in southern Europe on the north facing slopes of hills in Austria and Italy; 152 growth rings/inch with a diameter of 8 in suggests the tree was over 600 years old when it was felled.

The second conclusion is that the *Mary Rose* bows are substantially degraded. The strength of Al654 is reduced to 64 per cent and that of A1648 to 44 per cent of the predicted strength. The elastic stiffness of A1648 is about half of what it should be, while both A1654 and A3975 cracked across the sapwood. Roy King has commented that the sapwood of these bows 'is 50 per cent dead at least compared with new wood. It can be compressed with a fingernail and lacks substance'. If the sapwood made no contribution at all to the elastic stiffness of the bow and if we take one quarter of the thickness of the bow to be sapwood, the bending stiffness and also the strength of the bow would be as low as 42 per cent of the expected strength, like that in A1648. Probably both sapwood and heartwood have degraded, the sapwood more than the heartwood. This is confirmed by the darkening of the cell structure shown in the cross-section of the bow fragment.

What independent evidence is there for other longbows of this range of draw weights? In a footnote on page 45 Robert Hardy cites an archer using a 116 lb longbow to shoot arrows 350 yards. Howard Hill used a 172 lb longbow to shoot 391 yards. Little John's bow, which hung in the chancel of Hathersage church until 1729, is described by J.W. Hunter as 6 ft 7 in long with a girth at the centre of 5 in and a draw weight of 160 lb. This is the bow belonging to the Spencer Stanhope family described on page 54 which was on exhibition in the Wakefield City Museum. Finally the Type 16 arrow found in Westminster Abbey would have been suitable for a bow of 130–150 lb, as discussed in the Technical Appendix 2.

Range and efficiency of the *Mary Rose* bows

The range of arrows shot from the *Mary Rose* bows depends on the type and weight of the arrow and on the strength of the bow. The suggestions from our study are that two types of arrow were used, a heavy armour-piercing arrow with a long bodkin head weighing 58–73 g and a shorter, lighter arrow with a small barbed head weighing 33–42 g 'to harass the enemy at a distance'. Arrows of any weight much less than this, like modern target arrows or flight arrows, are too light to cause significant damage to their target, even if they were of sufficient spine to stand in such heavy bows. The maximum ranges estimated from Fig 4 in the Technical Appendix 2, *The Arrow*, are 320 yards for the bodkin and 350 yards for the lighter Type 16 arrow with the heaviest bows of 160–175 lb. With the lighter 100 lb bows the figures are 220 to 250 yards.

Our results also suggest that a bow made to come full compass, to work right into the centre of the bow, can be more efficient than the later sporting longbow. Klopsteg has claimed that the efficiency of the English longbow, the percentage of its stored energy transferred to the arrow, never exceeded 40 per cent. This may have been true of the Victorian longbow with a light target arrow, but Kooi has calculated an efficiency of 67 per cent for a 48 g arrow and 78 per cent for a 74 g arrow, released from A812. This demonstrates that the medieval longbow was significantly more efficient than previously believed and that the bowyer was remarkably skilful at matching the heavy arrow required in war to the heavy bow required to discharge it. Furthermore the medieval bowyer was able to make a heavy flatbow, like A1607, well before the American developments of the modern flatbow in the 1930s.

The medieval archer
What sort of a person would be capable of using the bows in battle?

★ ★ ★

What Sort of Men? Some Conclusions

There is a painting in the Christ Church, Oxford Ms Collection, an illumination of 1326, showing a castle defended by two women, one using a massive crossbow, the other an equally massive longbow. The moral of that is: training can make nearly all things possible. Does anything lead us to suppose that bows of the weights represented by the *Mary Rose* collection would be unusable? The answer must be no. If they were unusable they would not be there. So, since we admit them usable, what is there to suggest that the men who used them were specially selected, specially trained? The answer is: a very great deal. Those skeletons found in the *Mary Rose* which can undoubtedly be linked with archery tackle, and can be presumed archers, are large men, six-footers or so, and described by the Senior Consultant Anthropologist to the Trust who examined them as 'huge ... not *necessarily* tall, but massively boned'. They also exhibit changes to their shoulder blades which *could* be the result of working with heavy bows. The shipboard location of skeletons representing the highest percentage of bony changes attributable to the use of heavy bows occurs in the areas most associated with archery equipment. Even with the lighter bows we use for

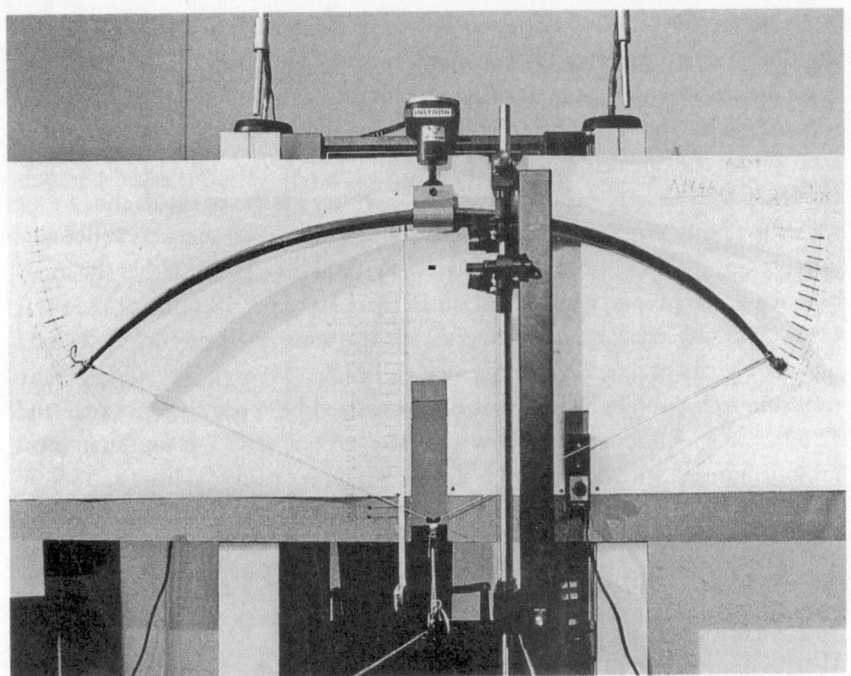

Mary Rose longbow number 1648 at 30 inch draw (76.2 centimetres) on the Instron test machine. This is the bow that was full drawn so many times successfully (BBC).

sport today it is in the shoulders, the upper arms and the elbows that things tend to go wrong. Absolute proof is difficult in this matter of the *Mary Rose* archers, but nothing has been found to suggest that the archers who would have wielded the bows found in the *Mary Rose* were anything but hefty fellows. Further than that, the lengths of the bows, from just over 6 ft to just under 7 ft, suggest men of some 5 ft 7 in to over 6 ft and the arrow measurements, with draw lengths of 28 in to 30 in, confirm these likely proportions.

Increasingly today there are to be found those who are teaching themselves to master bows of such great weights. As the facts are absorbed, the challenges that accompany those facts are taken up by enthusiasts in the archery world, especially among those aware of past history. I know of young archers who can handle weights well over a hundred pounds, as well as those who have trained themselves to shoot, with reliable accuracy, twenty and more arrows in a minute. Turn those few into thousands and one begins to get a genuine idea of the formidable power of our archer corps throughout the long years of its military ascendancy.

These modern shooters of great bows can never have had the early training nor constant practice of their predecessors. The earlier chapters of this book are thick with examples of how young they started and how hard they worked: 'They practised often with their longbows and shot with unerring aim', wrote the Spanish chronicler some 60 years before the *Mary Rose* sank. In Henry VIII's reign, Lord Herbert of Cherbury lamented the decline in the use of the bow 'proper for men of our strength'; in 1528 there were detailed instructions for the confiscation of crossbows and handguns, so that the longbow would be properly reinstated; in 1542 no archer of 24 years or over was to shoot at any mark of less than 220 yards distance; all physically fit men under 60 years of age, save clergymen and judges were to 'use and exercise shooting in longbows'; fathers of male children between the ages of 7 and 17 must provide them with a bow and two arrows; at 17 a young man must provide himself with a bow and four arrows, under pain of a fine; and so on, and so on. What sort of men could use the *Mary Rose* bows? Young, fit men in constant practice chosen for well-paid military service from a nation to whom the shooting of longbows had been second nature for 250 years at least.

Dr Rule, John Waller (centre) and Professor Levy (left) examining a crack that developed in a bow tip during testing at Instron (UK) in front of BBC Chronicle cameras (BBC).

It can be argued that by Henry VIII's reign, military archery was in decline. That is true, and for many reasons represented in the body of this book; nevertheless, can we infer that in these heavy bows rescued from the Solent we can see the kind of weapons that were used at Bosworth, at Towton, at Agincourt, Poitiers, and at Crécy 200 years before the *Mary Rose* archers sailed out of Portsmouth for the last time? I see no reason why we cannot be confident of that. If decline in the use of the weapon is going to change anything in the weapon itself, it will tend surely to diminish the strength of it, not increase it. We cannot but believe we now have available to see and to study nearly 140 bows that represent the great period of military archery. If anyone asks 'what did they use at Agincourt?' direct them to the Museum of the Mary Rose Trust in Portsmouth.

Three more questions: what was the function of the *Mary Rose* archers? By 1545 ship construction, sails and running rigging were so complicated that there was no room for the sort of massed archery that had been the order at Sluys or Cadsand. My own belief is that those archers who, or whose weapons, were found in action positions on the weather deck were there for two purposes: firstly as sharpshooters, forerunners of the men who shot from the tops at Trafalgar, one of whom killed Nelson. If an enemy came alongside to grapple and board, then archers, where they could clear crew and rigging, would be very useful at picking off the incautious. Secondly, they may have been used as range finders and windage indicators for the gunners. I think it likely there were many more archers on board than were there with bows at the ready. They would have been for land service, on the Isle of Wight perhaps, which the French had invaded, over the Channel, if we drove the French back to their home ports, or on this side of the Channel if the land troops had not dislodged all the French incursions along the South Coast. It is known the *Mary Rose* was carrying many more men than usual, making her dangerously overcrowded. The surplus was there to be put ashore for land fighting, and some of their bows were in boxes on the orlop deck, and we have them now, looking very much as they did on the morning of July 19th, 1545.

What is the future of the testing and conservation programme? The arrows are conserved and there is little now that can be done to further our knowledge. We know most of the materials used in their construction; we know their dimensions; we know how they were fletched, and how the fletchings were glued and bound on to the shaftment and how the nocks were cut and strengthened; we have largely to guess at the exact types and weights of their steel heads, because of their gross degradation. There they are, to be seen, but looking a little sad, set against the splendour of the bows.

The longbows of the *Mary Rose* are stable and safely conserved. They will be constantly re-examined to guard against deterioration. They may have more secrets to yield. Even psychometry was tried at one point: the practitioner confessed, after being in the storeroom for some time, to having heard the cries of men, breaking timber and crashing water; but it has to be said she knew what she sat among.

The conservation team will never let up in its efforts to reveal more, and of course when the full Trust reports are published there will be much more detailed evidence available to those who are interested; there are not a few who are impatient of the time that has already gone by, not perhaps realizing how slow and cautious the progress has to be. There will be no more testing of the bows themselves. Instead the making and testing of MRAs, *Mary Rose* Approximation bows will be given great attention.

So the last question concerns these bows. Three have already been made by Roy King, of strengths which are variously instanced in this chapter, but which after retillering now draw from 105 lb to 120 lb at 30 in. Recently Don Adams of Oregon, bowyer and timber expert, has generously given to the Trust two superb staves of Oregon yew, unjointed self staves of fine grain which should yield bows in the 150/160 lb range. Then it is a question of finding bowmen thoroughly capable of mastering such bows, not just of drawing them up; of matching arrows to the bows; and we're away! Shooting machines have been designed and tried, not successfully in my view because the rigid grip of a machine can never equal the elastic drawing, holding and loosing of an archer; so I believe the results to be quite untrustworthy. None of this programme is as easy as it may sound, but the auspices are good. The experimental Team has now been joined by a remarkable longbowman, Simon Stanley from Staffordshire, who can master bows of weights that defeat most archers. He has been shooting MRAs 1, 2 and 3 for hours together, untiring and in complete control. So far the results suggest very firmly that medieval military archers might expect to engage with the lighter types of arrow at 300 yards and over, and that 275 yards is not an exaggerated range for heavy war arrows shot from heavy war bows. Let us end with a few entries from Roy King's diary as he struggled with MRA 1: the first *Mary Rose* Approximation bow which did so much to validate our methods.

'Don Adams Oregon yew self stave 75¼ in long; as near as damn it, dead straight ...'

'Sapwood worked down slightly to remove small bumps. A good stave; no need to follow bumps ... '

'Sides worked down dead square to back, to what should be finished width, leaving just a little extra width at tips ... Slightly convex taper ...'

'Back generously radiused over full width, dropping a little extra at edges. Then there appears as if by magic the *Mary Rose* sharp edge ...'

'Coming to the conclusion that these bows were made almost by numbers. That given the quality of the wood ... a formula for manufacture could be closely followed ... a system that isn't used in Victorian bows ...'

'The bow is near totally inflexible as yet ...'

'The "Cunninge bowyer" takes over; "scratch and scrape" where gumption tells!'

'The bow held in a vice and each limb pulled back and held behind a stop at a 7 in brace for two minutes, for each limb.'

'The whole still bloody inflexible! With all my ten stone of muscle I managed to brace the bow in a swivel vice, but only just. The dacron string stretches as never before and settles to a 5 in bracing height ...'

'Left at 19 in draw for ten minutes ... draw weight at 65 lb ...'

'Left on tiller at 24 in draw. I'll beat the swine! Go for a cup of tea; return after half an hour ...'

'With all the strength I have I manage to brace the bow to 7½ in and leave it on the tiller at 26 in draw and retire for a cup of tea!!'

'After half an hour, bow registering 87 lb with a 26 in arrow.'

'Left at 26 in draw for one hour, then braced to 7 in and when drawn to 28 in gives 94 lb ... with little bits and scraping this is becoming quite a skilful job!'

'Left unbraced for five hours ... then drawn to 28 in ... 94 lb ... weight now static ...'

Finally it was crept from 28 in to 29 in and to 30 in, when it stabilized at 104–105 lb draw weight: 'horns fixed, giving 6 ft 2½ in between grooves'. He adds, 'I would regard this bow as the *ultimate* that could be expected of a yew bow of its excellent design, size and weight.'

Work is in progress.

APPENDICES

SOME TECHNICAL CONSIDERATIONS

1 The Design and Materials of the Bow,
by P. H. Blyth

The performance of medieval archers can be better gauged from the many surviving arrowheads than from the few bows. This study of the bow will therefore consider how far that performance depended upon the materials employed and upon the shape and size of the bow. It is not difficult to suggest imaginary improvements in the bow – for example making it short enough to be used on horseback, or making it of steel which would be less vulnerable than wood to changes in temperature and less vulnerable to humidity than a sinew-based composite. It is therefore of interest to consider how great the technical obstacles would have been to variation in different directions, and whether the traditional design represents an optimum.

The designer of a bow has essentially two problems: to match the bow to the archer, and to make it as efficient as possible in discharging the arrow. The first of these concerns quasi-static behaviour, in which the archer is trying to store as much energy as possible within the limits set by his strength and the length of his arm. The second concerns the dynamics of the release, some aspects of which are still obscure, but during which the bow must transfer as much of the stored energy as possible to the arrow, and retain as little as possible, while keeping the transfer smooth. Both phases are constrained by the materials employed, and they are closely connected because the same mechanism is used both to store the energy and to deliver it. It is therefore constantly necessary when considering one aspect of the bow to make cross reference to the others.

1 Energy storage vis-a-vis the archer

The energy stored in a bow is equivalent to the work done in drawing it, less the *hysteresis*, or energy left behind. The work done depends on the length of the draw and the average force applied. If the force on the string is plotted against the draw distance, as in fig 1, the total work will be equivalent to the area under the curve.

The shape of the curve will have a considerable effect on the work done and the energy stored; the more quickly the force builds up during the draw, the greater the average will be. The shape of the curve is determined partly by variations in the force exerted by the bow stave, and partly by the geometry of the system, as follows.

(i) When the bow is bent, it forms a spring whose resistance builds up from zero in a more or less linear manner. Over the total displacement, the average force on the tips will therefore be about half the maximum; however, the archer does not use the full displacement, because he bends the bow a certain amount before he strings or 'braces' it, and the average is therefore halfway between the maximum force and the force at the bracing displacement. If the stave can be arranged to bend a long way before it is braced, for instance by making a reflex stave in which the arms are initially pointing in the other direction from that in which they will be drawn, the average force over the draw will be higher, and it might seem that the distance through which the bow is bent before bracing should be made as great as possible.

However, the initial bending also stores energy, which is left unused, and all capacity to store energy has to be paid for by mass in the arms. The mass corresponding to unused energy will decrease the efficiency of the bow, because some of the energy stored in the draw will be required to accelerate it. Reflexing is therefore probably only worthwhile in composite bows made of materials such as sinew and horn which allow very large deflections, and in which the force builds up not in a linear manner but very slowly at first and then more steeply.

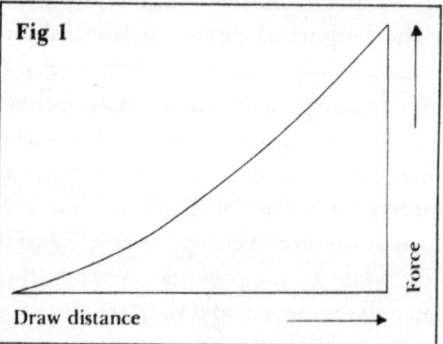

Fig 1

Draw force	Material	Actual weight with handle	Virtual weight
70 lb (319 N)	Yew	1 lb 8.5 oz (687 gm)	0.84 ± 0.28 oz (23.5 ± 7.8 gm)
72 lb (325 N)	Yew	1 lb 7 oz (645 gm)	0.90 ± 0.12 oz (25.2 ± 3.3 gm)
52 lb (235 N)	Yew	1 lb 4 oz (576 gm)	1.04 ± 0.13 oz (29.25 ± 3.6 gm)
73 lb (330 N)	Laminated wood	1 lb 8 oz (676 gm)	1.18 ± 0.1 oz (33.1 ± 2.8 gm)

NB: 1 lb = 4.448 Newtons which is abbreviated to N throughout the Appendices.

(ii) An analysis of the quasi-static geometry of the simple bow was published by C. N. Hickman under the title *The Dynamics of the Bow and Arrow* (1937). It gives a good account of the draw, although it is unsuccessful as a treatment of the dynamics of the release. When Hickman's equations are plotted graphically to estimate the energy storage in bows of different lengths, it is found that for a given draw-force, and a draw distance of 28 in (35 cm) from a straight stave, a 4 ft 6 in (135 cm) bow stores about 80%, and a 3 ft (90 cm) bow only about 60%, of the energy stored by a 6 ft (180 cm) bow. However it is only worth making the bow still longer if the draw distance is increased.

2 The efficiency of the release

When the bow is shot, the strain energy stored in the stave accelerates its arms and through them the string and the arrow. It is thus largely replaced by kinetic energy $K = \frac{1}{2}mV^2$, distributed in all three according to the mass m and the velocity V of each particle. The aim of the bowyer is to get as much of the energy as possible into the arrow. Two things are known empirically about this balance.

(i) So long as the velocity of the arms of the bow is increasing, the arms take up kinetic energy and reduce the amount available to the arrow. But in the second half of their travel they are slowed down by the string, and under certain circumstances some of the kinetic energy they lose is transferred to the arrow. This effect was demonstrated by C. N. Hickman (1929), but the amount is not very great. When the arms come to rest, a good deal of the kinetic energy they stored earlier is probably stored temporarily as strain in the string and in the arms themselves, to give rise to further vibrations after the arrow has left.

(ii) The efficiency of the bow, which is calculated as

$$\eta = \frac{\text{kinetic energy of the arrow}}{\text{work done in drawing the bow}}$$

rises with the weight of the arrow shot. The relationship was formulated by Klopsteg (1943) as

$$\eta = \frac{w}{k+w}$$

where w is the weight of the arrow and k is a constant for the bow. Klopsteg termed k the 'virtual weight' of the bow, as if the energy losses were regarded as an extra burden on the arrow, travelling beside it with the same velocity. The bows used in the tests described below proved to have virtual weight as table above.

Thus the first bow should be 50% efficient with an arrow of 0.84 oz (23.5 gm), 66% efficient with an arrow of 1.68 oz (47 gm) and 75% efficient with one of 2.52 oz (70.5 gm).

3 The materials of the bow

From the foregoing argument it appears that the arms of the bow should be as light as is consistent with their capacity to store energy. Table 1 gives an indication of the maximum energy, which could be stored as elastic strain in a unit weight, of some typical materials. *Strain* is proportional change in length, and it is taken that for small strains all the materials obey Hooke's law for elastic strain $\frac{\sigma}{\varepsilon}$ = constant, where σ is strain and ε is *stress*, or force per unit area —

although sinew and horn depart from Hooke's law to some extent. For simple tension and compression the constant defined is E, the Young's Modulus of elasticity of the material, a measure of its *stiffness*. Since stress is a force and strain a displacement, the strain energy per unit volume can be written:

$$U \quad \tfrac{1}{2}\sigma\varepsilon = \tfrac{1}{2}\frac{\sigma^2}{E} = \tfrac{1}{2}\varepsilon^2 E$$

and these expressions are divided by the density, ρ, to give the energy per unit weight.

As the squared term in the expression $\varepsilon^2 E$ implies, and as the table confirms, the most important variable for energy storage is the maximum strain, and for metal and sinew and horn that is determined by the *elastic limit*, the maximum strain from which the material will return to its original shape and size when unloaded. However, to accept a similar limit for wood in bending, which would allow strains of 0.5%–0.7%, would be unnecessarily restricting. The dimensions of actual bows imply strains of the order of 1.0%–1.1% in flat bows, and 1.2%–1.4% in 'D' sectioned longbows (and strains of that order have been directly measured). Since the wood is strained beyond the elastic

Table 1 **Factors affecting strain energy storage**

Material	Density lb/cu ft (gm/cc)	Elastic modulus E lb/in² (N/mm²)	Strength (see text) lb/in² (N/mm²)	Permissible strain (see text)	Energy/weight ft/lb per lb (Joules/gm)
Steel					
0.2% carbon quenched	475 (7.6)	30×10^6 (207,000)	11.2×10^4 (773)	0.187%	15.3 (0.046)
Spring steel, piano wire	475 (7.6)	30×10^6 (207,000)	45×10^4 (3,103)	0.75%	248 (0.746)
Sinew	81.25 (1.3)	0.18×10^6 (1,240)	1.5×10^4 (103)	4.1%	273 (0.822)
Buffalo horn	81.25 (1.3)	0.38×10^6 (2,650)	1.79×10^4 (124)	3%	303 (0.917)
Hardwoods					
Ash	37.5 (0.60)	1.73×10^6 (11,900)	1.68×10^4 (116)	0.97%	188 (0.565)
Elm	28.7 (0.46)	1.01×10^6 (7,000)	0.98×10^4 (68)	0.97%	109 (0.330)
Wych Elm	34.3 (0.55)	1.54×10^6 (10,600)	1.52×10^4 (105)	1.0%	171 (0.530)
Oak	38.1 (0.61)	1.55×10^6 (10,700)	1.52×10^4 (97)	0.97%	161 (0.485)
Softwoods					
Scots Pine	28.7 (0.46)	1.43×10^6 (9,900)	1.29×10^4 (89)	0.89%	131 (0.395)
Taxus Brevifolia (USA)	39.3 (0.63)	1.46×10^6 (10,000)	1.68×10^4 (116)	1.16%	219 (0.657)

limit it cannot recover completely, and when the bow is repeatedly loaded to the same high strain the first few cycles of loading and unloading will show serious energy losses. However, if carefully treated it will soon settle down, to provide a somewhat restricted recovery with little energy loss, as if the consequence of overloading had been to 'harden' the wood and render it fully elastic at higher strains, at the cost of some permanent deformation. (The cost also includes the formation of microscopic cracks which lead to immediate fracture if the stave is bent the other way, as can happen when the string breaks during a shot.)

The limit of recoverable strain attainable in this way has been estimated in the table by dividing the maximum possible strain, the 'modulus of rupture' or bending strength, by the elastic modulus. It must however be recognised in applying such formulae to wood that the properties of the material vary not only from piece to piece but according to the method of loading and the shape and size of the specimen tested. Both the bending strength and the elastic modulus of timber are conventionally measured by testing small specimens of square cross section, and must be adjusted for other shapes and sizes.

Taking the figures for timber as they stand, it appears that wood can store considerably more energy in bending than all but the very best steel and than lightly strained sinew and horn, for a given mass, but considerably less for a given volume, while it is overtaken on both counts by sinew and horn if the latter are fully strained.

The best timbers are those which combine a high specific bending strength, $\frac{B}{\rho}$, with a comparatively low specific modulus of elasticity, $\frac{E}{\rho}$. In general this criterion favours the hardwoods, such as ash and wych elm, because, while their specific modulus is much the same as that for the softwoods, their specific bending strength is higher, probably owing to their finer grain. Yew is an exception, being an unusually fine-grained softwood, and having an exceptionally high specific bending strength as well as a low specific elastic modulus. No reliable figures are at present available for *Taxus Baccata*, but it is probably comparable with *Taxus Brevifolia*.

The predictions in the last column of table 1 are subject to the following reservations. Firstly, wood is very variable, and these figures are averages. Secondly, the table does not show hysteresis, that is energy lost in the flexing, which probably emerges as heat. Hysteresis in the three yew bows mentioned above showed losses in a very slow draw and recovery between 1.9% and 6.5%, and many woods may be unsuitable for bows because their hysteresis is still higher. It is said that ash is one of them. Thirdly, the figures given refer to a temperature of 20°C. The strength and elasticity of yew wood vary from that by 1% per degree centigrade between 0°C and 50°C, and probably further in each direction; and that does not seem untypical of woods in general. Ascham remarks that a yew bow will snap if shot in freezing weather, and conversely that the handle can be weakened by the heat of the hand. This may account for

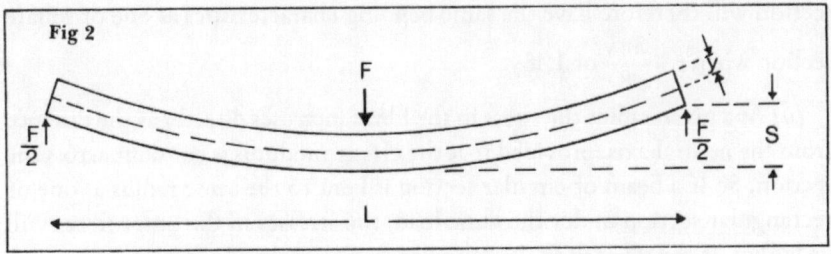

Fig 2

the greater success of the longbow in temperate climates, and the early development of the composite bow elsewhere.

4 The design of the stave

The actual energy storage in a bow will be less than the maximum found for the material in table 1, because it is impossible to stress all the material in a bent beam to the limit. We may consider the bow-stave as if it were a simply supported beam subjected to a load at the centre. The convex side or 'back' will be in tension; the concave, corresponding to the bow's 'belly', will be in compression with a so-called 'neutral axis', whose length is unchanged, lying between them (fig 2).

The stress in the fibres of the beam is then given by the general equation:

$$\frac{\sigma}{y} = \frac{M}{I} = \frac{E}{R}$$

where y is the perpendicular distance from the fibre to the neutral axis, M is the bending moment of the beam, I is the second moment of area (or Moment of Inertia) of its cross-section about the neutral axis, E is the elastic modulus of the material, and R is the radius of bending of the neutral axis.

From the equation we can deduce the following:

(i) The bending moment M describes the 'leverage' exerted by the ends of the bow in bending the segments nearer the centre. At any cross-section, M is the product of the force at the nearer end of the beam and the distance between that and the cross-section. The resistance of the beam to bending is not directly proportional to its area; the fibres at a distance from the neutral axis are not only more highly stressed than those closer in, but also have a greater effect on the bending, so that the shape of the cross-section must also be taken into account. Allowance for this is made in the second moment of area, I. For a rectangular section $I = \frac{by^3}{3}$ where b is the breadth and y half its depth, for a circular one $I = \frac{\pi r^1}{4}$ where r is the radius. A beam of circular

section will therefore have the same bending characteristics as one of square section when $r = \dfrac{3\pi y}{8}$ or $1.18y$.

(ii) At a given radius the stress in the fibres increases directly as the distance from the neutral axis, provided that the elastic modulus is constant across the section. So if a beam of circular section is bent to the same radius as one of rectangular section under the same load, the stresses in the outer fibres will be higher. If the stress in the rectangular section is already at a maximum, we might expect that it would be dangerous to exceed that stress in the circular beam, and so the latter would have to be content with a larger radius, a smaller deflection, and a smaller energy storage. However, tests have shown that it is in fact safe to treat solid beams of equal I as capable of equal deflection. The notional 'modulus of rupture' is increased by a 'form factor' which cancels the effect of the extra depth. The form factor was first discovered and used for the construction of wooden aircraft, but is now part of the standard building code. (For example the British Standard Code of Practice CP 112.) The traditional 'D' section is therefore not at a serious disadvantage *vis-a-vis* the flat bow. A positive advantage will be discussed below.

(iii) Since M increases towards the centre of the beam, in a beam of uniform cross-section in which both I and y are constant, the radius and the maximum strain will also increase towards the centre. In that case, the maximum deflection can be shown to be:

$$S = \dfrac{FL^3}{48EI}$$

which can be rearranged to give

$$S = \dfrac{L^2}{12R}$$

However, we can do better by arranging for I to vary directly with M, for example by narrowing the beam at each end. That will produce a constant radius R, so that the beam forms an arc of a circle and, by Pythagoras, the deflection will be approximately:

$$S = \dfrac{L^2}{8R}$$

and so

$$S = \dfrac{FL^3}{32EI}$$

Since the deflection is now 50% greater for the same applied force, the beam stores that much more energy, quite apart from the fact that a considerable amount of material has been removed. In the flat bow such trimming can be done entirely by removing material from the sides, so that the depth is uniform all along, and so is the maximum strain. With a rectangular cross-section the energy storage will be one-third that at maximum stress. Taking the energy storage of yew as given in table 1, a bow intended to store 66 ft/lb (90 J) (of which 59 ft/lb [80 J] would be used in the draw) would need to weigh $\frac{66}{219 \times 33}$ = 14.4 oz (408 gm) about 60% of the weight of the bows described above. Such trimming would also be exceptionally effective in reducing the dynamic 'effective mass' of the arms. The actual mass of a triangular bow arm affects its vibration as if it were a concentrated mass at the tip of about one-fifteenth, compared with a factor of just over one-quarter for a uniform stave. It is probably for that reason that the American wooden flat bows were so successful.

In a 'D' section bow such trimming cannot be done so effectively. However, it may well be that greater strains can be tolerated which more than make up the difference, because in a bow of even curvature the margin of safety given by the rounded belly is probably greater than the form factor allows. Since the energy required to propagate a crack increases with its breadth, it will perpetually be easier in a 'D' section for a new crack to start than for an existing one to grow, and the failure will spread all along the beam, rather than concentrate in one place. An old 'D' sectioned bow will frequently show compression creases spaced fairly evenly all along its length, and Ascham actually recommends relieving the stress on a crease by pricking the wood with a pin on either side towards the tip and the handle. Flat bows seem to have broken more frequently than 'D' sectioned ones in Neolithic times just as they do today, in spite of being less heavily stressed. Moreover the depth of 'D' sectioned bows frequently exceeds that of flat bows by 20%, instead of the 9% allowed by the form factor.

(iv) Finally, from the equations given above for deflection, it is clear that the stiffness of a beam must increase as the cube of its length, if a longer beam is to have the same deflection for the same force. Since the strain $\varepsilon = \frac{y}{R}$, the deflection can also be written:

$$S = \frac{L^2 \varepsilon}{8y}$$

In order to avoid making the bow too stiff, its breadth must be reduced by an even greater proportion – in fact it must vary as $1 - L^3$. There is therefore a sharp limit to the maximum length of a bow of given material and draw-force, and the longbow comes fairly close to it.

Conversely there are considerable difficulties in making a wooden cross-bow of high efficiency. As the bow gets shorter the radius required for a given deflection gets smaller; the depth must therefore be reduced and the breadth increased – or else the draw distance must be very short. In practice designers took the latter course, with draw distances as short as 5 in (12.5 cm). Theoretically a draw of 400 lb over that distance is as good as 100 lb over 20 in (50 cm), but the losses in the very short time allowed for the energy transfer are likely to have been much higher, and furthermore wood in large cross-sections is considerably weaker than in small ones. It is not surprising, therefore, that the crossbow was not a serious rival to the longbow until the development of a suitable steel, and as table 1 shows it requires a very good steel to beat wood.

There may be considerable gaps in our understanding, but there seems no reason to doubt that the traditional longbow, subject to the limitations of climate, represents something close to an optimum design.

Acknowledgement

For help in discussions during the writing of this paper, the author wishes to thank Professor J. E. Gordon, Professor W. D. Biggs, Dr A. J. Pretlove and Dr G. Jeronimides, of the University of Reading, and for information about wood, the Building Research Establishment, Prince's Risborough.

References

Hickman, C. N.	1929	Velocity and acceleration of arrows. Weight and efficiency as affected by backing of bow. Journ Franklin Inst 208, Oct 1929, pp 521–37.
Hickman, C. N.	1937	The dynamics of the bow and arrow. Journ App Phys, vol 8, June 1937, pp 404–409.
Klopsteg, P. E.	1943	The physics of bows and arrows. Am J Phys, vol 11, Aug 1943, pp 175–192.
British Standard Code of Practice CP112 part 2	1971	The Structural Use of Timber. The British Standards Institution, London. p29, § 3.12.1.4.

2 The Arrow, *by P. L. Pratt*

1 Introduction

'The successful discharge of an arrow depends entirely on its being correctly matched to the bow, and this is effected by the accurate combination of three factors, spine, weight and length.' In these few words E. G. Heath has summarised concisely the first of the three problems which the arrow-maker has to solve. After the discharge, the successful flight of an arrow depends upon minimising the aerodynamic drag, whilst retaining aerodynamic control, by correct design of the shaft, head and fletching. Finally, the successful attack of an arrow upon its target depends upon the velocity at impact, the weight of the arrow and the design and construction of the arrowhead. These three problems are interconnected, and any solution must represent a compromise. However, all solutions are constrained by the basic physics and applied mechanics of the problem, and by the materials available, and it is interesting to consider these in relation to the longbow.

2 The length and weight of arrows

One of the design parameters for the longbow fixed by the geometry of the human body is the length of the arrow. For most modern adult males an arrow length of 28–30 in (71.1–76.2 cm) is the greatest that can be drawn comfortably and safely to the arrowhead; a greater length than this adds weight to the arrow and reduces its velocity on leaving the bow. The energy available to the arrow from the bow is the product of the stored energy, E_s, and the efficiency, η. This transferred energy is converted into the kinetic energy of the arrow, so that for a mass, m, and initial velocity, V_0,

$$\eta E_s = \frac{1}{2} m v_0^2$$

and so

$$v_0 = \sqrt{\left(\frac{2\eta E_s}{m}\right)}$$

The weight of the arrow has two opposing effects; $v_0 \propto \frac{1}{\sqrt{m}}$ so that V_0 decreases as m increases; on the other hand, for a given bow, the weight of the arrow determines the efficiency η, because the energy stored in the bow is used to accelerate the bow-limbs and the bow-string as well as the arrow itself – the lighter the arrow the lower is the proportion of the stored energy transferred to it. Thus V_o, which is $\infty \sqrt{\eta}$, decreases as m decreases. Experimental results, in which these two opposing effects are combined, were obtained during 1975 at RARDE and at Imperial College, with John Waller shooting in the bow. The net result is a nearly linear reduction in V_0 with m, table 4, more or less confirming the earlier results of Hickman. The importance of keeping the arrow length to the minimum and thus keeping the weight small, so as to give the highest value of v_0, follows from the dependence of kinetic energy directly upon m but to the second power (v^2) upon v.

In practice, the weight of the arrow depends upon the purpose for which it is required. For maximum range, with a flight arrow, the weight is kept as small as possible to maximise v_0. For target arrows, where smoothness of discharge, efficiency and accuracy at shorter range are required, the weight increases with the draw-weight of the bow. Different manufacturers differ slightly in their recommendations, but typical figures are:

Table 2

Weight of bow	Weight of arrow			Spine GNAS
lb	N	oz	gm	units
30	133.5	0.57–0.69	16.16–19.56	81
40	178	0.69–0.91	19.56–25.80	61
50	222.5	0.91–1.14	25.80–32.32	49
60	267	1.14–1.37	32.32–38.84	40
70	311.5	1.37–1.60	38.84–45.36	35
100	445			25
150	667			16.5

For hunting arrows the required shape of the head determines its weight, and both the arrow-weight and the draw-weight of a hunting-bow are necessarily large.

3 The spine of an arrow and the Archer's Paradox

In the previous section the length and the weight of arrows were considered in general terms. In this section the need for the accurate combination of these factors with the spine, or the flexibility of the arrow for a successful discharge, is considered.

When the drawn bow-string is released, a sudden compressive load is applied through the nock down the length of the arrow. In addition, for a right-handed archer using the Mediterranean, or three-fingered release there is a force away from the fingers deflecting the rear of the arrow to the left. These two forces combine against the inertia of the arrow to buckle it and to set it vibrating laterally while it accelerates, as shown in Figure 3. Both the frequency and the amplitude of the vibration must be matched to the bow if the arrow is to avoid hitting the side of the bow repeatedly during its discharge.

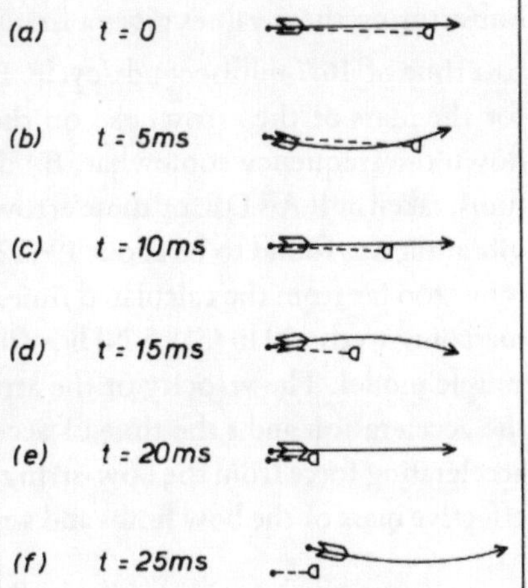

Fig 3 The Archer's Paradox

(a) t = 0

(b) t = 5ms

(c) t = 10ms

(d) t = 15ms

(e) t = 20ms

(f) t = 25ms

During the first half-cycle of this vibration, Figure 3a and 3b, the ideal shaft bends in such a way that the front part of the arrow is deflected by the bow, but only slightly. During the second half-cycle, Figure 3c and 3d, the mid-section of the arrow curves past the bow and finally, Figure 3e and 3f, the fletching clears the bow as the arrow straightens, leaves the string and reverses its bend again.

For the arrow to bend cleanly past the bow, its flexibility must match both its inertia and the thrust from the bow-string. The flexibility of the arrow, the spine, can be measured by supporting it at nock and head and determining the deflection at the mid-point, in hundredths of an inch, when a weight of 1½ lb is hung there. For example, a deflection of 0.5 in (50 hundredths) corresponds to a spine value of 50 GNAS units, and this would be suitable for a 50 lb (222.5 N) bow with a target arrow 28 in (71.1 cm) long weighing about 1.06 oz (30 gm). For heavier bows, as shown in Table 2, for heavier or longer arrows, a smaller spine value is needed. Other factors such as the speed of the bow, the quality of release, and the width of the bow-handle can influence the spine required in the arrow, and the final matching of the arrow to the bow must be confirmed by experiment.

We can calculate the frequency of flexural vibration of the ramin-wood arrows used in our experiments. The frequency

$$f = \sqrt{\left(\frac{C_1 E d^2}{l^4 \rho T}\right)}$$

where C_1 = constant, E = Young's Modulus = 14 GN/m², d = diameter of shaft = 0.375 in (9.37 mm), l = length of shaft = 28 in (71.1 cm), ρ = density = 657.0 kg/m³, and T is a correction factor involving the ratio $\frac{d}{l}$ and Poisson's ratio.

Substituting these values gives a frequency of 60.1 Hertz which corresponds to a time of 16.7 milliseconds/cycle. In this calculation no allowance is made for the mass of the arrowhead on the end of the shaft, and this must slow down the frequency somewhat. By direct measurement of high-speed cine films, taken at RARDE, of these arrows striking targets, the period of flexural vibration was found to be about 19–22 ms, with an average of 20.5 ms, which is not too far from the calculated time. We can estimate roughly the degree of matching to the 70 lb (311.5 N) bow used in these experiments by means of a simple model. The velocity of the arrow leaving the bow v_o = a.t, where a is the acceleration and t the time of acceleration. Now F = m a, where F is the accelerating force from the bow-string, and m is the mass of the arrow plus the effective mass of the bow limbs and string. Thus

$$t = \frac{v_0}{a} = \frac{m\,v_0}{F}$$

Substituting for F one-half of the draw-weight of the 70 lb bow, since the force falls from 70 lb to zero during acceleration, we find 21.7 ms as the time required for an arrow weighing 2.02 oz (57.4 gm) to leave the bow-string, as shown in Figure 3e. The time for the rear of the arrow to travel a distance equal to the bracing height at a velocity of 43.6 m/s is a further 3.75 ms, giving 25.45 ms in all. This corresponds to one and a quarter vibrations of the arrow, as shown in Figure 3 which should take (1.25 × 20.5) = 25.63 ms. By the time the rear of the arrow swings back on the next cycle, it will be well clear of the bow and travelling at maximum speed.

4   The design of arrow-shafts a Material and dimensions

The density and elastic modulus of woods used to make arrow-shafts are given in Table 3. As

Table 3   **Density and elastic modulus of wood**

	ρ kg/m^3	E GN/m^2	E/ρ
Ash	0.60 × 10^3	11.9 ± 2.17	19.83
Birch	0.60 × 10^3	13.3 ± 0.80	22.17
Beech	0.62 × 10^3	12.6 ± 1.22	20.32
Hornbeam	0.65 × 10^3	11.9 ± 0.85	18.31
Oak	0.61 × 10^3	10.1 ± 1.96	16.56
Ramin	0.59 × 10^3	14.0 ± 1.36	23.73

shown in the previous section, the most important design parameter is the frequency of lateral oscillation of the arrow-shaft, and both the density of the wood and the flexibility of the shaft enter into this calculation. However, the range of densities shown in Table 3 is limited, and the practical design of arrow-shafts is based essentially upon the necessary flexibility of the shaft. The lightest arrow for a given stiffness will be given by the material with the highest value of E/ρ.

The deflection of a thin cylindrical shaft supported at its ends under a central load is given by

$$y = \frac{C_2 \, P \, l^3}{E \, d^4}$$

where C_2 is a constant, P is the load, l is the length between supports, E is Young's Modulus, and d is the diameter. Given a fixed length of shaft, l, and a maximum deflection under a 1½ lb load that will match a particular bow, the design of the shaft is reduced to determining the diameter, d, which gives the required deflection with the value of E for the chosen wood. The spine values shown in Table 2 were obtained by a reciprocal extrapolation of the early data of Nagler and Rheingans, and it is not clear that these represent ideal values for the yew longbow. For the want of better, they have been used to check the design of our ramin-wood arrows, and to estimate the weight of bow suitable for the arrow in Westminster Abbey (section 7). Our 28 in (71.1 cm) arrows are made from ramin dowelling 0.375 in (0.95 cm) in diameter giving a shaft weight of 1.26 oz (35.86 gm) after fletching. Table 2 suggests that a spine of 35 GNAS units would be suitable for our 70 lb bow and the equivalent deflection of 0.35 in (0.89 cm) under a 1½ lb load compares rather well with the value of 0.348 in (0.88 cm) calculated from the formula for our ramin shafts. Nevertheless these arrows were found to be fractionally too stiff for the 70 lb bow, unless the loose was perfect.

b Shape

Most arrow-shafts today are of constant diameter along their length, because they are made from material in the form of tubes. To withstand the applied bending stresses, a shaft of constant diameter is not the most efficient shape, because the bending moment increases from zero at the ends to a maximum at the middle of the shaft. The resulting curvature increases in the same way from the ends to the middle. Uniform stressing, and a uniform curvature, can be obtained with a considerable saving of weight by tapering the shaft from the middle to the ends, that is by 'barrelling' the arrow. Wooden flight arrows and clout arrows are tapered so that a 5/16 in (0.79 cm) diameter shaft reduces to 3/16 in (0.48 cm) at the nock and the head. The most extreme tapering is shown in the Turkish flight arrows described by Payne-Gallwey, which reduce from 5/16 in (0.79 cm) at the centre to 3/16 in (0.48 cm) at the nock and to ⅛ in (0.32 cm) diameter at the head. The only drawback to the barrelled arrow, apart from the difficulty and the cost of making it, lies in its increased fragility. When striking a target at an oblique angle, the danger is that the tapered front section may break before the weight of the arrow has been applied to the target. It is interesting to see that the arrow in Westminster Abbey is tapered moderately at the rear, from 0.45–0.33 in (1.15–0.85 cm) but only slightly at the front to 0.41 in (1.05 cm); this must represent a good compromise for the shaft of a war arrow.

5 Design and construction of arrowheads

For flight arrows and for target arrows a simple ogival head, no larger in diameter than the shaft, offers the least resistance and the lowest weight. For hunting arrows, a wide cutting blade is the most effective against animals and types 13, 14 and 15 are representative of those broad-heads to be found in museums. In the Medieval Catalogue of the London Museum, Ward-Perkins suggests that the more developed types 14 and 15 were designed exclusively for hunting and there seems little reason to doubt this. The average weights of complete heads of these two types in the London Museum are 0.47 oz (13.2 gm) and 0.52 oz (14.8 gm) respectively and their shape is such as to offer considerable wind resistance.

Ascham says there were two kinds of arrowhead for war in Homer's day, the broad-heads or swallow-tails, and forked heads (type 6). The average weight of type 6 heads in the British Museum and the London Museum is 0.52 oz (14.8 gm), close to that for types 14 and 15. Ascham goes on, 'Our English heads be better in war than either – for the end being lighter they flee a great deal the faster, and so give a far sorer stripe – if the same little barbs they have were clere put away they should flie far better.' A recent survey of arrowheads reveals that the commonest by far is type 16 and this fits Ascham's description of a head with 'little barbs'. Furthermore, the average weight of 42 type 16 heads in the British Museum is 0.23 oz (6.45 gm) and of 21 in the London Museum 0.26 oz (7.51 gm), about half that of the broad-head or the forked head. Whereas Ward-Perkins suggests that type 16 should be included with types 14 and 15 as a hunting head, we believe that this represents the light medieval war-head designed, in the words of Payne-Gallwey, 'to harass an enemy especially his horses, at a distance beyond the reach of heavier war arrows'.

Following Ward-Perkins, the arrowhead designed to penetrate armour is the bodkin of types 7 or 8 with an average weight, depending upon length, of 0.35–0.71 oz (10–20 gm). This head couples low wind-resistance with the ability to wound even without full penetration of the armour, and the sharp point is less likely to bounce off than the blunter types 9 to 12.

Of the construction of arrowheads, Ascham says, 'I would wish the head makers of England should make their sheaf arrows more harder pointed than they be.' Successfully to attack armour the arrowhead should be harder than the armour, yet at the same time it must be tough so that the head does not break off on impact. This combination of properties must have been very difficult for the arrowsmith to reproduce in the 14th and 15th centuries. Only a limited metallurgical study of arrowheads has been made so far, but what has been done looks fascinating. By the kindness of J. Clark of the London Museum, a type 16 arrowhead was made available for microscopic examination. Briefly, this showed a composite structure with a hardened steel tip containing some 0.35% C with a very fine grain size, carried on a softer mild steel shank. The combination of forging and heat treatment of the tip has been so controlled that the steel is in a state of considerable hardness combined with great toughness, a good microstructure even by present-day standards. By water-quenching from 900°C the hardness could have been increased still further, as Peter Jones from RARDE has shown, but this would have been at the expense of much reduced toughness. A similar fine-grained tough microstructure has been found by Dr C. Brewer on the blade of a type 15 hunting head. This head was excavated recently at Hadleigh Castle, Essex.

It is clear that the medieval arrowsmith was able to design and manufacture highly sophisticated arrowheads suitable for the attack of both horse and man. The interesting questions now are how far could these arrows be shot and what damage could they do on arrival at their target?

6 The flight of the arrow

Shot in a vacuum, the final velocity of the arrow striking the target would be the same as the initial velocity on leaving the bow, and there would be no point in fletching the arrow since there would be no possibility, or necessity, of controlling the attitude of the arrow along its flight path. Shot through the air, the final velocity is reduced by the resistance of the air while control of the attitude, or steering, is achieved by the restoring pressure of the air on the vanes at the rear of the arrow when it yaws from its true course. The resistance of the air applies a drag force, D, to the arrow, which decelerates at a rate given by $a = \frac{D}{m}$. The drag on the arrow arises from (i) the head-on resistance which is proportional to the cross-sectional area of the arrow, (ii) the skin-friction of the head and of the shaft, and (iii) the skin-friction of the fletching. Two experimental methods have been used to estimate the drag force.

In the first method, the velocity of the arrow in flight is measured at two points a known distance apart. The work done by the drag force is equal to the loss of kinetic energy over this distance, so that $Dx = \frac{m}{2}(v_1^2 - v_2^2)$, where x is the distance. In this way English, in 1930, measured the drag on a target arrow; and his results were confirmed subsequently to better than 10% by the second method, in which the drag is measured directly in a wind tunnel. For any particular arrow $D = K v^2$, where K is the drag constant; English found a value of $K = 1.15 \times 10^{-6}$, where D is in lb, and v in ft/sec. The wind tunnel results, analysed by Rheingans, would give $K = 1.06 \times 10^{-6}$ for this arrow, and the 9% extra energy loss in free flight can perhaps be accounted for in terms of the energy needed to make the arrow spin and to steer the arrow on a straight path; the drag increases dramatically as the arrow yaws and is corrected by the vanes. By courtesy of Dr John Harvey of the Department of Aeronautics at Imperial College, we have been able to measure the drag of our ramin-shafted arrows in a wind-tunnel. Preliminary analysis of the results gives values of K between 2.1 and 3.5×10^{-6} depending upon the type of head. Part of the increased drag in our case is due to rather a large vane area totalling 12.35 sq in (79.68 sq cm) compared with English's 6.33 sq in (40.84 sq cm), and part to the shape of our arrowheads which are much less streamlined than target piles. The design of our fletching is developed from medieval pictures by John Waller, and the shape of the heads reproduced from those existing in museums by Ray Monnery.

For maximum range and maximum impact velocity the energy loss due to drag must be minimised, and the easiest way to do this is to reduce the size of the vanes. For a light target arrow three vanes, 2½–3 in (6.35–7.62 cm) long and ½ in (1.27 cm) high, are sufficient to give accurate control with a total surface area on both sides of 6.5–7.0 sq in (41.94–45.16 sq cm). The heavier the head the larger the total surface area of vane needed to balance the arrow in flight. For a heavy hunting broad-head the vanes are usually about 5 in (12.7 cm) long and ½ in (1.27 cm) high giving a total surface area of 12–13 sq in (77.4–83.9 sq cm). On the other hand for a light flight arrow a total area of 2–3 sq in (12.9–19.4 sq cm) is sufficient to maintain stable flight and this should give a K of about 0.5 0.65×10^{-6}. The feathers on a Turkish flight arrow are 2½ in (6.35 cm) long, as thin as paper, and only ¼ in (0.64 cm) high at the highest point, according to Payne-Gallwey. A further distinction of the Turkish arrow, released with a thumb-ring, is that the vanes are placed close to the nock, since the usual space of 1¼–1½ in (3.18–3.81 cm) for the fingers in the European release is not required. In this way about 10% greater control leverage is obtained for the same area, so that less energy is lost in control and the arrow will fly farther and steadier.

The choice of material for the vanes is important. Feathers are not weatherproof although they are resilient and not easily damaged by contact with the bow. Smooth on top and rough below, three feathers from the same side of the bird, properly mounted, make the arrow rotate in flight, and so reduce the planing effect of the wings of a hunting arrowhead. Sadly, the stability gained by this rotation comes from the kinetic energy of the arrow, so that the faster the rotation the shorter is the range. Spirally mounted feathers are needed to control very large hunting blades and they reduce the velocity and range of the arrow very considerably. Smoother materials like paper or plastic have much lower drag coefficients than feather, by as much as a half, and so they give greater range and impact velocity. Paper is fragile, although Payne-Gallwey claims an increased range for a Turkish flight arrow of at least 30 yards (27.43 metres) in 360 yards (329.18 metres) using parchment instead of feathers. Plastic vanes are in many ways ideal, being waterproof, tough and with low drag, but they were not available to the medieval fletcher nor are they permitted to the modern longbowman. Also they can be dislodged should they strike the bow during a bad loose, or if they are struck on the target by another arrow. They would seem an ideal choice for a war arrow.

7 The range of an arrow

The ranging power of an arrow is given by the ballistic coefficient $C_o = \frac{Am}{K}$, where A is a constant, m the weight of the arrow, and K the drag constant from the previous section. Once this is known, the simplest method of determining the range and impact velocity in terms of the initial velocity and angle of launch is to use Ingalls' ballistic tables. In this section the maximum range of arrows shot from longbows is considered and compared with the range measured experimentally.

The results from the previous section are

Table 4 **Ranges of arrows with 70 lb longbow**

	Weight			Velocity	Energy	Efficiency	Max range
	gm	$K \cdot 10^{-6}$	C_o	v_o m/sec	Joules	η	yd
Flight	16.2	0.5	0.10	(64.1)	(33.3)	(0.41)	310
Target	24.0	1.15	0.078	(60.1)	(43.3)	(0.53)	270
Lozenge bodkin	47.9	2.1	0.071	46.5	51.8	0.63	180
Long bodkin	57.4	2.78	0.066	43.6	54.6	0.67	170
Broadhead	73.6	3.5	0.066	38.7	55.1	0.67	150

assembled in the first five columns of Table 4, together with some extrapolations and interpolations for the 70 lb yew self-bow used in our experiments so far.

Maximum ranges for these arrows have been obtained from Ingalls' tables, and the range as a function of arrow-weight is shown in Figure 4 and in the final column of Table 4. Experimental points for the 70 lb (311.5 N) bow, shown as crosses in Figure 4, were obtained by measuring the maximum range for a number of arrows, up-wind and down-wind, and taking the average. These lie below the curves drawn smoothly through the theoretical points, which are shown as filled circles. The position of the curve for the lighter arrows with the 70 lb bow has been fixed using the experimental range of 300 yards (274.32 metres) for a flight arrow given by C. J. Longman in the Badminton volume on archery, and using the range of 280 yards (256.03 metres) obtained by Robert Hardy with arrows weighing 0.75 oz (21.26 gm). Longman quotes H. Ford as believing 'that, with practice, 300 yards is fairly attainable by many archers of the present day', while Ford himself reached 308 yards with a 68 lb (302.46 N) yew self-bow.

From the curve for the 70 lb bow (311.5 N) a family of curves has been generated for bows weighing 100, 150 and 200 lb (444.8, 667.2 and 889.6 N). The assumptions involved in doing this are that the draw-curves for the bows are linear, that the length, drag and weight of the arrows is the same for all the bows, but that the efficiency of the bow decreases as the draw-weight increases especially for the lighter arrows. For the 150 and 200 lb bows these assumptions might appear extreme, but they suggest that Howard Hill's shot of 391 yards (357.53 metres) with a 172 lb (765.06 N) bow was made with an arrow weighing 0.71–0.85 oz (20–24 gm), which is not unlikely. A recalculation of the spine of the Westminster Abbey arrow now suggests a figure of 19 GNAS units, with 17 units as the lower limit. This would be suitable for a bow of 130 lb (578.24 N), with 147 lb (653.85 N) as the upper limit, and the range shown in Figure 4 is about 280–290 yards (256–265 metres) for the 130 lb bow. This is the same as Double's 'forehand shaft' range of 14 to 14½ score, and supports Payne-Gallwey's similar figure for the light medieval war-head. By 1590 Sir Roger Williams was complaining that 'out of 5,000 archers not 500 will make any strong shootes', and 'few or none do anie great hurte 12 or 14 score off'. Payne-Gallwey's claim of 250 yards for a heavy war arrow, weighing some 2.1 oz (60 gm), could be achieved with the bow weighing 130 lb also, but not with a 70 lb bow. Finally, Payne-Gallwey's figure of 500–515 yards (457–471 metres) for a longbow flight arrow shot from a powerful crossbow is clearly outside the range of any longbow that could be drawn by an archer.

The ranges shown in Figure 4 seem to represent what can be done now, or what could have been done with the medieval longbow. The exact position of

Fig 4 The range and weight of arrows

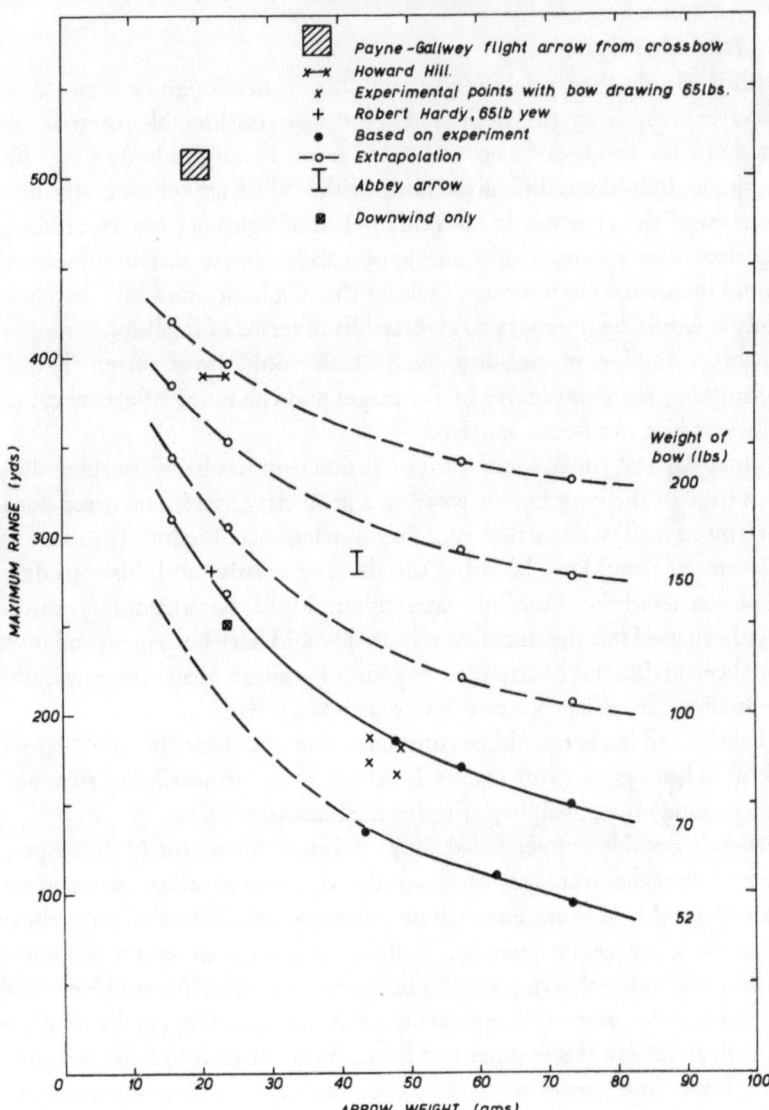

the lines especially for the heavier bows needs to be confirmed by experiment, and it is important to realise that these represent the likely performance of a well-trained competent archer with a good bow and with arrows having drag coefficients as shown in Table 4. A better, more efficient bow, an exceptional archer, or arrows of the same weight but with reduced drag undoubtedly could, and sometimes have, exceeded these ranges.

3 The Target, *by Peter Jones*

1 Introduction

Without doubt the steel-headed arrow shot from a longbow is an extremely effective weapon against flesh. There must be considerable internal rupture caused by the passage of a broad-headed arrow through a body either human or equine. Indeed it is difficult to imagine that a 'hit' anywhere on the body by an arrow of this type would not result in loss of fighting capacity, although, as a general observation, combatants appear to be able to sustain injuries which would in normal circumstances render them at least immobile. In a detailed study it would be necessary to classify hits in terms of fatalities, loss of attack capability and loss of mobility. Such a task would prove extremely difficult considering the complexity of the target and the range of arrow types and velocities likely to be encountered.

Since earliest times some form of protection has been considered useful even though the very fact of wearing a protective system incurred penalties in terms of load to be carried, mobility restriction and reduced vision. Clearly any armour should not be either too thick or massive and this consideration alone restricted the choice of materials employed. For example, wood could have been used but the thickness required would have been great and it would have been as difficult to articulate as a suit of armour. However, it was suitable for an unarticulated protective device such as a shield.

Fabrics and leather could be considered and undoubtedly would have been useful as linings, but soft materials do not cause projectiles to ricochet and they preclude the possibility of projectile break-up.

Metals, notably copper-based alloys and steel during the Middle Ages, had none of these disadvantages. Steel, an alloy of iron and carbon, is amenable to a wide range of heat treatments resulting in improved mechanical properties after fabrication, whereas copper-based alloys, except for softening, are not. Low carbon steels are relatively easy to hammer-weld which would have enabled the armourer to start with several pieces of material. Copper-based alloys are difficult to join in this manner and hence more difficult to fabricate. Further, hard bronze and brasses are likely to be susceptible to season cracking, manifested as fracture, which is caused by the conjoint action of residual stresses from mechanical working and atmospheric contamination, especially from ammonia.

For these reasons steel was the favoured material. Its use resulted in the development of long narrow arrowheads, for reasons which will be discussed.

Of equal importance to the material used is the actual construction of the armour system.

Characterisation of a target of the medieval period must lack precision for two reasons; firstly because very little original material survives intact and secondly because the period was one of transition from mail armour to plate armour, and most soldiers would have worn a combination of the two. The task is made no easier by the probable wide range of equipment, from the complete or partial plate armours of the nobility to the variably lesser protection of the lower orders.

Notwithstanding the difficulties of defining the target in terms of either sensitivity or protection, without doubt the two most difficult armour systems to be defeated would have been plate and mail. If the longbow arrow could defeat these then it would defeat the others and it is to these two, especially plate, that this discussion will be confined.

2 The microstructure and mechanical properties of armour

The resistance to penetration of armour depends upon the strength, toughness, ductility and thickness of the material. As was mentioned earlier, steel can be heat-treated and mechanically worked to produce a variety of mechanical properties. The action of carbon is to strengthen the steel and to make the material amenable to heat treatment. Thus, typical properties for a 0.2% carbon steel are:

Condition	Tensile strength	Elongation
Cold worked	44 tons/sq in (69 kgf/mm^2)	12%
Quenched	51 tons/sq in (80 kgf/mm^2)	14%
Slow cooled	30 tons/sq in (47 kgf/mm^2)	35%

The tensile strength is the maximum load the material will sustain in a simple tension and the elongation is the deformation that is required to produce fracture.

Clearly it is impossible to measure the strength and ductilities of these components directly by the manufacture of test pieces because of the rarity of the pieces of armour. Therefore the materials have to be assessed indirectly by means of microstructural examination and hardness measurements which involve no spoiling of the object.

Microstructural examination by optical microscopy reveals the arrangement of iron crystals within the sample and this can be related to the processing route and the mechanical properties. Iron-carbon alloys have several atomic arrangements depending upon temperature and pressure. At room temperature it is a body-centred cubic arrangement, ferrite, and it is relatively soft although it can be hardened by mechanical work such as forging or rolling. However, heating a 0.2% carbon steel to 900°C transforms the structure to a face-centred cubic arrangement, austenite. This structure is softer than ferrite and is suitable for easy forming. Slow cooling of this material will give the room temperature allotrope, that is ferrite together with carbon as iron carbide. However, rapid cooling, quenching, does not give sufficient time for these changes to occur and a metastable allotrope martensite is formed by an internal shear mechanism. It would be inappropriate to discuss the martensite reaction here but it can be likened to an internal mechanical working process which gives an increase in strength. It will give higher strengths for the same levels of ductility than the cold-worked material and is to be preferred as a hardening mechanism.

Slowly cooled material shows a polygonal crystal structure, and the grains are elongated by mechanical work. Martensite is revealed as a fine needle-like product. By these means the common treatment processes in steel can be deduced.

Very little material being available for examination, our knowledge of the manufacturing techniques known to the armourer is limited. However Professor Pratt has examined, using this method, an English jack plate, circa 1570, and this shows a ferrite-iron carbide structure with a hammered surface. The hardness was 120 DPN (tensile strength in tons/sq in = DPN/5) with a carbon content of about 0.1% to 0.2%. However a fine German gauntlet of 1500 showed a martensitic microstructure with a hardness of 350 DPN indicating that the part must have been quenched.

As an alternative to hammer-hardening the surface, carburising has been found in an Italian breastplate, circa 1520, although the plate had not been further hardened by quenching. All the examples quoted are rather later than the period of interest; however it is probably safe to assume that hardening by quenching was in use by 1400 because it was then known to sword makers.

It is therefore possible to conclude that all the simple treatments of steel were known to the armourers of the period. The question of interest is why all armour was not quenched as was the fine German gauntlet. The probable answer is that the austenite martensite reaction not only involves distortion due to differential stresses engendered during quenching but also a volume expansion. This means quenching can result in a considerable loss of shape integrity in thin sections, especially if the quench rate is not controlled and the method of immersing the component is performed incorrectly. Further, attempts to straighten a hardened steel sheet with the distortions caused by quenching can prove difficult and give rise to cracking. During the heating of the material, prior to quenching, there is a risk of decarburisation. This is the surface oxidation of carbon which reduces the hardening response of the steel. However, it can be largely eliminated by taking proper precautions. For these reasons the manipulation of hardened thin sheets may have been restricted to only the most skilful armourers.

Mail armour, it is thought, was manufactured by drawing the steel to wire, wrapping it around a stick and cutting down. This left open rings of steel which could be forged or riveted together. A sample of 15th century mail was examined by Professor Pratt and shown to have a carbon content of 0.16%C to 0.22%C.

It should be said that the steel of this period is 'dirtier' than modern steels; that is, it contained more non-metallic inclusions, usually iron oxides, than do modern materials. This is a consequence of the direct reduction process employed at the time. Inclusions have a deleterious effect on mechanical properties because they assist in the processes of crack initiation and propagation. It is expected that the elongation of early steel will be lower than the figures mentioned previously. It is impossible at this stage to say how far ductility will have been degraded until a systematic survey has been completed.

Summarising, it can be said that steels of 0.1 to 0.2% Carbon were employed and that a strength range from 20 tons/sq in (31 kgf/mm^2) to 70 tons/sq in (110 kgf/mm^2) was available. Further that face hardening by hammering and carburising was known. However the ductilities and hence the toughness of the steels would be lower than those manufactured today.

3 The thickness of armour

Of equal importance to its material properties is the thickness of armour because the thicker the material the more energy it is capable of absorbing.

Recent work by the authors on material in the Tower of London, using calipers and an ultrasonic thickness meter, can be summarised as follows. All figures are in thousandths of an inch:

Bascinet, 1380, German
Thickest: (top front) — 150 thou (3.81 mm)
Thinnest: (visor snout) — 60 thou (1.52 mm)

Bascinet, 1370–80, German
Thickest: (top front) — 96 thou (2.44 mm)
Thinnest: (side) — 50 thou (1.27 mm)

Bascinet, 1370–80, Italian
Thickest: (top front) — 120 thou (3.05 mm)
Thinnest: (back) — 60 thou (1.52 mm)

Bascinet, 1370–80, Italian
Thickest: (top front) — 180 thou (4.57 mm)
Thinnest: (side and back) — 100 thou (2.54 mm)

Pair of cuisses, 1390, Italian
70–50 thou
(1.78–1.27 mm)

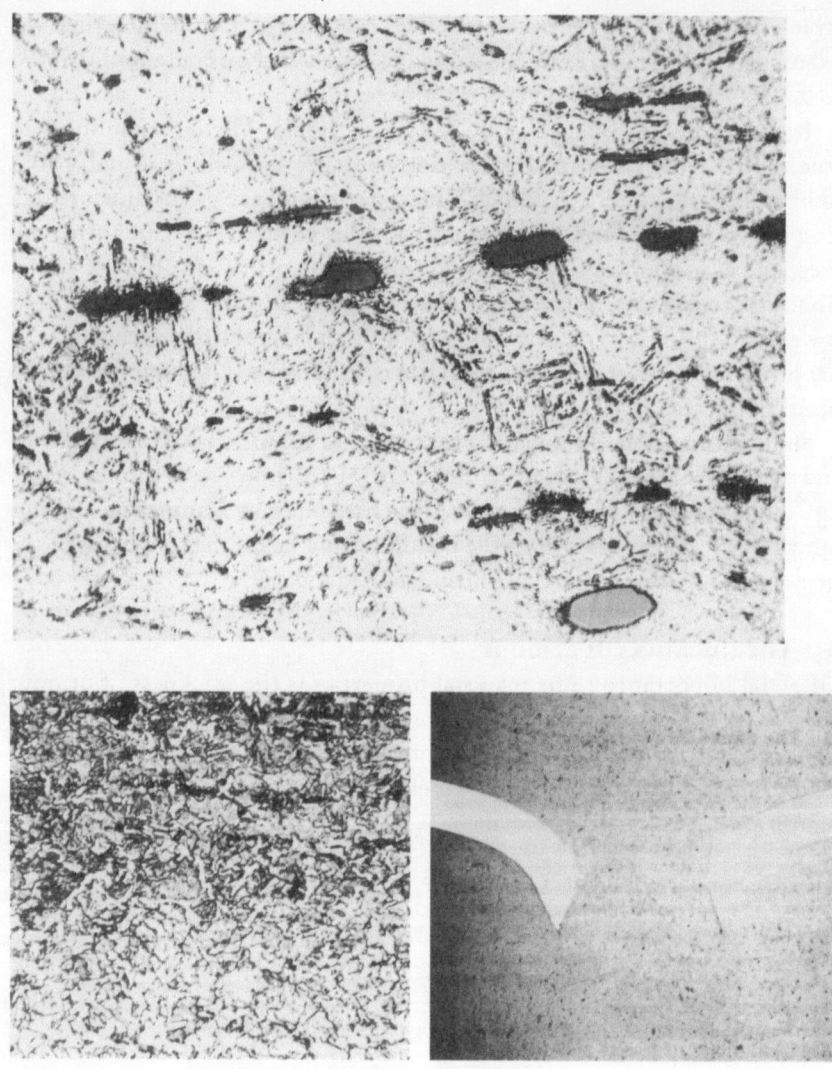

Above top *Section from German gauntlet, × 500, showing a quenched martensitic structure and non-metallic inclusions.*

Above left *Section from an Italian breastplate, × 75, showing a carburised surface.*

Above right *Section of an arrow perforation, × 5. The arrow struck normally; it shows back flow of material and final rupture was by ductile failure.*

No breastplate circa 1400 was available but a composite half armour, Italian, circa 1470, gave measurements between 80–110 thou (2.03–2.79 mm).

Thus it can be seen that the top of the head was the most heavily armoured portion of the body (100–180 thou: 2.54–4.57 mm) followed, probably, by the chest (80–110 thou: 2.03–2.79 mm), and then the legs (50–70 thou: 1.27–1.77 mm). At first sight, the level of protection is proportional to the sensitivity of the target area and the probability of a direct rather than a glancing arrow attack.

4 Penetration mechanisms

Armour functions by either absorbing or dissipating the projectile energy. The energy of the missile is absorbed by elastic and plastic deformation of the target and in extreme cases actual fracture. Some energy will be lost as an elastic wave passing through the target plate, and some as sound. Energy dissipation is accomplished either by spreading the high local loads engendered by the point contact of the arrow or by the fracturing of the arrowhead so that high loads are not sustained. In real situations, if it has insufficient energy to penetrate, the projectile is usually defeated by some combination of these mechanisms.

Most of the energy transmitted to the plate is absorbed by plastic deformation of the material. The material is caused to flow around the arrow point, sideways, forwards and sometimes backwards, and a simple case of hole enlargement has been considered by W. T. Thompson who developed the equation for a conical projectile:

$$W = \pi R^2 h_o \left[\tfrac{1}{2} \sigma y + \rho \frac{(VR)^2}{L} \right]$$

where W = work done by the projectile, 2R = diameter of the projectile, h_o = the plate thickness, σy = the yield strength of the material, ρ = density, V = depth of penetration, and R/L = tan semi-angle of the projectile head.

From this analysis it is evident that the work needed to penetrate a plate is proportional to its thickness and strength. Further the model indicates that the work required can be minimised by using a long narrow arrowhead, that is one with minimum values of diameter R and R/L. This effect can be seen simply in terms of the volume of armour to be displaced; that is, the amount of armour steel to be pushed aside during penetration will depend upon the cross-sectional area of the penetrator, and further that the applied load will be over a smaller area and so the stress level will be higher.

It must, however, be recognised that the equation takes no account of the backward flow of material, of coronet formation, or the frictional losses caused by the movement of the arrowhead through the material. It might also be expected that brittle materials would fracture with lower amounts of deformation and hence be perforated with lower energies than is suggested by the equation which has no criteria of fracture built into it. It is further limited because in its present form it is only suitable for penetrations that are circular in section.

Arrows were not generally circular in section, but either leaf shaped or square in section. Round sections existed but were not common probably because it is difficult to grind a point on them. Of more direct use is the experimentally determined 'critical velocity'; this is the velocity at which the plate is just penetrated to a sufficient degree to be damaging. Unfortunately it is unlikely that arrows of the long bodkin type would penetrate completely and clearly some arbitrary defeat criteria would have to be selected in any series of tests. It is important to realise that very few arrow strikes were likely to be 'normal', at 90° to the armour. Arrows were likely to reach the target at an inclination determined by the trajectory which could be anything between 90° and 0° to the ground. Further, allowing for the curves in armour plate which delineate the shape of the body, and for deliberate angling of the armour, the number of normal strikes would represent a proportion only of the number of arrows discharged.

The degradation of ballistic performance with strike angle is therefore of extreme interest. Analogous to critical velocity is critical angle, that is the angle at which the projectile just fails to penetrate. These variables are related in the empirical De Marre equation:

$$W \frac{(V \cos \theta)^2}{d^3} = C \left(\frac{t}{d}\right)^\alpha$$

where W = weight of projectile, d = diameter of projectile, V = velocity of projectile, θ = angle of attack, t = thickness of plate, and C and α are experimentally determined constants. So that, when the value of the left-hand side of the equation exceeds that of the right, penetration occurs. This shows that as the obliquity (the deviation from normal) increases then the velocity must also increase. Further, as with the Thompson equation, small diameter projectiles require lower strike velocities than larger ones. As has been shown by tests a broad-head is a less effective penetrator of armour than a bodkin and this is to be expected from shape considerations, because a long narrow shape has to deform less material than a broad one. Of course a broad-head has a much more devastating effect against unprotected flesh. As is frequently found two incompatibles must be 'traded off': in this case

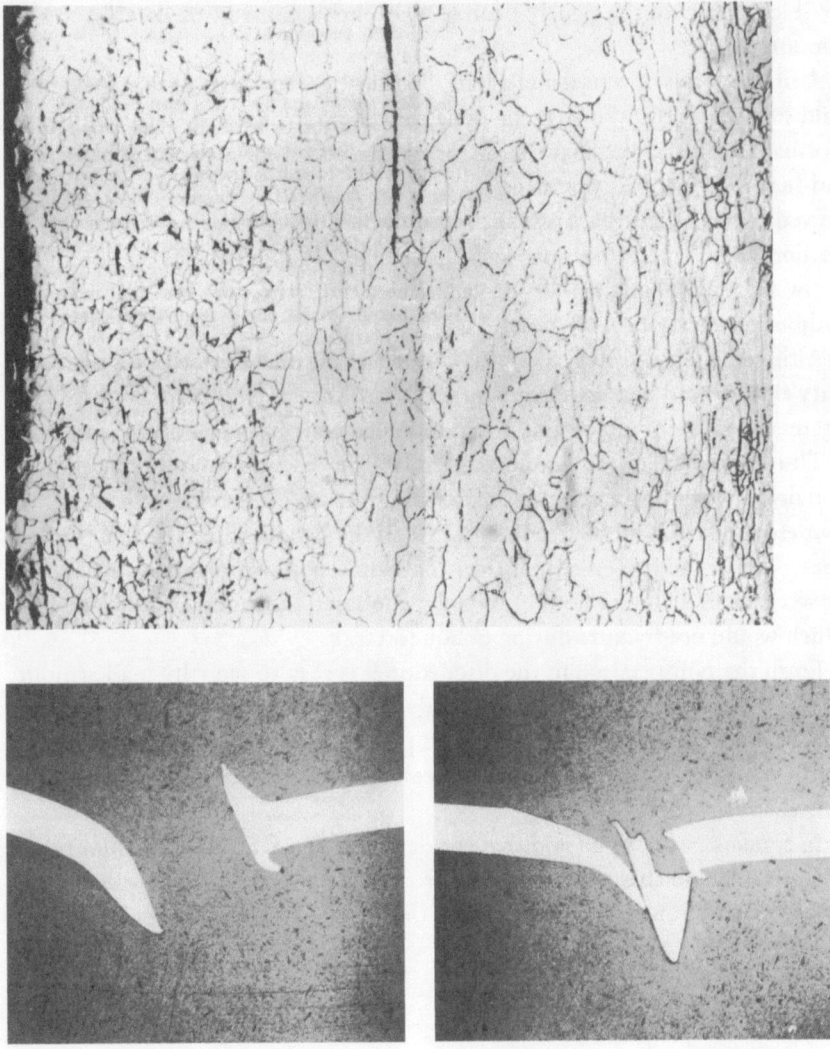

Above *Section from an English jack plate, × 200, showing a slow-cooled ferritic structure and hammer hardening.*

Above Left *Section of an arrow perforation, × 5. The arrow struck at 45°; it shows build up on the front face, and a smaller hole than in the photograph at the bottom of the facing page.*

Above right *Section of an arrow perforation, × 5. The arrow struck at 60°; it shows the fractured arrow tip stuck in the plate.*

terrible damage to the lightly protected targets, against some damage to the armoured ones.

A simple series of tests using a long-bodkin-headed arrow against a 1½ mm mild steel sheet showed that the penetration mode changed with obliquity. Normal attack showed material deformation both forwards and backwards, and that the steel was fractured by ductile failure. At 45° the material had flowed forward, but backwards only on one side, and the perforation was smaller. At 60° the arrow made a hole but the bending stresses fractured the arrow tip and the main body ricocheted. At 70° the projectile ricocheted without penetration.

Although the mild steel used was a modern one probably with higher ductility than would be expected in an early steel, these tests demonstrated the interesting range of effects that could be expected.

Thus it can be seen that as the attack angle increases the amount of penetration decreases until the arrowhead actually fractures. It is evident that smaller diameter bodkins will be less able to sustain the bending stresses than thicker ones. This leads to the conclusion that some compromise was necessary between small diameter bodkins which penetrated easily and larger diameters which would not fracture during oblique attacks.

From the points raised in the discussion it is easy to see why mail armour was less effective than plate. Firstly, only one or two rings had to be broken, probably across the weld, so much less plastic deformation was required to defeat mail than plate. Secondly, there was little chance of richochet and arrowhead fracture.

In conclusion it can be said that the design of arrowheads and armour was inter-related and that a series of compromises between damage to the body, ease of penetration and the integrity of the penetrator had to be reconciled.

BIBLIOGRAPHY

A selection of some works consulted but not mentioned in the text

Some contemporary sources
Archaeologia Cambrensis, 3rd Series, 1866
Chronicon Galfridi le Baker de Swynbroke
Cambrensis, Giraldus, *Opera and Expugnatio Hibernica*
Cymmrodorion, *Transactions of the Society of Ministers' Accounts for West Wales, 1277–1306*

Contemporary chronicles of the Hundred Years' War
Jean le Bel; St Denys; Pierre Fénin; Jean Froissart; Gesta Henrici Quinti; Letters and Despatches of Henry of Monmouth; Le Laboureur; Titus Livius; Enguerrand de Monstrelet; Christine de Pisan; Pistolesi; Le Récit de Tramecourt; St Remy; Jean Juvenal des Ursins; Villani
Gaston de Foix, *Le Roi Modus*, 1486
Edward, second Duke of York, *The Master of the Game*, 1406

Archery, The Badminton Library
Beith, Maj Gen J. H., *The Royal Company of Archers*, Blackwood
de Belleval, Marquis R., *La Bataille d'Azincourt*, Dumoulin 1865; *La Grande Guerre*, Durand, 1862
Belloc, H., *British Battles: Crécy*, Swift
Blackmore, H. L., *Hunting Weapons*, Barrie & Jenkins
Boynton, L., *The Elizabethan Militia*, David & Charles
Bryant, Sir A., *The Age of Chivalry*, Collins

Burke, E., *The History of Archery*, Heinemann
Burne, A. H., *The Crécy War*, Eyre & Spottiswood; *The Agincourt War*, Eyre & Spottiswood
Cambridge Ancient History, Cambridge University Press
Chandler, D., *The Art of Warfare on Land*, Hamlyn; *Battlefields of Europe*, Evelyn
Chenevix Trench, C., *A History of Marksmanship*, Longman
Churchill, Sir W. L. S., *History of the English Speaking Peoples*, Cassell
Contamine, P., *Azincourt*, Collections Archives
Cruickshank, C. G., *Army Royal*, Clarendon; *Elizabeth's Army*, Oxford University Press
Dreyer, C., *Med Bue og Pil*, Steen Hasselbachs
Duff, J., *Bows and Arrows*, Macmillan, New York
Edwards, C. B., *The Tox Story*, Royal Tox Society
Elmer, R. and Smart, C. A., *The Book of the Longbow*, Doubleday
Featherstone, D., *Bowmen of England*, New English Library
Gallice, H., *L'Art d'Archerie*, Renouard
Gordon, L. L., *Military Origins*, Kaye & Ward
Grimley, G., *The Book of the Bow*, Putnam
Hansard, G. A., *Book of Archery*, Longman
Hare, K., *The Archer's Chronicle*, Williams & Norgate
Hargrove, A. E., *Anecdotes of Archery*, Hargrove's Library
Herrigel, E., *Zen in the Art of Archery*, Routledge, Kegan Paul
Hewitt, H. J., *The Black Prince's Expedition of 1355*, Manchester Univ Press; *The Organisation of War under Edward III*, Manchester Univ Press
Hibbert, C., *Agincourt*, Batsford
Hodgkin, A. E., *The Archer's Craft*, Faber
Hollister, C. W., *The Military Organisation of Norman England*, Clarendon
Jacob, E. F., *Henry V and the Invasion of France*, Hodder & Stoughton
Keen, M., *Outlaws of Medieval Legend*, Routledge, Kegan Paul; *History of Medieval Europe*, Routledge, Kegan Paul
Le Livre d'Or de la Gendarmerie Nationale
Lhote, H., *The Search for the Tassili Frescoes*, Hutchinson
Lipson, E., *Economic History of England*, Black
Lowe, J., *The Trees of Great Britain*, Macmillan
Mann, Sir J., *An Outline of Arms & Armour in England*, HMSO
Mazas, A., *Vies des Grands Capitaines*, Lecoffre
Morris, J., *The Welsh Wars of Edward I*, Clarendon
Moseley, W. M., *An Essay on Archery*, 1792
Myatt, F., *The Soldier's Trade*, Macdonald & Jane's
Newhall, R. A., *The English Conquest of Normandy*, Russell & Russell; *Muster and Review*, Harvard University Press
Nicholas, Sir H., *History of the Battle of Agincourt*, Johnson
Nicholl, J., *An Houre Glasse of Indian Newes*, 1607
Oakeshott, R. E., *The Archaeology of Weapons*, Lutterworth Press
Oman, Sir C., *History of the Art of War*, Methuen

Bibliography

Oxley, J. E., *The Fletchers & Longbowstringmakers of London*, Worshipful Company of Fletchers
Poole, A. L., *Oxford History of England*, Clarendon
Powicke, M., *Military Obligation in Medieval England*, Clarendon
Renfrew, C., *Before Civilization*, Cape
Roberts, T., *The English Bowman*, 1801
Schuyler, K. C., *Archery*, Barnes
Smith, R., *A Treatise on Martial Discipline*, temp Eliz I
Stein, H., *Archers d'autrefois, archers d'aujourdhui*, Longuet
Swinehart, R., *In Africa*, Boyertown
Tansley, A. G., *The British Isles and their Vegetation*, Cambridge Univ Press
Trevelyan, Sir G. M., *English Social History*, Longman
Vedel, H. and Lange, J., *Trees & Bushes*, Eyre & Methuen
Waring, T., *Treatise on Archery*, 1877
Winlock, H. E., *The Slain Soldiers of Neb-Hepet-Re*, Metropolitan Museum of Art
Winton, R. K., *The Forest and Man*, Vantage Press Wrottesley, Gen Hon G., *Crécy and Calais*, Harrison
Young, Brig P. and Adair, J., *Hastings to Culloden*, G. Bell & Sons
Ziegler, P., *The Black Death*, Collins

Articles

Archer Antiquaries, Journals of the Society of 1958–1976
The Archer's Register, 1864–1914
The British Archer, 1955–1976
Clark, J. G. D., *Prehistoric Longbows*, Proceedings of the Prehistoric Society, 1963
Glover, R., *English Warfare in 1066*, Eng Hist Rev, 1952
Hadfield, M., *Trees in Anglo-Saxon England*, Quarterly Journal of Forestry, 55 (3)
Notes and Queries, 1891, 1899
Newhall, R. A., *Discipline in an English Army*, Military Historian, 1917
Sherborne, J. W., *Indentured Retinues and English Expeditions to France, 1369–1380*, Eng Hist Review, Vol 79, 1964
The Sporting Magazine, 1798–1799
Willard, J. F., *The Use of Carts in the Fourteenth Century*, History, Vol 17, 1932

INDEX

A

Abbeville 88, 93, 149
Abingdon Chronicle 42
acacia 31, 32
Accles & Pollock bow 214
Acheux 86, 139
Acocanthera tree 35
adder 227
Africa 16, 20, 21, 28, 31, 33, 209, 218, 225, 228, 229, 230
Agache, Gobin 86
Agincourt 57, 78, 91, 101, 102, 103, 105, 116, 128, 140, *142*, *144*, 147, *148*
Ahlefeldt-Laurvig-Bille, Count 27, 36, 216, 218, 219, 222, 229, 231, 236
aim, aiming 59, 60, 192, 220
alder 18
ale 61, 111, 129
Alençon, Comte d' 72, 143, 153
Algeria 16
aluminium 214
American Nat Archery Assoc 203
American Round 208

ammunition 81, 115, 164
Andaman Islands, islanders 19, 21
Anglo-Saxon 43
Anglo-Saxon Chronicle 44
arbaletriers 154
Arc, Joan of 148, 149, 150, 154, 155, 163
Archers' Hall 184
Archers' Paradox *222*, 228, *289*
Armagnac 151
armour – *passim*
armourer 61, 73, 135
armoury 88, 255, 229
array, arrayers 64, 104, 106, 107, 116, 158
arrow-plate 242
artiller, artillery 113, 114, 156, 157, 163, 187, 189
Artois 127
Arundel, Earl of 89, 98
Arundel, M. 216
Ascham, Roger 40, 71, 149, 169, 170, 172, 173, 174, 178, 179, 185, 192
ash 24, 71, 160, 171, 208, 218, 222

Index

Ashcott Heath bow *22*, 24
Asia 18, 27, 28, 31, 197, 214, 231
aspen 25, 171
Atholl, David of 70, 71
Auray 128
Azincourt – see Agincourt

B
bacon 61, 111
baggage train 84, 88, 113, 119, 120, 123, 136, 138, 141, 143, 146
Baliol, Edward 70, 71
ballista 50
bamboo 209, 215, 219, 229, 231, 233
banner *93*, 95, 99, 126, 139, 143, 144, 158
bannerets 82, 107, 142
Bannockburn 53, 67, 68
Bar, princes of 142
barbs *59*, 73, 74, 92, 158, 172, 182, 207, 220, 225, 233
Barnet 156
Barnsdale, Yorks and R. 53, 55
barons 57, 58, 63, 67, 82, 135, 142
Barwick, Humphrey 177
bascinet 61, 63, 144
'battle' 68, 69, 71, 81, 84, 90, 97, 115, 117, 126, 131, 139, 140, 141, 144, 145, 146, 168
battleaxe 68, 127, 146, 154, 156
Bayeux Tapestry *44*, 45, 46
beans 111
bear 201, 202, 209, 219, 229
Beaugé 149
Beaumont, Henry de 70, 71
Beaumont, Jean de 139
beaver 63, 144
beech 72, 171, 213
beef 111, 118
beefwood 234
Belgians, Belgium 49, 139
Berwick 70
Béthencourt 138
Béthune 86, 88
billets 235, *236*, *239*
billmen 154, 158, 159
birch 25, 72, 172

Black Army 104
Black Death 102, 129
blackthorn 72, 171
Blanchetaque 87, 88, 89, 137
Blangy 139
boar, wild 216
bodkin 73, 74, 92, 98, 201, 231
Boehm, Lt Col Hugo 187
Bohemia, king of 86, 94, 99
Bohun, de 57, 67
bois d'arc 234
Boleyn, Ann 168
bolts 60, 64, 87
Bonhomme, Jacques 131
boots 76, 159
Bordeaux 81, 102, 117, 120, 123, 125, 155
Bosworth 157, 162, 163
bowcases 61, 76, 173, 174, 179
bowls 166, 188
bowyers 73, 76, 112, 135, 161, 168, 170, 178, 180, 181, 184, 185, 197, 198, 203, 206, 210, 212, 214, 218, 222, 231, 233, 235, 241, 270–4
Bowyers, Worshipful Co of 233
bowmarks 147, 162, 168, 172, 182, 183, 251, 258, 260, *261*
bowstrings – *passim*
bracer 162, 173, 174, 179, 243, 244
Braganza Shield 183
Braose, William de 50, 56
Brasell, Brazilwood 71, 172, 234
Brazil 32, *35*, 71
bread 61, 110, 122, 128
breaking 171, 213, 214, 221, 238
breastplate 125, 132, 159, 168, 211, 303
Brecknock 103, 150
brigandine 174
Bristol 49, 107, 111, 112, 135
Britain 40, 41, 42, 51, 54, 76, 197, 214, 230
British 139, 211, 226, 228
British Long Bow Society 23, 55, 195, 229, 233
Britons 40, 227
Brittany 80, 128, 163

311

broadheads 52, *59*, 73, 168, 172, 220, 223, 227, *229*, 230, 243
Bronze Age 27, 28
Bruce, David 102
Bruce, Robert 67, 68, 69
Buchanan 191, 205, 241
Buckinghamshire 109, 177, 185
buckler 150, 243
buffalo 226, *229*, 230, 241
bullet 177, 179, 211, 224, 228, 229
Burgundy 150
butts, 76, 107, 144, 147, 150, 160, 178, 182

C

Cadsand 79
Caen 84, 85
Caesar, Julius 228
Calais 102, 106, 107, 116, 130, 136, 138, 140, 141, 149, 156
California 205, 212
caliver 164, 175, 177
Cambrensis, Giraldus 49, 50, 51, 56, 57, 71
camels 32, 68
cannon 88, 97, 156, 164
Canute, king 47, 177
canvas 61, 76, 112
cap 76, 148, 150, 156
captains 62, 125, 130, 150, 151, 152, 157, 166, 174, 175
Captal de Buch 117, 126, 128, 131
Carcassonne 119, *119*, 120
casks 61
cast 170, 212, 215, 222
Castillon 155, 156
cattle 111, 119
cavalry – *passim*
cedar 234
Celtic 105
centenar 61, 63
Chandos, Sir John 128
Channel, the 79, 81, 105, 116, 122, 152, 228
Charles I of England 178
Charles II of England 180
Charles V of France 128

Charles V of Spain 166
Charles VI of France 134, 139
charters 47, 106, 168
Chaucer 243
Chauvigny 126
cheese 61, 111
cherrywood 231
Cheshire 103, 107, 108, 109, 111, 122, 164
Chester 53, 55, 56, 64, 82, 107, 113, 122
chevauchée 118, 120, 122, 123, 129, 136
Chicago 203, *214*
China, Chinese 20, 21, 48, 231
Chronicle of Breteyne 54
chrysals 171, 212
Churchill, Sir Winston 133
City, the 133, 178, 184
Civil War 132, 165, 179, 180
Civil War (American) 200
Clarke, F. H. 225, 226
cloak 76, 150, 211
clothes 61, 76, 107, 109, 146
clout 182, 187, 193, 203
coal 61, 122
cockfeather 173, 207, 243
Commissioners of Array 52, 82, 107, 158
Commons, House of 161, 231
compass, come round 171, 241
composite *17*, *190*, 197, 209, 213, 215, 231
compression 171, 212, 237
Conquest, the 44, 53
Constable of France 84, 128, 155
corn 118, 122, 205
cornelwood 234
Cornish, Cornwall 90, 111
corselets 174
cowhorn 241
crabapple 234
Crécy – *passim*; 78–99
cresting 207, 243
criminals 62, 69
crowberry 25
Crusades, the 48, 49, 57
cuirass 141
culverins 155

Index

D
dagame 234
dagger 52, 59, 76, 90, 115, 154, 174, 244
Dancaster, John 130
Danes 24, 41, 42, 43
Danish 40, 223
Dauphin 126, 134, 136
deer 52, 201, 206, 208, 209, 215, 219, 224, 229, 231
Denmark 23, 27, 29, 40, 218, 222, 225
Derby, Earl of 79, 81, 82
Derbyshire 53, 58, 62, 64, 66, 103
Despenser, Hugh 87, 88
destriers 58
des Ursins, Jean Juvenal 131
dice 140, 166
Dorset 62, 112
Double-Armed Man *176*, 178
Douglas, Earl of 132, 149
Dover 104, 110, 130, 150
Dowsen 181
drawing 59, 97, 169, 171, 173, 189, 192, 201, 215, 216, 223
draw weight 73, 74, 189, 209, 215, 223, 246, 254, 255
Dreyer, Carl 219, 223
duck 201, 202, 215
Duff, James 205, 219
du Guesclin, Bertrand 128
Dunois 154
Dupplin Muir 69, 71
dysentery 136, 149, 163

E
eagle 200, 207, 215
ebony 31
Edgecote Field 159
Edington Burtle 25
Edmund, king 42, *42*
Edward I 56, 57, 60, 62, 63, 64, 66, 67, 68, 71, 104, 132
Edward II 54, 56, 67, 69, 70
Edward III 48, 67, 70, 71, 73, 76, 78, 79, 81, *84*, 85, 86, 87, 88, 89, 93, 98, 100, 102, 103, 104, 107, 114, 120, 128, 129, 132, 133, 137, 175, 217
Edward IV 159, 160, 161, 168
Edward VI 40, 168
Edward the Confessor 57
Edward, Prince 64, 84, 86, 103, 104, 107, 117, 118, 119, 121, 123, 124, 126, 127, 128, 142
Edwards, C. B. 190
Edwin, king of Northumbria 41
Egeskov 218, 228
eggs 61
Egypt 17, *31*, *32*
elder 171
elephant 33, 36, 226, 230, 243
Elizabeth I, Queen 40, 169, 174, 177, 181
Elizabeth II, Queen 182
elm 23, 25, 50, 71, 168, 219, 222, 234
Elmer, Dr Robert 203, 211, 212, 215
Elmham, Thomas 104, 106, 135, 137, 140, 148
Elyot, Sir Thomas 40, 174
embarkation 107, 109, 111
Erpingham, Sir Thomas 143
esquires 82, 107, 135, 142
Ettrick 66, 164
Eu 137, 142
Evesham 57

F
Falkirk 58, 63, 66, 67, 68, 164
farriers 135
farmers 178, 183
'Fast' 168
feathers 76, 112, 166, 172, 198, 200, 207, 208, 220, 228, 243, 244
Fergie 181, 205
Ferguson, James 178
Field archery 183, 203, 243
Field of the Cloth of Gold 162
'Fighting Man, The' 43, 46
films 52, 54, 148, 228
Finch, Hon D. 185, 187
Finsbury Archers 182, 183, *191*
firwood 29, 210
fish, fishing 111, 122, 199, 205, 208, 220
fishtail joint 236, *239*

313

'fistmele' *242*, 244
Flanders 79, 80, 81, 88, 116, 168
fletchers 112, 135, 166, 178, 183, 198
Fletchers' Co 166, 178
fletching 31, 37, 73, 76, 92, 132, 172, 200, 207, 220, 222, 243
flint arrowheads 16, *20*, 27, 30, 208
Flodden 163, 164
Florida 201, 211
Flynn, Errol 52, 53
Ford, Horace 189, *191*, 192, 193, 195, 197, 202, 205
Formigny 154
François I of France 162
francs archiers 155, 156
Frederic IX of Denmark 218
fret 170
Froissart 79, 81, 93, 94, 98, 118, 121, 125, 127, 131
Frost, Nicholas 235
furs 106, 127, 197
fustic 234
fyrd 43, 47, 61

G

gambling 178, 180
game 40, 201, 208, 209, 216, 218, 220, 221, 224, 225, 228
garrison 61, 62, 63, 79, 80, 85, 130, 136, 150, 152
Gascons, Gascony 64, 67, 69, 80, 83, 106, 111, 117, 120, 123, 126, 155
Gaunt, John of 79, 122
Genoese 78, 81, 87, 94, 95, 97, 100, 101
George IV 182
George VI 182
German 28, 38, 40, 94, 139, 148, 160, 174, 222
Germany 22, 24, 41
Ghent 79, 83
Glacial period 16, 206
Gloucestershire 82, 109
gloves 63, 168, 174, 179, 232
glue 76, 198, 206, 208, 220, 236, 242
Glyndwr, Owain 132
Goch, Iolo 132

gold 106, 127, 138
gold, the 182, 187, 201
goose 112, 162, 172, 220
Goose Prize 182
gorgets 63, 159
grain 76, 170, 235, 236, 238, 240
Grand National Archery Meeting 192, *193*, 200
greaves 159
Greece 28
Greeks 213
greenheart 213, 233
Grimley, Gordon 37
Grinnell-Milne, Duncan 48
grounds, archery 178, 183, 200
grouse 201, 202, 215
Guienne 126, 156
Guild of St George 168
Guines 130, 151
gunpowder 157, 163
guns, gunners 78, 113, 135, 139, 141, 154, 156, 157, 160, 163, 164, 168, 174, 177, 178, 201, 220, 230
Gwent 56, 58, 60, 63

H

Hainault, John of 83, 94, 99
Halidon Hill 69, 71
handgun 102, 156, 157, 162, 166, 168, 177
handle 74, 75, 206, 213, 221, 228, 233, 235, 237, 238, 239
hardbeam, hornbeam 72, 172
hares 220, 225
Harfleur 114, 135, 136
Harold, king 43, 45, 46, 93
harquebusiers 168, 174, 177
Harris, Valentine 55
Harrison, William 174
Harrow School 182
Hastings 42, 43, 45, 47, 48, 58, 93, 104
hauberk 62, 132
hay, 117, 122
hazel 25, 160, 219
Heath, E. G. 32
hedge 124, 125, 126, 149, 175

Index

helmets 61, 63, 76, 95, 104, 125, 126, 133, 141, 142, 146, 156, 162, 168
Henry I 49, 168
Henry II 49, 78, 199
Henry III 52, 56
Henry IV 131, 132
Henry V 57, 58, 71, 78, 85, 102, 103, 114, 118, 133, 134, 135, 136, 138, 139, 140, 143, 147, 148, 149, 154, 162, 235
Henry VI 151
Henry VII 162
Henry VIII 29, 114, 162, 163, *165*, 165, *167*, 169, 178
heralds 138, 139, 147
Herbert, Lord 164, 178
herce formation 91
Hereford 57, 64, 111, 113
Hereford, Ralph Earl of 42
heron 200, 202
Hesdin 94, 149
Hickman, Dr Clarence 212, 213
hickory 213, 234
hide 198, 209, 222
Hill, Howard 73, 203, 215, 221, 229
hobelars 67, 69, 82
holding 173
holly, white 200
Holman, Dennis 33
Holmegaard 23
Homer 21
Homer, Stuart *240*
Homildon Hill 132
honey 122
Honfleur 151, 152
Honourable Artillery Co 182, 183, 184
Hood, Robin 52, 53, 55, 56, 58, 219
horn 17, 18, 50, 52, 198, 219, 241, 252
horses – *passim*
horseshoes 166
hose 76
housecarls 43, 45, 47
humidity 160
Hundred Years' War 19, 74, 85, 128, 132, 156
Hungerford, Sir Edward 180, 183
Hungerford, Sir Walter 103, 140, 153
hunter, hunting – *passim*; 219–43
Huntingdon, Earl of 79
Hyde Park 180

I

Iceland 19, 228
Imperial State Crown 142
Independence, American 216
India 19, 71
Indians, Brazilian 33
Indians, American 21, *198*, *202*, 203, 209, *210*, 211, 215, 219
infantry 57, 59, 60, 64, 67, 68, 69, 81, 83, 84, 87, 102, 106, 141, 142, 149, 180
inflation 165
Ireland 27, 40, 49, 57, 231
Iron Age 29
ironwood 210, 234
Ishi 205, 206, 207, 209, *210*
Italians 97, 191
Italy 49, 63, 122, 160
ivory 50, 52

J

jack, jacket 76, 104, 137, 148
Jacquerie, the 131
James I of Scotland 162
James IV of Scotland 162, 164
Japan, Japanese 231, *232*, 233
Jauderel, Jodrell 103, 104, 122
John, king of France 122, 123, 124, 126, 127, 128
juniper 25, 206
Jutland 26, 219

K

Kamba 33, *34*, 36
Keasey, Gilman 212
Kent 82, 110, 111
Kenya 18, 221, 222, 226
King, John 182
Klopsteg, Dr Paul 212, 215
knights 66, 70, 82, 84, 97, 98, 107, 124, 125, 127, 130, 131, 135, 141
Kyriel, Sir Thomas 154

L

Laboureur 139
laburnum 160, 234
laminates 213, 215, 219, 229, 231
Lancashire 139, 164, 183
Lancaster 157
Lancaster, John of 123, 129
Lancastrians 156, 159, 160, 161
lances 58, 61, 67, 80, 97, 102, 107, 112, 116, 124, 129, 139, 141, 144, 150, 152, 155
lancewood 234
Langland, William 53, 54
Latimer, Bishop 168, 174
Le Baker, Geoffrey 93, 130
Le Bel, Jean 62, 93
Le Crotoy 88, 93
Leicester, Earl of 177, 181
Leicester House 184, 185
lemonwood 203, 219
leopard 226, 230
Lever, Sir Ashton 184, 185
levies 43, 45, 49, 59, 60, 62, 64, 82, 86, 94, 95, 100, 128, 152, 158
Liangulu 33, *34*, *35*, 36
lime trees 17, 25
Lincolnshire 122
linen 95, 211
lion 226, 227, 230
livery 158, 166, 184
Llantrisant 104
Llewelyn, Prince 57, 60, 64
London 44, 60, 79, 107, 122, 178, 183, 205, 219
London Museum 73, *75*
loosing 60, 168, 173, 193, 201, 211, 220, 223, 225
loot 106, 118, 131, 135
Louis XI of France 161
Low Countries 79, 180

M

Macclesfield 58, 63, 104, 109
Maior, John 54
Maisoncelles 140, 144
Marches, marcher lords 49, 56, 58, 62, 64, 83
Markham, Gervase 178
Marshal of England 110
Mary Rose the 29, 165, 74, 76, Chapter 11 *passim*
Mascy, Hamo de 109, 116, 121
mason 61, 135
meal 61, 205
Meare Heath bow *22*, 24, 30
meat 61, 111, 118, 138, 164, 198
medieval 71, 74, 75, 76, 85, 89, 91, 124, 139, 162, 199, 212, 228, 233, 235
men-at-arms 69, 79, 80, 82, 84, 87, 89, 90, 92, 95, 98, 99, 102, 103, 106, 115, 119, 124, 126, 127, 129, 133, 135, 136, 141, 143, 151, 152, 154, 158, 163
mercenaries 44, 57, 64, 79, 94, 122, 129, 141, 152, 162, 165
Meriden 187
Mesolithic 17, *20*, 21, 25
militia 179
moisture 160, 243
Monmouth 49, 57
Montfort, Simon de 57
moose 215, 219, 229
Morley, Sir Robert 79
Muir, Peter 181, 191
mulberry 231, 234
Munro, Robert 181
Muntz, Hope 46
muscle 189, 190, 192, 193
musket 76, 162, 175, 177, 178, 180, 188, 211, 217
muster 105, 111, 122, 150, 151, 152, 174
mutton 111, 118

N

Nairobi 223, 225
Najera 128
Napoleon, Prince Louis 92
Neade, William 178
Neb-Hepet-Re 32
Neolithic *18*, *22*, 23, 25
Neville's Cross 102, 106
Nicopolis 147

Index

nobility, the 94, 97, 106, 110, 122, 131, 144, 147, 159
nocking, nocks 29, 74, 76, 97, 132, 166, 172, 173, 174, 206, 207, 211, 219, 222, 238, 240, *241*, 252, *258*
Normandy 40, 43, 57, 78, 83, 84, 122, 124, 137, 149, 152
Normans 41, 44, 46, 48, 49, 56, 58, 78, 105, 151, 228
Norman-French 45
Norse 40, 43
Northampton 69, 107, 156
Northampton, Earl of 79, 81, 84, 86, 88, 90, 97, 98, 107
Northumberland, Earl of 133
Norway 221
Norwegian 39, 40, 227
Nottinghamshire 58, 62, 116
Nydam *26*, 29, 30

O

oak 17, 25, 27, 72, 104, 132, 172
oats 61, 111, 117
obsidian 207, 209
Oman, Sir Charles 133, 143
Ordnance 179
Oregon 234
Orewin Bridge 64
Oriflamme 94, 126
Orleans, Duke of 124, 142
osage 209, 213, 219, 234
outlaw 52, 53, 58, 62, 76, 105
Oxford 111, 179
Oxfordshire 82, 130, 178

P

pack animals 83, 101, 111, 115, 136
'painted' bows 112
Paleolithic 16, *20*
panther 201
Paris 83, 86, 87, 136, 139, 154
Parliament 133, 179
Parthians 68
Pass, the 103, *105*
Patay 154, 163

pay 58, 59, 60, 63, 69, 106, 109, 110, 150, 151, 152
Payne-Gallwey, Sir Ralph 95, 101
peasant 44, 52, 59, 87, 103, 122, 132
Pedro of Portugal 128, 142
Pembroke 60
penetration 175, *261*, *262*
pennons 126
Percy, Henry, Lord 132
Péronne 138, 139
Philip IV of France 78, 79, 80–3, 85, 87, 93, 95, 99
Philip, son of king of France 127
Picardy 78, 102
pike, pikemen 154, 162, 168, 174, 178, 187, 188, 217
pike, to 50, 238
pillage 85, 152
'pinch' 27, 170
pine 17, 21, 25, 219
Pinkie 168
pins 75, 170, 220, 234, 238, 239
pioneers 86, 135
ploughman 129
plumber 61
Plymouth 104, 109, 110, 116, 122
poachers 36, 54, 62, 76, 224
poison 36, 41, 157
Poitiers 101, 103, 104, 105, 123, 124, 128, 131, 141
Pope, the 48, 57, 123, 174
Pope, Saxton T. 205, 208, 209, 211, 219, 220, 228
pork 111, 122
Portsmouth 69, 83, 84, 89, 104, 110
practice 58, 60, *62*, 73, 76, 110, 178, 191
Pre-Glacial period 25
Preservation 246, 250, 251, 252
prices 60, 129
'prick', pricking 172, 182
Prince Madoc 199
Prince Regent, the 187
Prince of Wales 57, 82, 89, 90, 97, 98, 100, 103, 104, 107, 111, 113, 127, 132
Prince's colours 243

Prince's Lengths 187
Prince's Reckoning 187, 208
prisoner 84, 87, 123, 127, 128, 130, 133, 137, 139, 140, 146, 147, 152
Privy Council 177
proclamations 129, 166, 169

Q
Quai de Caux 135
quail 202, 208
quarrel 60
quarry 197, 199, 243
Queen's Bodyguard in Scotland 181
querbole 76, 104, 148
Quintus Fabius Maximus 128
quiver 60, 76, 87, 97, 115, 156, 179, 182, 208, 220, 243

R
rabbits 215, 220, 230, 243
ransom 106, 128, 130, 132
Rausing, Prof Gad 50
rations 136
rearguard 88, 118, 154
recruitment 63, 106, 117, 158
recurve 197, 215, 233
retinues 82, 107, 110, 116, 149, 162
Rhine 28, 38, 160
rhinoceros 225, 230
Rich, Barnaby 175
Richard I 48, 54, 57
Richard II 131, 132
Richard III 162, 163
rifle 76, 92, 101, 193, 202, 224, 227, 228, 229, 230
Roberts, Thomas 184, 188
rock carvings 18, 20, 22, 27
roebuck 224, 225
Roman arch 74, 212
Romans 28, 38, 41, 48, 68, 94, 129
Roses, Wars of the 132, 156, *161*
rosewood 234
Rouen 86, 139, 141
rovers 182, 194, 203
Royal Artillery Co 181

Royal British Bowmen 188
Royal Co of Archers *181*, 182, 187, 195, 205
Royal Kentish Bowmen *186*, 187
Royal St Leonards Bowmen 188
Royal Toxophilite Soc 183, 187, 188, 195, *237*
Rupert, Prince 179
Russia 20, 21
Rutland 82, 107, 111

S
saddle 211
safari 227, 229
Sahara 16, 17, 32
SS Crispin and Crispinian 141
St Denis 95, 126
St George 62, 80, 126, 143, 182
St George's Fields 182
Salisbury 109
Salisbury, Earl of 124, 158
salt 61, 62, 209
saltpetre 113, 216
sand 200, 206
Sandwich 104, 107, 109, 110, 117
sappers 135
sapwood 74, 75, 112, 206, 209, 213, 236, 238
sassafras 234
Saxon 40, *41*, 42, 43, 45, 46, 53
Scandinavia 25, 27, 32, 38, 50, 228
Scandinavians 42, 234
schiltrons 58, 66, 67, 68, 132
Scorton Arrow 186
Scottish, Scotland – *passim*
seasoning 74, 170, 206, 207, 221, 235
Second World War 148, 187
Seine, river 85, 135
shaft 171, 172, 173, 188, 191, 200, 206, 207, 219, 222, 228, *241*, *242*
sheaf 60, 72, 92, 98, 109, 112, 113, 114, 162, 166, 168, 174, 243
sheriffs 82, 111, 112, 128, 160
ship burials 29
ships 69, 78, 79, 80, 83, 86, 111, 112, 116, 117, 135, 160
Shrewsbury 57, 107, 132, *134*, 157

Shrewsbury, Earl of 151, 154, 155, 156
Shropshire 63, 64, 109
Siberia 21, 22
siege 61, 111, 118, 119, *121*, 129, 135, *153*, 156, 164, 168
silk 200, 243
sinew 18, 198, 206, 207, 209, 213
Sluys 79, 80, 104
smiths 135, 219
Smythe, Sir John 157, 177
snakewood 234
snow 158, 159
Somerset 24, 25, 112
Somme, river 86, 87, 88, 94, 137
Southampton 83, 104, 109, 110
Spain 22, 128, 163, 234
Spaniards 128, 211
Spanish 175, 198, 234
spear, spearmen 15, 38, 56, 59, 63, 66, 67, 68, 83, 90, 91, 102, 106, 112, 144, 145, 159, 205, 209
Spencer-Stanhope 74, 76
spine 73, 222, 228, 243
Spurs, battle of the 163
Staffordshire 82, 109
stag 154, 224, 242
stakes, archers' 137, 144, 154, 166, 187
stalking 198, 225, 227, 230
Stamford Bridge 43
stance 60, 189
standard 110, 146
standing 173
statutes 128, 159, 160, 166, 177
stave 74, 75, 76, 83, 112, 160, 161, 162, 166, 170, 178, 197, 213, 221, 222, 235, 236, *237*, 240
steel bow 214
Stephen, king 49
Stirling 65, 67
Stone Age 23, 27, 205
stones 200, 206
stopperwood 233
suger cheste 171
sulphur 113
sumac 231

surcoat 62, 133
swan 135, 200
Sweden 27, 38, 214, 221
Swinehart, Bob *229*, 230
Switzerland 25, 160
sword 76, 80, 111, 113, 126, 136, 145, 146, 151, 154, 187, 188

T

tab 243
tackle 170, 180, 209, 219, 231, 243
Talbot, Sir Gilbert 180
Tambeskjelve, Einar 39, 219, 231
taper 219, 221, 238, 239, 240, 241
target 52, 192, 194, 195, 197, 200, 203, 208, 209, 213, 215, 216, 220, 222, 223, 225, 228, 243
Tartar 233
television 54, 103
Tell, William 219, 225
tension 212, 213, 237
tests 209, 260–3, 265–71
Tewkesbury 109, 156, 161
Thames, river 79, 111, 148
Thompson brothers 200, 203, *204*, 211, 214
Thompson, Ned 242
thong 206, 207
thumb 207, 240
tillering 170, 235, *240*
tips 76, 206, 219, 240, 241
Tournay 188
Tower of London 74, 88, 112, 113, 135, 166, 177, 178
Towton 156, 159
tracking 198
trained men, bands 175, 177
trajectory 192, 216
Tramecourt 141
transport 112, 113, 114, 122, 158
Treasury, Treasurer 150, 151, 152
Trevelyan, G. M. 49
Troels-Smith, Dr 27
True Temper bow 214
Tweed, river 69, 132
Tyler, Wat 131

U

Ullrich, Earl 212, 234
uniform 61, 76, 107, 109, 150, 155, 182, 202
United Bowmen of Philadelphia 199
Umphraville, Gilbert 70, 71
United States 197, 205, 220
University of California 205

V

Vachan, Vaughan 103, 121
varnish 112, 221, 242
vaward, vanguard 83, 87, 94, 118, 124, 140, 141
Veddahs 21, *33*
Venice 160, 161, 165
venison 52, 161, 205
Verneuil 149
Viborg 26, *28*
Victoria, Victorian 181, 188
Viking 38, 39, 50, 219, 228, 229
Villani 93
Vimose 29
'vintaines' 109, 110
vintenar 60
visor 133
Vortiger(n) king 40, 174
Voyennes 138

W

wages 61, 69, 82, 103, 106, 109, 121, 129, 151, 152, 165
wagons 81, *85*, 88, 89, 98, 112, *113*, 114, 117, 122, 124, 125, 136, 140, 148
Wallace 65, 66, 67
walnut, black 234
Waring, Thomas 184, 185, 187, 188
Warwick, Earl of 98, 124
Waterloo 162, 188
wax 221, 242
wedge formation 70, 91, 124, 141, 154
Westminster, Dean and Chapter 72

Westminster, Palace of 129, 160
wheat 61, 111, 129
'white' bows 112
Wilcox, Russell 213, 215
William, Duke of Normandy 43, 44, 46, 57
William Rufus, king 48
willow 25, 72, 206
Winchelsea 63, 116
Winchester, Statute of 64
wine 61, 106, 109, 117, 118, 119, 120, 122, 130, 137, 140, 162, 163
Wirral 109
Wood, Sir William 183
Woodley, Frank *34*, 35, 36
Woodmen of Arden 187, 195, 243
woodpecker 201
wool trade 78
Worcester, Earl of 132
wych elm 71, 160, 234
wych hazel 207
Wye, river 49
Wylie, J. H. 136, 144, 147

Y

Yarmouth 111, 116
yeomen 52, 82, 122, 162, 164, 180, 243
yew 23, 24, 25, 28, 29, 41, 50, 52, 71, 74, 75, 112, 132, 149, 160, 161, 162, 166, 190, 194, 196, 197, 209, 212, 213, 214, 218, 219, 221, 222, 225, 227, 229, 233, 239, 240, 254
York 44, 157
York, Duke of 139, 141, 146
Yorkist 159, 160
York Round 187, 192, 196, 203, 205
Yorkshire 49, 53, 62, 116, 181, 183
Young, Arthur 208, 219, 228

Z

Zealand 23